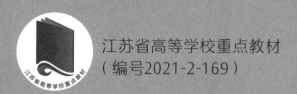

江苏省高等学校重点教材
（编号2021-2-169）

Fundamentals of Photophysics

光物理基础

张秀娟　揭建胜　编著

苏州大学出版社
Soochow University Press

图书在版编目(CIP)数据

光物理基础 = Fundamentals of Photophysics：英文 / 张秀娟，揭建胜编著. —苏州：苏州大学出版社，2022.9
ISBN 978-7-5672-4043-8

Ⅰ.①光… Ⅱ.①张…②揭… Ⅲ.①物理光学-英文 Ⅳ.①O436

中国版本图书馆 CIP 数据核字(2022)第 144764 号

书　　名：	Fundamentals of Photophysics 光物理基础
编　　著：	张秀娟　揭建胜
责任编辑：	万才兰　王　娅
封面设计：	吴　钰
出版发行：	苏州大学出版社(Soochow University Press)
社　　址：	苏州市十梓街 1 号　邮编：215006
印　　装：	镇江文苑制版印刷有限责任公司
网　　址：	www.sudapress.com
邮　　箱：	sdcbs@suda.edu.cn
邮购热线：	0512-67480030
销售热线：	0512-67481020
开　　本：	787 mm×1 092 mm　1/16　印张：21　字数：473 千
版　　次：	2022 年 9 月第 1 版
印　　次：	2022 年 9 月第 1 次印刷
书　　号：	ISBN 978-7-5672-4043-8
定　　价：	98.00 元

凡购本社图书发现印装错误，请与本社联系调换。服务热线：0512-67481020

>>> Contents

CONTENTS

Chapter 1 Light Absorption

1.1   Introduction to Photochemistry and Photophysics   / 1
    1.1.1   Photochemistry and Photophysics in Nature   / 1
    1.1.2   The Nature of Light   / 2
    1.1.3   Definition of Photophysics and Photochemistry   / 5
    1.1.4   Development of Photophysics and Photochemistry   / 7

1.2   Light Absorption   / 9
    1.2.1   Basic Concepts of Light Absorption   / 9
    1.2.2   Electronically Excited States   / 10
    1.2.3   Fundamental Principles of Light Absorption   / 12
    1.2.4   Spin Multiplicity   / 15
    1.2.5   Selection Rules for Light Absorption   / 16
    1.2.6   Absorption Spectrum   / 19

References   / 20

Chapter 2 Deactivation of Excited States

2.1   Brief Introduction to Deactivation of Excited States   / 22

2.2   The Jablonski Diagram   / 24

2.3   Intramolecular Processes   / 28
    2.3.1   Non-radiative Transitions   / 28
    2.3.2   Radiative Transitions   / 35

2.4   Intermolecular Processes   / 47
    2.4.1   Vibrational Relaxation   / 47
    2.4.2   Energy Transfer   / 49
    2.4.3   Electron Transfer   / 51

Fundamentals of Photophysics

2.5　Dynamics of Excited State Deactivation　/ 52

References　/ 56

Chapter 3　Fluorescence and Phosphorescence

3.1　Introduction to Fluorescence　/ 57

　3.1.1　Luminescence in Nature　/ 57

　3.1.2　History of Fluorescence　/ 58

　3.1.3　Fluorescence Processes　/ 60

　3.1.4　Types of Fluorescence　/ 61

3.2　Fluorescence Spectroscopy　/ 62

　3.2.1　Absorption Spectrum　/ 63

　3.2.2　Excitation Spectrum　/ 64

　3.2.3　Emission Spectrum　/ 64

3.3　Several Rules about Fluorescence　/ 66

　3.3.1　Kasha's Rule　/ 66

　3.3.2　Mirror-image Rule　/ 69

　3.3.3　The Stokes Shift　/ 72

3.4　Fluorescence Features　/ 76

　3.4.1　Fluorescence Lifetime　/ 76

　3.4.2　Quantum Yield　/ 80

　3.4.3　Fluorescence Intensity　/ 83

　3.4.4　Fluorescence Polarization　/ 84

3.5　Factors Affecting Fluorescence　/ 88

　3.5.1　Effect of the Molecular Structure　/ 88

　3.5.2　Effect of the Environment　/ 91

3.6　Phosphorescence　/ 97

　3.6.1　Definition of Phosphorescence　/ 97

　3.6.2　Phosphorescence Features and Measurement　/ 98

　3.6.3　Low-temperature and Room-temperature Phosphorescence　/ 99

References　/ 100

Chapter 4　Supramolecules in Photophysics

4.1　Introduction to Supramolecular Chemistry　/ 102

　4.1.1　Definition of Supramolecule　/ 102

　4.1.2　Difference Between Molecular Chemistry and Supramolecular Chemistry　/ 104

　4.1.3　The Importance of Supramolecular Chemistry and Its Applications in Nature　/ 108

4.1.4 The Development of Supramolecular Chemistry / 115
4.2 Supramolecules Formation / 123
　　4.2.1 Gibbs Free Energy / 123
　　4.2.2 Enthalpy Reduction Factor / 124
　　4.2.3 Entropy Increasing Factor / 142
4.3 Typical Molecules / 150
　　4.3.1 Crown Ethers / 150
　　4.3.2 Cyclodextrin / 152
　　4.3.3 Calixarene / 155
　　4.3.4 Dendrimer / 158
　　4.3.5 Metal Organic Framework / 161
　　4.3.6 Carbon Nanotube / 164
4.4 Basic Functions of Supramolecules / 167
　　4.4.1 Molecular Recognition / 167
　　4.4.2 Molecular Transportation / 169
　　4.4.3 Chemical Reaction / 171
References / 177

Chapter 5　Fluorescence Quenching, Energy Transfer, and Electron Transfer

5.1 Fluorescence Quenching / 179
　　5.1.1 Definition and Theory / 179
　　5.1.2 Classification of Fluorescence Quenching / 181
　　5.1.3 Quencher / 183
　　5.1.4 Quenching Process / 187
5.2 Collision Process / 188
　　5.2.1 Collision with a Paramagnetic Species / 188
　　5.2.2 Collision with a Heavy Atom / 188
5.3 Excimer Formation / 189
5.4 Exciplex Formation / 192
5.5 Energy Transfer / 194
　　5.5.1 Introduction to Energy Transfer / 194
　　5.5.2 Radiative Transfer / 197
　　5.5.3 Non-radiative Transfer / 198
5.6 Electron Transfer / 205
　　5.6.1 Definition and Classification of Electron Transfer / 205
　　5.6.2 Mechanism and Characteristics of Photoinduced Electron Transfer / 207

Fundamentals of Photophysics

    5.6.3 Classification of Photoinduced Electron Transfer　/ 208

    5.6.4 Comparison Between Electron Transfer and Electron-exchange Energy Transfer　/ 210

    5.6.5 Applications of Photoinduced Electron Transfer　/ 211

References　/ 214

Chapter 6   Fluorescence Chemical Sensors

6.1 Chemical Sensors and Fluorescence Chemical Sensors　/ 216

    6.1.1 Chemical Sensors　/ 217

    6.1.2 3R and 3S　/ 218

    6.1.3 Fluorescence Chemical Sensors　/ 220

    6.1.4 Design of Fluorescence Chemical Sensors　/ 223

6.2 Recognize　/ 224

    6.2.1 Electrostatic　/ 225

    6.2.2 Hydrogen Bonding　/ 227

    6.2.3 π-π Stacking Interaction　/ 228

    6.2.4 Cation-π Interaction　/ 229

    6.2.5 van der Waals Force　/ 230

    6.2.6 Hydrophobic Effect　/ 231

6.3 Relay　/ 232

    6.3.1 Photoinduced Electron Transfer Mechanism　/ 232

    6.3.2 Photoinduced Charge Transfer Mechanism　/ 237

    6.3.3 Fluorescence Resonance Energy Transfer Mechanism　/ 242

    6.3.4 Excimer Formation Mechanism　/ 248

    6.3.5 Configuration Changes or Conformation Transition Mechanism　/ 250

    6.3.6 Excited-state Intramolecular Proton Transfer Mechanism　/ 252

6.4 Report　/ 254

    6.4.1 Polycyclic Aromatic Compounds　/ 255

    6.4.2 Intramolecular Conjugated Charge Transfer Compounds　/ 259

    6.4.3 Conjugated Polymers　/ 264

6.5 Fluorescence Chemical Sensor Application　/ 268

    6.5.1 Transition Metal Sensors　/ 268

    6.5.2 Fluorescent Sensors for Biological Analytes　/ 270

    6.5.3 Water Quality Monitoring　/ 272

References　/ 276

Chapter 7　Photophysics in Life

7.1　Photosynthesis　／277

　　7.1.1　Natural Photosynthesis　／277

　　7.1.2　Artificial Photosynthesis　／280

7.2　Phototherapy　／282

　　7.2.1　Types of Treatment　／283

　　7.2.2　Photo-controlled Delivery　／285

7.3　DNA Technology　／286

　　7.3.1　DNA Sequencing　／286

　　7.3.2　High-sensitivity DNA Stains　／289

　　7.3.3　DNA Hybridization　／291

7.4　Coffee Ring Effect　／294

　　7.4.1　Mechanism of the Coffee Ring Effect　／294

　　7.4.2　Suppression of the Coffee Ring Effect　／295

　　7.4.3　Example: Organic Crystal Growth　／297

7.5　Sunscreen　／298

　　7.5.1　Effects of Sunscreen　／298

　　7.5.2　Classification of Sunscreen　／299

　　7.5.3　Precautions for Sunscreen Use　／301

7.6　Organic Photovoltaics　／302

　　7.6.1　Device Structure of an Organic Photovoltaic　／303

　　7.6.2　Working Mechanism of Organic Photovoltaics　／304

　　7.6.3　Advantages and Disadvantages of Organic Photovoltaics　／307

　　7.6.4　The Future of Organic Photovoltaics　／308

7.7　Organic Light-emitting Diodes　／310

　　7.7.1　Device Structure of an Organic Light-emitting Diode　／310

　　7.7.2　Working Mechanism of Organic Light-emitting Diodes　／311

　　7.7.3　Advantages and Disadvantages of Organic Light-emitting Diodes　／312

　　7.7.4　The Future of Organic Light-emitting Diodes　／313

7.8　Glow Sticks　／314

　　7.8.1　Mechanism of Glow Sticks　／315

　　7.8.2　History of Glow Sticks　／316

　　7.8.3　Dangers of Opening Glow Sticks　／317

7.9　Infrared Detection　／318

　　7.9.1　Principles of Infrared Detection　／318

Fundamentals of Photophysics

   7.9.2 Applications of Infrared Detection / 320
 7.10 Bionic Navigation Sensors / 322
   7.10.1 Mechanism of Polarization Navigation / 322
   7.10.2 Structure of Compound Eyes of Insects / 324
   7.10.3 Artificial Celestial Compass / 325
References / 327

Chapter 1

Light Absorption

1.1 Introduction to Photochemistry and Photophysics

1.1.1 Photochemistry and Photophysics in Nature

Photochemical and photophysical processes have been intimately related to the development of human and his environment even before his appearance on the planet. Ever since the first morning of Creation, life has not been merely a chemical process but one in which light from the sun played an important role. Initially, the sun's rays produced organic molecules of the earth's primordial atmosphere through simple photochemical reactions. Subsequently, a sophisticated series of photosynthesis processes made it possible for simple cells to become autotrophic, provided the necessities of life, stored solar energy in the form of fossil fuels, and still supply us with practically all our food.

From the point of view of living matter, photochemistry is not only a means of harnessing light energy but also plays an important role in determining the composition of matter in interstellar space and the formation of atmospheric pollutants. Of course, photophysical processes also occur in nature. Suffice it to say that the world would not be colored if sunlight were completely absorbed or completely reflected by the objects that surround us, and we would not be able to enjoy fireflies or other beautiful scenes without bioluminescence. Typical photophysical and photochemical processes are shown in Figure 1.1.

Photophysics and photochemistry are also important from an artificial viewpoint. It's impact in the chemical, physical, biological, and medical sciences and technologies, including nanotechnology, is being felt increasing in a spectacular manner. For example, photochemical and photophysical concepts are at the basis of important applications such as protection of dyes and plastics (and also human skin) from the damaging effect of sunlight, waste water cleaning,

Fundamentals of Photophysics

design of fluorescent compounds for a variety of sensing applications (wind galleries, security, optical brighteners, pollutant detectors, display devices, molecular switches and logic gates, biological markers, cellular properties, and functions), creation of photochromic materials used in sunglasses, fashion clothes and optical memories, and development of laser devices and of light-powered molecular machines. Not only that, other interesting fields concern photomedicine, multiphotonic processes, solar-powered green synthesis, molecular photovoltaics, and solar energy conversion by water photodissociation.

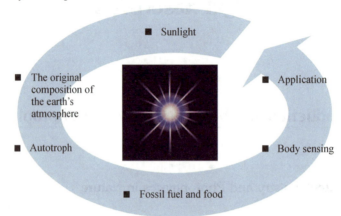

Figure 1.1　Typical photophysical and photochemical processes in nature

1.1.2　The Nature of Light

Light exhibits both wave-like and particle-like properties, a concept known as wave-particle duality (as shown in Figure 1.2). The theory was based on the research of many scientists. Physicist Robert Hooke proposed that light should be a longitudinal wave. In 1665, Hooke published his *Micrographia*, in which he described observations under the microscope of the "colors of thin blades", explained today as a phenomenon of interference and diffraction. In 1672, Newton described an experiment on the dispersion of light. In his view, the mixing and decomposition of light is like the mixing and decomposition of particles of different colors. He expounded the color theory of light with the corpuscular theory. Different from his view, Dutch physicist Christiaan Huygens is the systematizers of the wave theory of light. He inherited and refined Hooke's ideas. Huygens believed that light is a mechanical wave and light wave is a longitudinal wave propagated by material carrier. The two theoretical factions have engaged in a long dispute. In the early 20th century, Planck and Einstein proposed the quantum theory of light. In March 1905, Einstein published a paper entitled "A Speculative View on the Generation and Transformation of Light" on the German *Annals of Physics*. For instantaneous values of time, light behaves as a particle. This is the first time in history to reveal the unity of wave particles in micro-objects, that is, the duality of wave particles. This scientific theory finally gained wide acceptance in academic circles.

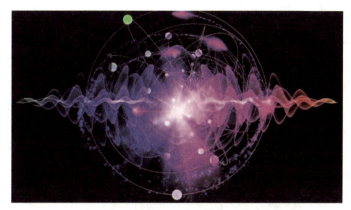

Figure 1.2 Wave-particle duality of light

In the wave model, the characteristics of electromagnetic radiation are wavelength λ, frequency ν, and velocity c.

$$\lambda \nu = c \tag{1-1}$$

The value of c is constant (2.998×10^8 m·s^{-1} in vacuum), whereas λ (and ν) may cover a wide range of values (Figure 1.3). The units for λ and ν are the meter (m) and the hertz (Hz), respectively. In some cases, the wavenumber $\bar{\nu}$ (defined as the number of waves per centimeter) is also used to characterize electromagnetic radiation. The electromagnetic spectrum encompasses a variety of types of radiation from γ-rays to radio waves, distinguished by their wavelengths (or frequencies, or wavenumbers).

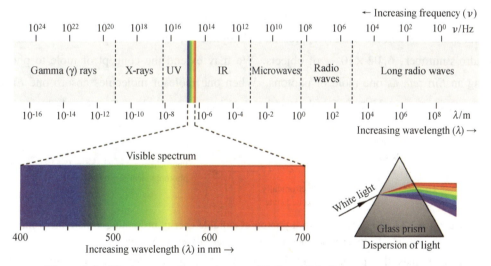

Figure 1.3 Electromagnetic spectrum, with the visible light spectrum enlarged

When dealing with photochemistry and photophysics, i.e. electronically excited states, the term "light" refers to electromagnetic radiation capable of causing electronically excited states or participating in molecular electronic deactivation which means the portion of the electromagnetic spectrum ranging from 200 to 1,000 nm (1.5×10^{15} to 3×10^{14} Hz or 5×10^4 to 1×10^4 cm^{-1}), which includes the near-UV, visible, and near infrared regions (Figure 1.3).

Fundamentals of Photophysics

In the quantum model, a beam of radiation is considered as a series of photons, or quanta. Each photon has a specific energy E, and its relation with frequency ν is:

$$E = h\nu \tag{1-2}$$

where h is the Planck's constant (6.63×10^{-34} J·s), and this equation demonstrates the important properties relating to the energy of photons. This picture of light as made up of individual photons is essential to photophysics. Photons are at the same time energy quanta and information bits. The two most important processes occurring in the biological world: photosynthesis and vision provide a living example of the two-sided nature of light. All the natural phenomena related to the interaction between light and matter and the great number of applications of photochemistry and photophysics can ultimately be traced back to these two aspects of photons.

The interaction of light with molecular systems is generally an interaction between one molecule and one photon, as shown in Figure 1.4. For a molecule A in its ground state which absorbs a photon to produce an electronically excited molecule A^*, we may write as:

$$A + h\nu \rightarrow A^* \tag{1-3}$$

where A denotes the ground state molecule, $h\nu$ the absorbed photon, and A^* the molecule in an electronically excited state. As the equation implies, the excited molecule A^* is the molecule A with an extra energy $h\nu$. From Equations 1-1 and 1-2, it follows that the photon energy is 9.95×10^{-19} J and 1.99×10^{-19} J, respectively, for light of 200 and 1,000 nm. In order to better understand the magnitude of photon energy, we must compare it with the energies of chemical bonds, which are usually expressed in kilojoules or kilocalories per mole. A mole is an Avogadro's number, 6.02×10^{23}, of objects. We may extend the concept of mole to photons, defining an *Einstein* as one mole of photons. When one mole of molecules absorb one *Einstein* of photons, it is equivalent to one photon absorbed by one molecule (Equation 1-3).

Figure 1.4 Light excitation with a photon of suitable energy promoting a molecule from its ground state to an electronically excited state

Table 1.1 lists the properties of light with different wavelengths. The energy of one *Einstein* of photons at 200 nm is 599 kJ (143 kcal), and that of one *Einstein* of photons at 1,000 nm is 119.8 kJ (28.6 kcal). These energy values are of the same order of magnitude of those required to break chemical bonds (e.g. 190 kJ·mol^{-1} for the Br-Br bond of Br_2; 416 kJ·mol^{-1} for the C-H bond of CH_4). Therefore, the energy gained by the molecule when it absorbs the photon is not negligible. For example, absorption of a 300 nm photon by naphthalene corresponds to an energy of 400 kJ · mol^{-1}, comparable with the energy that would be taken up by naphthalene [$C_{p,m}$(g) = 136 J · K^{-1} · mol^{-1}] if it were immersed in a heat bath at 3,000 K. The availability of this additional energy is the reason why an excited molecule must be considered a new chemical species. It has its own chemical and physical properties, which are usually very different from those of ground state molecules. Whether light absorption causes bond breaking as would be expected on the basis of the energy argument is another story. As we will see later, it depends on the competition among various deactivating processes.

Table 1.1 Properties of light with different wavelengths

Colour	λ/nm	ν/(10^{14} Hz)	$\bar{\nu}$/(10^4 cm^{-1})	E/(kJ · mol^{-1})
Red	700	4.3	1.4	170
Orange	620	4.8	1.6	193
Yellow	580	5.2	1.7	206
Green	530	5.7	1.9	226
Blue	470	6.4	2.1	254
Violet	420	7.1	2.4	285
Ultraviolet	<300	>10.0	>3.3	>400

1.1.3 Definition of Photophysics and Photochemistry

Photophysics and photochemistry both deal with the impact of energy in the form of photons on materials. Photochemistry focuses on the chemistry involved as a material is impacted by photons, whereas photophysics deals with physical changes that result from the impact of photons.

Many scientists have also explained photochemistry and photophysics. According to C. H. Wells, photochemistry and photophysics are the chemical and physical processes experienced by molecules that absorb ultraviolet or visible light. N. J. Turro believed that photochemistry studies the chemical behavior and physical processes of molecules in the excited states of electrons. Since electron excited states are usually formed by absorbing ultraviolet or visible light, the essence of the two statements is the same.

Basic photophysics in the framework of photobiology is concerned with processes that occur when sunlight, filtered through the earth's atmosphere, interacts with matter (atoms and molecules) present on the earth. The spectrum of solar radiation striking the earth (Figure 1.5) spans from 100 nm to 10^6 nm (1 nm = 10^{-9} m) and can be divided into the ultraviolet (UV) range (100 nm to 400 nm), visible range (400 nm to 700 nm) and infrared (IR) range (700 nm to 10^6 nm). UV radiation has both damaging and beneficial effects on living matter. UV radiation also causes photochemical reactions that create a protective ozone layer in the atmosphere. As the name suggests, the visible part of the spectrum is the light that human eyes

Fundamentals of Photophysics

can detect. An important part of electromagnetic radiation reaching the earth is IR radiation.

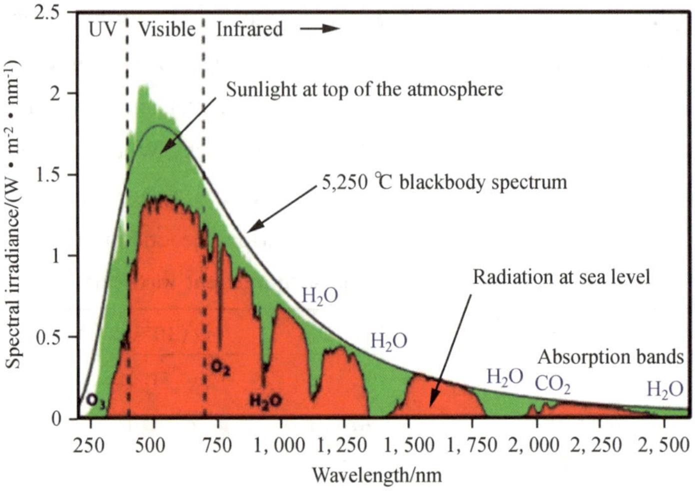

Figure 1.5 Solar radiation spectrum above the atmosphere, and at the surface of the earth

In many ways, there is a great similarity between a material's behavior when struck by photons, whether the material is small or macromolecular. Where are the differences? Differences are related to the size and the ability of polymers to transfer the effects of radiation from one site to another within the chain or macromolecular complex.

Photophysics involves the absorption, transfer, movement, and emission of electromagnetic, light, energy without chemical reactions. By comparison, photochemistry involves the interaction of electromagnetic energy that results in chemical reactions. Let's briefly review the two main types of spectroscopy about light. In absorption, a detector is placed in the direction of the incident light and the transmitted light is measured. In emission studies, the detector is placed at an angle away from the incident light, usually 90 degrees. When absorption of light occurs, the resulting polymer, P^*, contains excess energy and is said to be excited.

$$P + h\nu \rightarrow P^* \tag{1-4}$$

The light can be simply reemitted.

$$P^* \rightarrow P + h\nu \tag{1-5}$$

Bimolecular occurrences can occur, leading to an electronic relaxation called quenching. In this approach, P^* finds another molecule or part of the same chain, A, transferring the energy to A.

$$P^* + A \rightarrow P + A^* \tag{1-6}$$

Generally, the quenching molecule or site is initially in its ground state. The details of quenching will be covered in Chapter 5.

1.1.4 Development of Photophysics and Photochemistry

Figure 1.6 lists the important scientists in the history of photophysics and their contributions. A brief history of the development of photophysics and photochemistry is shown in Figure 1.7. The first investigation was made in 1777 by the Swedish chemist Carl W. Scheele, who observed that violet light was the most effective in darkening silver chloride. But it was only in 1817 that Theodor von Grotthuss established that only the light absorbed is effective in producing photochemical change. This first general principle of photochemistry passed unnoticed until 1841, when it was restated by John W. Draper and, as a consequence, is now termed the Grotthuss-Draper law. Photochemistry arose out of its empirical phase when modern physics established that light is radiated in the form of discrete quanta (called photons), the energy of which is proportional to the frequency of the light, and absorption is equivalent to the capture of photons by atoms or molecules. With this concept in mind, Johannes Stark and Albert Einstein between 1908 and 1913 independently formulated the photo equivalence law that essentially states that there should be a 1∶1 equivalence between the number of molecules decomposed and the number of photons absorbed. However, experiments showed that usually this 1∶1 ratio is not observed, indicating that the Stark-Einstein law is not sufficient to characterize a photochemical process and that absorption of a photon can be followed by other processes. This makes it possible to distinguish between the light-initiated reaction (photochemical primary process) and any subsequent chemical reactions (photochemical secondary processes). In some cases, such secondary reactions can proceed by a chain mechanism, which explains why one photon can decompose a great number of molecules. An obvious reason why the number of molecules dissociated in the photofragmentation reaction is less than the number of absorbed photons may be the efficient recombination of the primary products. It was soon realized, however, that even for other types of photoreactions, for example, photoisomerization, the number of reacted

Figure 1.6 Important scientists in the history of photochemistry and photophysics and their contributions

Fundamentals of Photophysics

molecules can be much less than the number of absorbed photons. It was thus clear that absorption of a photon is a necessary, but not sufficient condition, to cause a photoreaction and that light energy can be used by a molecule for other purposes. However, it was found, indeed, that in some cases photoexcitation does not cause any reaction but leads to emission of light, that is, a photophysical process, and that in other cases neither a chemical change nor light emission is observed.

Before the second decade of the twentieth century, an important limitation in the development of photochemistry was the lack of sufficient light sources and analytical techniques. In fact, the only light source used by the early pioneers, such as Lemoine in Paris and Ciamician in Bologna, was the sun.

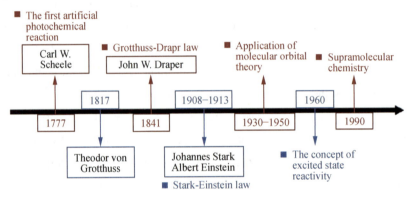

Figure 1.7　A brief history of the development of photophysics and photochemistry

After World War Ⅰ, photochemistry became a territory of physical chemists who were particularly interested in the photolysis of small molecules in the gas phase. The notion of competition among photochemical and photophysical processes for electronically excited-state decay was gradually recognized. Between 1930 and 1950, the development of molecular orbital theory allowed the interpretation of electronic absorption spectra of organic molecules and rationalized trends in a range of related molecules. Some years later, the main lines to interpret the absorption and emission spectra of metal complexes became available. Since 1960 the concepts to understand the reactivity of electronically excited states emerged and correlations between structure and photochemical reactivity or photoluminescence were developed, first for organic molecules and then for metal complexes. Within a few years, the tight link between photochemistry and photophysics was established. It is clear that photochemistry (a term that often includes photophysics) is a distinct and separate part of chemistry because it does not involve the ground state of a molecule, but rather a new species: the electronically excited state. And focused photochemical experiments, improved spectroscopic techniques, and computational methods began to provide adequate characterization of electronically excited states of several classes of molecules. Around 1990, investigations were extended to supramolecular species and photochemistry and photophysics began to play an important role in organic chemistry, as well as in novel scientific ventures such as information processing at the molecular level and creation of molecular devices and machines. In recent years, tremendous advances in technology have allowed the time windows for studying photochemical and photophysical properties of molecules to be shortened to those allowed by the uncertainty principle, and at the single-molecule level.

1.2 Light Absorption

1.2.1 Basic Concepts of Light Absorption

Light is composed of particles known as photons, each of which has the energy of Planck's quantum, hc/λ, where h is Plank's constant, c is the velocity of light, and λ is the wavelength of the radiation. Light has dualistic properties of both waves and particles; ejection of electrons from an atom as a result of light bombardment is due to the particle behavior, whereas the observed light diffraction at gratings is attributed to the wave properties. When a beam of light interacts with a molecule, some interesting photophysical processes have taken place. The different processes related to light interactions with molecules can be represented as in Figure 1.8.

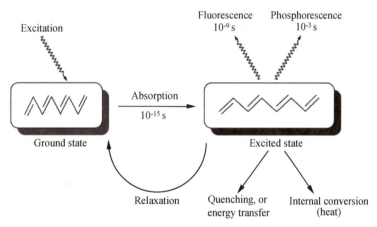

Figure 1.8 Different processes associated with light interaction with a molecule

The absorption of light by materials produces physical and chemical changes. The changes have both negative and positive aspects. On the negative side, such absorption can lead to discoloration generally as a response to unwanted changes in the material's structure. Absorption also can lead to a loss in physical properties, such as strength. In the biological world, it is responsible for a multitude of problems, including skin cancer. It is one of the chief modes of weathering by materials. Our focus here is on the positive changes effected by the absorption of light. Typical examples in semiconductor manufacturing field are photoresist, including negative-lithographic resists and positive-lithographic resists. Absorption of light has intentionally resulted in polymer cross-linking and associated insolubilization, which forms the basis for coatings and negative-lithographic resists. Light-induced chain breakage is the basis for positive-lithographic resists. Photoconductivity forms the basis for photocopying, and photovoltaic effects form the basis for solar cells being developed to harvest light energy. It is important to remember that the basic laws governing small and large molecules are the same.

Light absorption can be summarized as follows. The absorption of light is a prerequisite for it to be able to cause a chemical change. This can be viewed as another expression of the laws of energy conservation. Most molecules have all their electrons paired in their ground states, and the simplest (but not the only) effect of light absorption is the promotion of an electron from the highest occupied molecular orbital (HOMO) to the lowest unoccupied molecular orbital (LUMO) (Figure 1.9). Broadly speaking, light absorption is the process in which light is absorbed and converted into energy. When electrons absorb energy, they become "excited" and move to higher energy levels. Molecular oxygen and stable (or persistent) free radicals are exceptions of molecules that have unpaired electrons in their ground states.

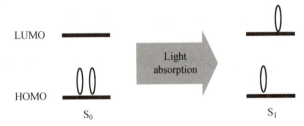

Figure 1.9 Promotion of an electron from the HOMO to the LUMO

1.2.2 Electronically Excited States

During light absorption in semiconductor, a photon having energy equal to the energy difference between two electronic states can use its energy to move an electron from the lower energy level to the upper one, producing an electronically excited state. The photon is completely destroyed in the process, and its energy then becomes part of the total energy of the absorbing species.

As illustrated in Figure 1.10, under normal conditions, the molecule is at the lowest energy level, and the electron moves in the orbital closest to the nucleus. This stationary state is called the ground state, which is the steady state of the electron. At absolute zero, all particles are in the lowest possible state of energy, that is, all particles are in the ground state. When ground state molecules are excited by external energy (such as thermal energy, light energy, etc.), their outer electrons absorb certain energy and jump to different energy states (such energy states are called the electronically excited state). Therefore, molecules may have different excited states. All photochemical reactions are chemical reactions that occur when molecules are elevated to excited states.

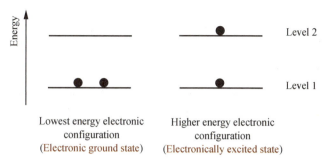

Figure 1.10 Schematics of electronic ground state and electronically excited state

Absorption of ultraviolet and visible light by molecules results in electronic transitions in which changes in both electronic and vibrational states occur. Such transitions are called vibronic transitions. At thermal equilibrium, the population of any series of energy levels obeys the Boltzmann distribution law. If N_0 molecules are in the ground state then the number N_1 in any higher energy level can be given by the equation $N_1/N_0 = \exp(-\Delta E/RT)$, where exp refers to the exponential function (ex on calculators), ΔE is the energy difference between the two energy levels, R is the gas constant (which has a value of 8.314 J·K^{-1}·mol^{-1}) and T is the absolute temperature. According to the equation above, electron distribution at the high energy level is closely related to the different energy levels when the temperature is a content.

Calculations based on the Boltzmann distribution law show that, at room temperature, most molecules will be in the $\nu = 0$ vibrational state of the electronic ground state and so absorption almost always occurs from $S_0(\nu = 0)$ as shown in Figure 1.11.

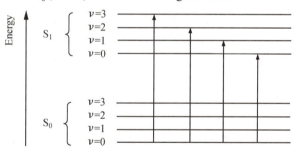

Figure 1.11 Schematic diagram of the electronic ground state and the first excited electronic state, with their associated quantized vibrational energy levels, for an organic molecule
(The vertical arrows show vibronic transitions due to the absorption of photons.)

Electronically excited states of molecules are at the heart of photochemistry, photophysics, as well as photobiology and also play a role in material science. The excited state describes an atom, ion or molecule with an electron in a higher normal energy level than its ground state. The length time of a particle spends in the excited state before falling to a lower energy state varies. Short duration excitation usually results in release of a quantum of energy, in the form of a photon or phonon.

1.2.3 Fundamental Principles of Light Absorption

Fundamental principles relating to light absorption are the basis for understanding photochemical transformations. Four fundamental principles of light absorption, the Grotthuss-Draper law, the Stark-Einstein law, the Franck-Condon principle and the Lambert-Beer law, will be discussed below.

i. Grotthuss-Draper law

The Grotthuss-Draper law (also called the principle of photochemical activation) states that only light which is absorbed by a chemical entity can bring about photochemical change.

Materials such as dyes and phosphors must be able to absorb "light" at optical frequencies. This law provides a basis for fluorescence and phosphorescence. The law was first proposed in 1871 by Theodor Grotthuss and in 1842, independently, by John William Draper.

ii. Stark-Einstein law

The Stark-Einstein law (the second law of photochemistry) is named after German-born physicists Johannes Stark and Albert Einstein, who independently formulated the law between 1908 and 1913. It is also known as the photochemical equivalence law or photo-equivalence law. If a species absorbs radiation, then one particle (molecule, ion, atom, etc.) is excited for each quantum of radiation (photon) that is absorbed. The photochemical equivalence law applies to the part of a light-induced reaction that is referred to as the primary process.

The Stark-Einstein law states that the primary act of light absorption by a molecule is a one-quantum process. For each photon absorbed only one molecule is excited. This law is obeyed in the vast majority of cases. Exceptions occur when very intense light sources such as lasers are used for irradiation of a sample. In these cases, concurrent or sequential absorption of two or more photons may occur.

iii. Franck-Condon principle

The Franck-Condon principle states that when a molecule is undergoing an electronic transition, the nuclear configuration of the molecule experiences no significant change.

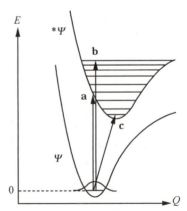

This process is illustrated in Figure 1.12. From a classical viewpoint, this principle states that no changes in nuclear position or nuclear kinetic energy occurs during the transition. This means that the transition may only be represented by a vertical arrow, the process of **a** in Figure 1.12, connecting the two potential energy surfaces. A

Figure 1.12 Schematic illustration of the Franck-Condon principle from a classical viewpoint

transition represented by arrow, the process of **b** in Figure 1.12, would imply a change in the kinetic energy, and a transition represented by arrow, the process of **c** in Figure 1.12, would entail a change of the nuclear positions.

Classically, the Franck-Condon principle is the approximation that an electronic transition is most likely to occur without changes in the position of the nuclei in the molecular entity and its environment.

Expressed in quantum mechanical terms, the Franck-Condon principle states that the most probable transitions between electronic states occur when the wave function of the initial vibrational state most closely resembles the wave function of the final vibrational state (Figure 1.13).

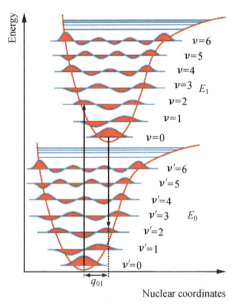

Figure 1.13　Schematic diagram of the Franck-Condon principle

iv. Lambert-Beer law

The extent of absorption of light varies a great deal from one substance to another, with the probability of absorption being indicated by the molar absorption coefficient (ε). As light is absorbed, the intensity of light entering the substance, I_{in}, is greater than the intensity of the emerging light, I_{out}. Generally, there is an exponential relationship between the relative absorption (I_{out}/I_{in}) and the concentration (c) and the path length (l) of the absorbing substance, the exponential relationship can be described as follows:

$$I_{out}/I_{in} = 10^{-\varepsilon cl} \tag{1-7}$$

In order to insight the change in the order of magnitude, we usually take logarithms to the base 10, and then we can give out the transformational relationship:

$$\lg(I_{out}/I_{in}) = -\varepsilon cl \tag{1-8}$$

Move the minus into the logarithms, and thus we can get:

Fundamentals of Photophysics

$$\lg(I_{in}/I_{out}) = \varepsilon cl \tag{1-9}$$

The left-hand-side quantity is defined as the absorbance, represented by the letter A, and the linear relationship between absorbance, molar absorption coefficient, concentration and path length is known as the Lambert-Beer law:

$$A = \varepsilon cl \tag{1-10}$$

The Lambert-Beer law can generally be applied, except where very high-intensity light beams such as lasers are used. Because, in such cases, a considerable proportion of the irradiated species will be in the excited state and not in the ground state, therefore the result of absorbance in such cases cannot represent the intrinsic characteristics of the substance.

The units of ε require some explanation here as they are generally expressed as non-SI units for historic reasons, having been used in spectroscopy for many years. Units of other variables are described as follows:

- Concentration, c, traditionally has units of moles per liter, $mol \cdot L^{-1}$.
- Path length, l, traditionally has units of centimeters, cm.
- A has no units since it is a logarithmic quantity, the value of A indicates the change of relative intensity.

According to the Lambert-Beer law:

$$\varepsilon = A/cl \tag{1-11}$$

Therefore, the units of ε usually given are: $cm^{-1} \times (mol \cdot L^{-1})^{-1} = L \cdot mol^{-1} \cdot cm^{-1}$.

For a given substance, the molar absorption coefficient varies with the wavelength of the light used. A plot of ε (or $\lg \varepsilon$) against wavelength (or wavenumber) is called the absorption spectrum of the substance just as shown in Figure 1.14. The principal use of absorption spectra from the photochemists' point of view is that they provide information as to what wavelength (λ_{max}) a compound has at its maximum value of the molar absorption coefficient (ε_{max}). Thus, irradiation of the compound at λ_{max} allows optimum photoexcitation of the compound to be carried out. In addition, if we know the molar absorption coefficient of one substance, we could design the corresponding experiments and get the relationship between absorption and concentration based on the Lambert-Beer law. Therefore, the Lambert-Beer law is frequently used in the analytical determination of concentrations from absorbance measurements.

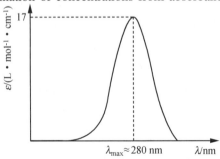

Figure 1.14 Absorption spectrum of propanone (acetone)

1.2.4 Spin Multiplicity

In quantum mechanical model, three quantum numbers are used to describe an atomic orbital: principal quantum number n, orbital angular-momentum quantum number l and magnetic quantum number m_l.

The principal quantum number, n, can have integral values of 1, 2, 3, etc. As n increases, the atomic orbital is associated with higher energy.

The orbital angular-momentum quantum number, l, defines the shape of the atomic orbital. For example, s-orbitals have a spherical boundary surface, while p-orbitals are represented by a two-lobed shaped boundary surface. l can have integral values from 0 to $(n-1)$ for each value of n. The value of l for a particular orbital is designated by the letters s, p, d and f, corresponding to l values of 0, 1, 2 and 3 respectively (Table 1.2).

Table 1.2 Values of principal and angular-momentum quantum numbers

N	l
1	0(1s)
2	0(2s)1(2p)
3	1(3s)1(3p)2(3d)
4	1(4s)1(4p)1(4d)3(4f)

The magnetic quantum number, m_l, describes the orientation of the atomic orbital in space and has integral values between -1 and $+1$ through 0 (Table 1.3).

Table 1.3 Values of angular-momentum and magnetic quantum numbers

l	Orbital	m_l	Representing
0	s	0	an s orbital
1	p	$-1, 0, +1$	3 equal-energy p orbitals
2	d	$-2, -1, 0, +1, +2$	5 equal-energy d orbitals
3	f	$-3, -2, -1, 0, +1, +2, +3$	7 equal-energy f orbitals

In order to understand how electrons of many-electron atoms arrange themselves into the available orbitals, the fourth quantum number called spin quantum number m_s, is necessary to be defined. It can have two possible values, $+1/2$ or $-1/2$. These are interpreted as indicating the two opposite directions in which the electron can spin, up ($\uparrow$) and down ($\downarrow$).

The total spin, S, of a number of electrons can be determined simply as the sum of the spin quantum numbers of the electrons involved and a state can be specified by its spin multiplicity. Spin multiplicity is defined as: $2S+1$. According to the different of S, the spin of the system could be described to singlet state ($S=0$), triplet state ($S=1$) and double state ($S=1/2$) (Figure 1.15).

Fundamentals of Photophysics

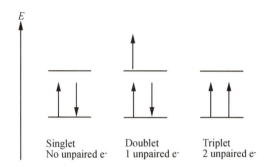

Figure 1.15 Comparison of singlet state, doublet state and triplet state

- Singlet state: When the spins of a pair of electrons are opposite, the net spin S is 0, and $2S+1=1$. Its energy level is not affected by the external magnetic field to produce fission. It is a diamagnetic species, and most of the organic molecules are resistant to magnetic. They have a singlet ground state.

- Triplet state: When the spins of a pair of electrons are the same, the net spin S is 1, $2S+1=3$. Triplet states are more common in the excited states of luminescent dyes. Species with triplet ground states are relatively rare, except for oxygen. There are some diradical molecules with triplet ground states.

- Double state: If there is an unpaired electron with a net spin of $1/2$, $2S+1=2$. The most typical example of a double state is an organic single radical.

Take the spin multiplicity of helium atoms for example. In the lowest excited-state helium atom there are two possible spin configurations:

- The two electrons have opposite spins: Total spin $S = 1/2 - 1/2 = 0$. Spin multiplicity = $2S+1=1$. A species such as this is referred to as an excited singlet state.

- The two electrons have parallel spins: Total spin $S = 1/2 + 1/2 = 1$. Spin multiplicity = $2S+1=3$. In this case, the species is referred to as an excited triplet state.

We can discuss the terms singlet state and triplet state as facts about minute systems such as atoms. The key difference between singlet state and triplet state is that singlet state shows only one spectral line whereas triplet state shows the threefold splitting of spectral lines.

1.2.5 Selection Rules for Light Absorption

In molecular systems, certain conditions, called selection rules, need to be met before light absorption can occur. There are two major selection rules for absorption transition: spin-forbidden transitions and symmetry-forbidden transitions.

Spin-forbidden transitions mean transitions between states of different multiplicities are forbidden, i.e. singlet→singlet and triplet→triplet transitions are allowed, but singlet→triplet and triplet→singlet transitions are forbidden. However, there is always a weak interaction between the wave functions of different multiplicities via spin-orbit coupling.

Symmetry-forbidden transitions mean a transition can be forbidden for symmetry reasons. It is important to note that a symmetry-forbidden transition can be observed because the molecular vibrations cause some departure from perfect symmetry (vibronic coupling).

i. Spin-forbidden transitions

The effect of electron spin on transition intensities is given by the factor in the transition moment expression. When the transition occurs with no changes in multiplicity, singlet→singlet and triplet→triplet transitions are spin allowed. As the ground state of molecules is usually a singlet, singlet→singlet transitions are very common. In fact, they account for most of the absorption bands observed for molecules that have a singlet ground state. Another transition, triplet→triplet transitions are very important in transient spectroscopy, when a pulse of light generates triplet excited states that can be examined by absorption spectroscopy before they decay.

Because of the orthogonality of spin wave functions, the integral vanishes whenever the initial and final states have different spin multiplicity, and the corresponding electronic transitions are called spin forbidden. Consequently, singlet→triplet transitions are spin forbidden.

This spin selection rule is valid to the extent to which spin and orbital functions can be separated rigorously. Departures from this approximation can be dealt with in terms of a perturbation called spin-orbit coupling, by which states of different spin multiplicity can be mixed. This perturbation increases as the fourth power of the atomic number of the atoms involved. Thus, spin-forbidden transitions of typical organic molecules are actually almost unobservable ($\varepsilon_{max} < 1$ L·mol^{-1}·cm^{-1}). However, spin-forbidden transitions of metal complexes can reach quite sizable intensities (e.g. 10^2 L·mol^{-1}·cm^{-1} for 5d metal complexes). The spin selection rule plays a fundamental role for molecules that do not contain heavy atoms which could break the rule of spin-forbidden transitions from internal or external, in other words, for the vast majority of organic molecules whose absorption spectrum is dominated by spin-allowed transitions.

ii. Symmetry-forbidden transitions

According to the quantum theory, the intensity of absorption by molecules is explained by considering the wave functions of the initial and final states (ψ and ψ^*, respectively). An electronic transition will proceed most rapidly when ψ and ψ^* most closely resemble each other; that is, when the coupling between the initial and final states is the strongest. Since, when the rate of absorption is the greatest, the molar absorption coefficient ε is the greatest when the electron transition is most probable. The greatest values for ε also occur when the wave functions ψ and ψ^* most closely resemble each other. Compared to the $\pi\rightarrow\pi^*$ transition, the $n\rightarrow\pi^*$ transition is a weak absorption, which is a consequence of the orbital symmetry selection rule. Transitions involving a large change in the region of space the electron occupies are forbidden. The orbital overlap between the ground state and the excited state should be as large as possible for an allowed transition. π- and π^*-orbitals occupy the same regions of space, so overlap between them is large (Figure 1.16). The orbital overlap between n- and π^*-orbitals is very much smaller, as these orbitals lie perpendicular to each other (Figure 1.17).

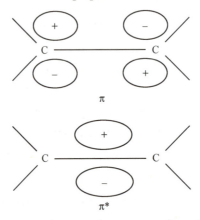

Figure 1.16 π and π^* molecule orbitals associated with the $>C=C<$ chromophore
(Both the π- and π^*-orbitals lie in the plane of the paper.)

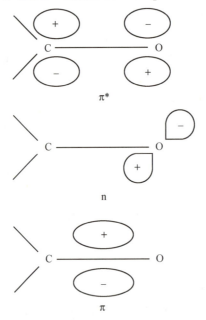

Figure 1.17 Molecular orbitals associated with the $>C=O$ chromophore (The π- and π^*-orbitals lie in the plane of the paper but the n-orbitals are perpendicular to the plane of the paper.)

The phasing of the molecular orbitals (shown as +/−) is a result of the wave functions describing the orbitals, where + shows that the wave function is positive in a particular region in space and − shows that the wave function is negative.

Hence, according to the symmetry selection rule, $\pi \rightarrow \pi^*$ transitions are allowed while $n \rightarrow \pi^*$ transitions are forbidden. However, in practice the $n \rightarrow \pi^*$ transition is weakly allowed due to the coupling of vibrational and electronic motions in the molecule, which is called vibronic coupling, and is a result of the breakdown of the Born-Oppenheimer approximation.

1.2.6 Absorption Spectrum

Absorption spectrum is an electromagnetic spectrum in which a decrease in intensity of radiation at specific wavelengths or ranges of wavelengths characteristic of an absorbing substance is manifested especially as a pattern of dark lines or bands. The measurement of an absorption spectrum is based on the Lambert-Beer law, and shows the ability of the investigated sample to absorb light at different wavelengths. The measurement process of absorption spectrum in detail is displayed in schematic diagram shown in Figure 1.18.

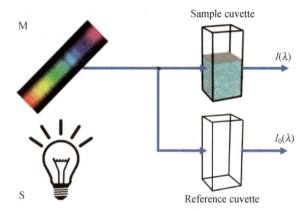

Figure 1.18 Measurement principle of light absorption spectra

As light absorption occurs almost always from the lowest vibrational level of the electronic ground state, the absorption spectrum characterizes the energetic structures of the electronic excited states of an aromatic molecule.

The UV-visible spectrum of a very dilute solution of anthracene in benzene, which clearly shows small fingers superimposed on a broader band (or envelop). These fingers are called the vibrational fine structure and we can see that each finger corresponds to a transition from the $v = 0$ of the ground electronic state to the $v = 0, 1, 2, 3$, etc. vibrational level of the electronically excited state. The absorption spectra of rigid hydrocarbons in nonpolar solvents may show vibrational fine structure (Figure 1.19). The spectrum shows that many vibronic transitions are allowed, and the intensities of the different vibronic transitions vary.

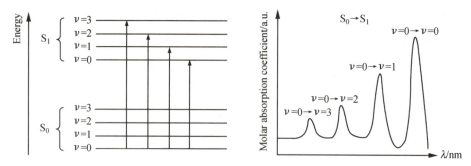

Figure 1.19 Diagram of fine structure of light absorption

Absorption spectra of other organic molecules in solution tend to be broad, featureless bands with little or no vibrational structure (Figure 1.20). This is due to the very large number of vibrational levels in organic molecules and to blurring of any fine structure due to interaction between organic molecules and solvent molecules. The hypothetical spectrum in Figure 1.19 shows the vibrational structure hidden by the enveloping absorption spectrum, and the peak of the absorption curve does not correspond to the 0 – 0 band because the most probable vibronic transition here is the 0 – 4 transition.

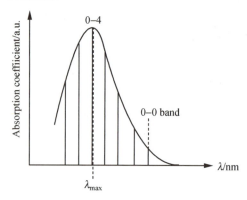

Figure 1.20 Broad, featureless absorption spectrum of the solution of an organic compound.

References

[1] Fabio, W. O. Da Silva. The evolution of the wave theory of light and textbooks [J]. *Rev. Bras. Ensino. Fis.*, 2007, 29(1): 149 – 159.

[2] Westermayr, J. & Marquetand, P. Machine Learning for electronically excited states of molecules [J]. *Chem. Rev.*, 2021, 121(16): 9873 – 9926.

[3] Liu, T. & Sullivan, J. P. *Pressure and Temperature Sensitive Paints* [M]. 2nd ed. Heidelberg: Springer, 2021: 15 – 31.

[4] Suppan, P. *Chemistry and Light* [M]. Cambridge: The Royal Society of Chemistry,

1994: 11 – 26.

[5] Jaffe, H. H. & Orchin, M. *Theory and Applications of Ultraviolet Spectroscopy* [M]. Hoboken: John Wiley & Sons, Inc., 1962: 85 – 93.

[6] Balzani, V., Credi, A. & Venturi, M. *Molecular Devices and Machines: Concepts and Perspectives for the Nanoworld* [M]. 2nd ed. Weinheim: Wiley-VCH, 2008: 1 – 21.

[7] Vincenzo, B., Paola, C. & Alberto J. *Photochemistry and Photophysics: Concepts, Research, Applications* [M]. Weinheim: Wiley-VCH, 2014: 1 – 10.

[8] Liliana, M. *Green Chemistry and Computational Chemistry: Shared Lessons in Sustainability* [M]. Amsterdam: Elsevier, 2021: 1 – 39.

[9] Turro, N. J., Ramamurthy, V. & Scaiano, J. C. *Modern Molecular Photochemistry of Organic Molecules* [M]. Sausalito: University Science Books, 2010: 3 – 38.

[10] Willock, D. J. *Molecular Symmetry* [M]. Chichester: John Wiley & Sons, Ltd, 2009: 6 – 18.

Chapter 2

Deactivation of Excited States

In this chapter, we shall focus on an overview of the physical deactivation processes relating to organic molecules, along with a simple kinetic analysis of these processes. Physical deactivation processes may be classified as:

1. Intramolecular processes

• Radiative transitions, which involve the emission of electromagnetic radiation as the excited molecule relaxes to the ground state. Fluorescence and phosphorescence are known collectively as luminescence, and they will be described in detail in Chapter 3.

• Non-radiative transitions, where no electromagnetic radiation emits during the deactivation process.

2. Intermolecular processes

• Vibrational relaxation, where molecules with excess vibrational energy undergo rapid collision between themselves and solvent molecules to make it in the lowest vibrational level of a particular electronic energy level.

• Energy transfer, where the electronically excited state of the donor molecule is deactivated to a lower electronic state by transferring energy to the acceptor molecule, and then the acceptor molecule is promoted to a higher electronic state. In this process, the acceptor is also called quencher and the donor is called sensitizer.

• Electron transfer, considered as a photophysical process, involves donor molecule with an excited state interacting with a ground state acceptor molecule. In this process, an ion pair is formed, which may undergo back electron transfer, resulting in quenching of the excited donor.

2.1 Brief Introduction to Deactivation of Excited States

The excited state describes an atom, ion or molecule with an electron in a higher than normal energy level than its ground state. Electronically excited states of molecules are endowed with excess energy due to their formation by photon absorption. These short-lived excited states will lose their excess energy within a very short period of time through a variety of deactivation processes (Figure 2.1) and return to a ground state configuration. In this regard, if the excited molecule returns to its original ground state, then the dissipative process is a physical process. However, if a new molecular species is formed, then the dissipative process is accompanied by a chemical change.

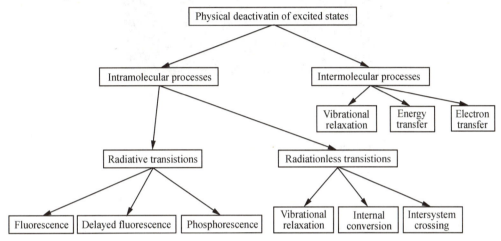

Figure 2.1　Physical deactivation of excited states of organic molecules

The time that a particle spends in the excited state before falling to a lower energy state is different. Short duration excitation usually results in form of a photon or phonon due to release of a quantum of energy. This process of return to a lower energy state is called decay. Fluorescence is a fast decay process, while phosphorescence needs a much longer decay time. In other words, decay is the inverse process of excitation. Besides, an excited state that lasts a long time is called a metastable state. Singlet oxygen and nuclear isomers are typical examples of metastable states. Singlet oxygen is produced by the reaction that occurs between a photosensitizer molecule and ground state oxygen. The photosensitizer molecular triplet states are chemically reactive due to their long decay times and the presence of unpaired valence electrons. And, reactivity with ground state oxygen ($^3O_2$) will generate singlet oxygen ($^1O_2$) and ground state photosensitizer (Figure 2.2). Nuclear isomers are atoms with the same mass number and atomic number, but with different states of excitation in the atomic nucleus (Figure 2.3). The metastable states will be formed due to the need for a larger nuclear spin change in order for them to return to the ground state.

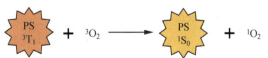

Figure 2.2　A photosensitizer molecular triplet state ($PS\text{-}^3T_1$) reacting with ground state oxygen ($^3O_2$) to generate ground state photosensitizer ($PS\text{-}^1S_0$) and singlet oxygen ($^1O_2$)

Figure 2.3　Schematic diagram of nuclear isomers

2.2　The Jablonski Diagram

Electronically excited states have high-energy content and therefore must undergo deactivation within a short period of time. Deactivation can take place through states of the original molecule leading back to the ground state (photophysical processes) or with formation of other species (photochemical processes). In reality, the extensively used distinction between photophysical and photochemical processes has a weak meaning because an electronically excited state, owing to its peculiar physical and chemical properties, may already be considered as a new chemical species compared with the ground state. Furthermore, light excitation can transform a molecule into an isomer, which is at the same time an excited state of the original molecule and, from a chemical viewpoint, a different chemical species.

The Jablonski diagram is named after Aleksander Jabłoński (left of Figure 2.4), a Polish physicist, who is regarded as the father of fluorescence spectroscopy due to his outstanding contributions. In his work "On the Influence of the Change of Wavelengths of Excitation Light on the Fluorescence Spectra", he experimentally demonstrated that the fluorescence spectrum is independent to the wavelength of the excitation light. Besides, the other most notable contributions to fluorescence spectroscopy were furthering the understanding of the theory of fluorescence polarization in solutions; the concept of concentration quenching; and the development of the famous diagram, which explains the spectra and kinetics of fluorescence, delayed fluorescence and phosphorescence.

Then, the diagram was further corrected by French physicists Jean Baptist Perrin and his son Francis Perrin (center and right of Figure 2.4), who both contributed greatly to the theory of fluorescence in the 1920s and 1930s. And the more correctly diagram called the Perrin-Jablonski diagram recognizes their important contributions. Jean Baptiste Perrin introduced the concept of resonant energy transfer between molecules and developed a theory to explain thermally activated delayed fluorescence. Francis Perrin developed the active sphere model for quenching, established the relationship between fluorescence quantum yield and lifetime and the theory for fluorescence polarization.

Figure 2.4   Aleksander Jabłoński (left), 1898 – 1980, Jean Baptiste Perrin (center), 1870 – 1942, and Francis Perrin (right), 1901 – 1992

A Jablonski diagram is an energy on a vertical axis. Energy levels in most of these diagrams are schematically diagrams, though energy levels can be quantitatively denoted. The rest of the diagrams are established by columns. Every column usually acts for a specific spin multiplicity for a particular species. However, some diagrams divide energy levels within the same spin multiplicity into different columns. Within each column, horizontal lines correspond to eigenstates for that particular molecule. Bold horizontal lines stand for the limits of electronic energy states. Within each electronic energy state are multiple vibrational energy states. They may be coupled with the electronic state. Usually only a portion of these vibrational eigenstates are exhibited, because there are a large number of possible vibrations in a molecule. Each of these vibrational energy states can be further segmented into rotational energy levels. However, typical Jablonski diagrams omit such intense levels of detail.

A photon with a particular energy is absorbed by the molecule of interest as the first transition in most Jablonski diagrams. This is showed by an upward straight arrow (Figure 2.5). One method that can make an electron to excite from a lower energy level to a higher energy level is by absorbance. Then, the energy of the photon is transferred to a particular electron. That electron then transitions to a different eigenstate corresponding to the amount of energy transferred. Only certain wavelengths of light are possible for absorbance, that is, wavelengths that have energies that correspond to the energy difference between two different eigenstates of the particular molecule. Absorbance is a very fast transition, on the order of 10^{-15} seconds. However, most Jablonski diagrams do not exhibit a time scale for the phenomenon being indicated. This transition will usually occur from the lowest (ground) electronic state due to the statistical mechanical issue of most electrons occupying a low-lying state at reasonable temperatures. There is a Boltzmann distribution of electrons within this low-lying level, based on the energy available to the molecules. This energy available is a function of the Boltzmann's constant and the temperature of the system. These low-lying electrons will transition to an excited electronic state as well as some excited vibrational state.

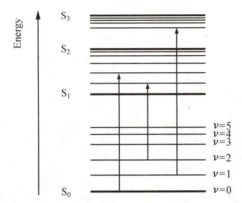

Figure 2.5 Three possible absorption transitions represented

Besides, the properties of excited states and their relaxation processes are conveniently represented by a Jablonski diagram (Figure 2.6).

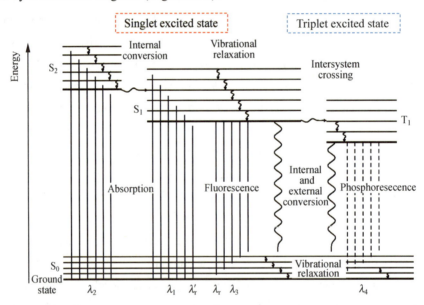

Figure 2.6 The Jablonski diagram for an organic molecule, illustrating excited state photophysical processes

The Jablonski diagram shows:

- The electronic states of the molecule and their relative energies.

Singlet electronic states are denoted by S_0, S_1, S_2, etc. and triplet electronic states as T_1, T_2, etc.

- Vibrational levels associated with each state are denoted as $v=0$, $v=1$, $v=2$, etc. in order of increasing energy.

- Radiative and non-radiative transitions are drawn as straight arrows and wavy arrows, respectively.

- If an electronically excited state is formed as a "vibrationally-hot" excited molecule (with $v>0$), then it will undergo vibrational relaxation within that electronic energy level until it reaches the $v=0$ level. The vibrational relaxation within each electronically excited state is drawn as a vertical wavy arrow.
- Non-radiative transitions (internal conversion and intersystem crossing) between electronic states are isoenergetic processes and are drawn as wavy arrows from the $v=0$ level of the initial state to a "vibrationally-hot" ($v>0$) level of the final state.

The photophysical processes are summarized in Table 2.1.

Table 2.1 Summary of the photophysical processes shown in Figure 2.6

Relaxation process	Details
Vibrational relaxation	Involves transitions between a vibrationally excited state and the $v=0$ state within a given electronic state when excited molecules collide with other species such as solvent molecules, e.g. $S_2(v=3) \leadsto S_2(v=0)$. The excess vibrational energy is dissipated as heat.
Internal conversion	Involves non-radiative transitions between vibronic states of the same total energy (isoenergetic states) and the same multiplicity. Internal conversion between excited states, e.g. $S_2 \leadsto S_1$ is much faster than internal conversion between S_1 and S_0.
Intersystem crossing	Intramolecular spin-forbidden non-radiative transitions between isoenergetic states of different multiplicity, e.g. $S_1 \leadsto T_1$.
Fluorescence	Photon emission. Fluorescence involves a radiative transition between states of the same multiplicity (spin allowed), usually from the lowest vibrational level of the lowest excited singlet state, S_1. $S_1(v=0) \rightarrow S_0 + hv$.
Phosphorescence	Photon emission. Phosphorescence involves a spin-forbidden radiative transition between states of different multiplicity, usually from the lowest vibrational level of the lowest excited triplet state, T_1. $T_1(v=0) \rightarrow S_0 + hv$.

The Jablonski diagram involves energy changes in molecules and the time of these processes. In this regard, the following points should be considered.

Energy:

- The energy difference between the ground and the lowest excited state is usually much higher than the energy difference between two successive excited states.
- Singlet and triplet excited states with the same subscripts (i.e. S_1 and T_1) usually belong to the same excited configuration and their energy separation may be small or large, which depends on the repulsive strength between electrons.
- According to the Franck-Condon principle and excited state distortion, light absorption in molecules usually generates the excited state with a high vibrational level. For the same reason, light emission also generates the ground state with a high vibrational level.

Time:

Fundamentals of Photophysics

• Light absorption takes place in a very short time scale compared with the time scale of nuclear movements.

• The rate constants in different deactivation processes are determined by the particular pathway.

It should be noted that the Jablonski diagram shows what sorts of transitions that can possibly happen in a particular molecule. The time scales of each transition dictate each of these possibilities. The faster the transition, the more likely it is to happen as determined by selection rules. Therefore, understanding the time scales of each process is essential to understanding if the process may happen. The average time scales for basic radiative and non-radiative processes are shown in the table as following:

Table 2.2 Average timescales for radiative and non-radiative processes

Process	Transition	Timescale/s
Light absorption (Excitation)	$S_0 \to S_n$	ca. 10^{-15} (instantaneous)
Internal conversion	$S_n \to S_1$	10^{-14} to 10^{-11}
Vibrational relaxation	$S_n^* \to S_n$	10^{-12} to 10^{-10}
Intersystem crossing	$S_1 \to T_1$	10^{-11} to 10^{-6}
Fluorescence	$S_1 \to S_0$	10^{-9} to 10^{-6}
Phosphorescence	$T_1 \to S_0$	10^{-3} to 100
Non-radiative decay	$S_1 \to S_0$	10^{-7} to 10^{-5}
	$T_1 \to S_0$	10^{-3} to 100

Each process listed above can be combined with a single Jablonski diagram for a particular molecule. It gives an overall picture of possible results of perturbation of a molecule by light energy. Jablonski diagrams are used to easily visualize the complex inner workings of how electrons change eigenstates in different conditions. Through this simple model, specific quantum mechanical phenomena are easily communicated.

2.3 Intramolecular Processes

2.3.1 Non-radiative Transitions

Non-radiative transitions occur between isoenergetic vibrational levels of different electronic states. As the total energy of the system does not change, there is no photon emitted and the process is represented by a horizontal wavy line in the Jablonski diagram, although downward wavy arrows are usually used to indicate it. Non-radiative transitions are essentially irreversible processes because they are related to an entropy increase (higher density of vibrational levels in the lower energy excited state) and the vibrational relaxation within the lower excited state is very fast.

i. Internal conversion (IC)

In case that an electron is excited, there are many ways that energy may be dissipated. One of them is vibrational relaxation, a non-radiative process. This is labeled on the Jablonski diagram as a curved arrow between vibrational levels (blue arrow in Figure 2.7). However, if vibrational energy levels overlap electronic energy levels intensively, the condition could happen that the excited electron can transition from a vibration level in one electronic state to another vibration level in a lower electronic state. This process is called internal conversion and is mechanistically the same as vibrational relaxation. It is also marked as a curved line on a Jablonski diagram, between two vibrational levels in different electronic states (green arrow in Figure 2.7). Internal conversion appears due to the overlap of vibrational and electronic energy states. Along with the energy increase, the manifold of vibrational and electronic eigenstates becomes ever closer distributed. When the energy levels are higher than the first excited state, the manifold of vibrational energy levels overlaps with the electronic levels strongly. This overlap gives a larger degree of probability that the electron can transition between vibrational levels that will lower the electronic state. Internal conversion occurs in the same time along with vibrational relaxation, therefore, it is a very possible way for molecules to dissipate energy from light perturbation. However, because of a lack of vibrational and electronic energy state overlap and a large energy difference between the ground state and the first excited state, internal conversion is very slow for an electron to return to the ground state. This slow return to the ground state allows other transitive processes to compete with internal conversion at the first electronically excited state. Both vibrational relaxation and internal conversion occur in most perturbations, yet are seldom the final transition.

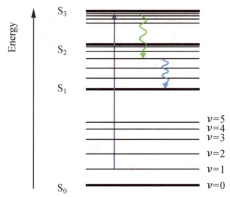

Figure 2.7 Possible scenario with absorption (purple arrow), internal conversion (green arrow), and vibrational relaxation (blue arrow) processes shown

Internal conversion refers to intramolecular non-radiative transitions between vibronic states of the same total energy (isoenergetic states) and the same multiplicity. Relaxation from an upper excited electronic state such as S_2, S_3, etc. to a lower electronic excited state with the same multiplicity happens quickly by the non-radiative process of internal conversion. Because the difference in energy of these upper excited states is relatively small, there is a high probability of the $v=0$ level of, say, S_2 being very close in energy to a high vibrational level of S_1, allowing rapid energy transfer between the two electronic levels to occur. Due to the rapid rate of internal conversion between excited states, other radiative and non-radiative transitions do not universally occur from upper electronically excited states as they are unable to compete with internal conversion.

Internal conversion involves intramolecular non-radiative transitions between vibronic states of the same total energy (isoenergetic states) and the same multiplicity, for instance, $S_2(v=0)$ ⇝ $S_1(v=n)$ and $T_2(v=0)$ ⇝ $T_1(v=n)$. Representative timescales are of the order of 10^{-14} – 10^{-11} s (internal conversion between excited states) and 10^{-9} – 10^{-7} s (internal conversion between S_1 and S_0).

The fact that there is a much larger energy difference existing between S_1 and S_0 than that between any successive excited states means internal conversion between S_1 and S_0 occurs more slowly than that between excited states in general. Thus, whichever upper excited state is initially created by photon absorption, prompt internal conversion and vibrational relaxation processes mean that the excited-state molecule rapidly relaxes to the $S_1(v=0)$ state from which fluorescence and intersystem crossing compete effectively with internal conversion from S_1. This is the basis of Kasha's rule, which states that because of the very rapid rate of deactivation to the lowest vibrational level of S_1 (or T_1), luminescence emission and chemical reaction by excited molecules will always stem from the lowest vibrational level of S_1 or T_1.

The distinction between singlet and triplet states is because photon induced excitation always results in a state of the same multiplicity, i.e. singlet to singlet or triplet to triplet. Since most ground states are singlets, it is indicated that the excited states, initially formed by absorption of light, must also be singlets. Internal conversion of excited states to reduce energy states of the same multiplicity happens quickly with loss of heat energy (relaxation). Alternatively, an excited state may get back to the ground state by emitting a photon (radiative decay). In the study of acetone, about 80% of the excited singlet states lose energy by internal conversion and approximately 3% by fluorescence. Conversion of a singlet state to a lower energy triplet state, or vice versa, is called intersystem crossing and is slower than internal conversion. Radiative decay from a triplet state is called phosphorescence and is universally, enormously slow. The non-radiative decay recorded in the last row may occur through transferring intermolecular energy to a different molecule. This collisional process is named quenching if the focus is on the initially excited species, or sensitization if the newly produced excited state is of interest. Photochemical sensitization popularly appears by a $T_1 + S_0 \rightarrow S_0 + T_1$ reaction. The new triplet excited state may then undergo characteristic reactions of its own.

Alkene isomerization is a case of photochemical reaction involving the internal conversion. A photochemical reaction takes place when the internal conversion and relaxation of an excited state causes a ground state isomer of the initial substrate molecule, or when an excited state undergoes an intermolecular addition to another reactant molecule in the ground state. The cis-trans photochemical isomerization of stilbene is a reaction of the first kind, as exhibited in (Figure 2.8). Both cis- and trans-stilbene undergo $\pi \rightarrow \pi^*$ electron excitation by absorption of UV light. However, isolated double bonds need light with 180 nm wavelength for such excitation, and conjugation with the phenyl substituents lowers the transition energy to about 300 nm, a more easily achieved source. The molar absorptivity of the cis-isomer is less than that of the trans-isomer because steric crowding of the ortho sites leads to the phenyl groups to twist slightly out of coplanarity.

The stability of the stereoisomers of stilbene originates from a 62 kcal·mol^{-1} barrier to rotation about the double bond produced by the π-bond. This bonding is absent in the $\pi \rightarrow \pi^*$ excited state (magenta curve in Figure 2.8). Both the initial S_1 states formed from the cis and trans- ground states are slightly twisted (the cis by 25° & the trans by 13°) with the C = C double bond being lengthened by about 4.5%. These local S_1 states rapidly relax to a general lower energy twisted configuration ($\theta \cong 90°$). Non-radiative internal conversion of this S_1 twisted state results in the transition state region of S_0, which decays equally to the ground states of the cis- and trans-isomers. This simple configurational isomerization about a single double bond is referred to as a One-Bond-Flip (OBF) event.

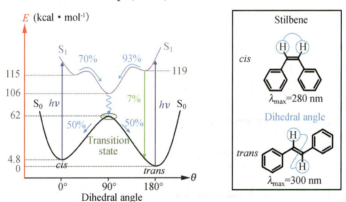

Figure 2.8 Schematic illustration of alkene isomerization (Internal conversion of the S_1 twisted state leads to the transition state region of S_0.)

A small proportion (6%) of the trans-S_1 state fluoresces back to the trans-isomer, but there is less than 0.1% fluorescence from the cis-S_1 state. Triplet states (not shown) may also be formed, but there is no phosphorescence observation in solution, and non-radiative decay from such states results in similar cis-trans isomerization. In viscous solvents, the quantum yield of trans- to cis-isomerization decreases, accompanied by an increase in fluorescent decay. Freezing the solutions to a rigid glass could prevent isomerization and maximize fluorescence efficiency.

Inspection of the cis-S_1 local minimum near $\theta = 0°$ shows that only 70% of the molecules in this state relax to the lower energy twisted S_1 state. The remaining 30% undergo an electrocyclic rearrangement to an isomeric 4a, 4b-dihydrophenanthrene (DHP) S_1 state, as illustrated on the diagram. Molecules occupying this new excited state then relax to either DHP or cis-stilbene ground states. Over time, DHP accumulates and may be converted to phenanthrene by mild oxidation with air or iodine. Therefore, substituted phenanthrenes is available from trans-stilbene precursors prepared by Wittig or Grignard procedures.

ii. Intersystem crossing

Another non-radiative process that can take place is known as intersystem crossing (ISC) from a singlet to a triplet or a triplet to a singlet state. Intersystem crossing involves intramolecular spin-forbidden non-radiative transitions between isoenergetic states of different multiplicity (Figure 2.9). This process is very rapid for metal-containing compounds. This process can take place on a time scale of $10^{-8} - 10^{-6}$ s for an organic molecule, while for organometallics it is about 10^{-11} s. The enhancement of rate is due to spin-orbit coupling present in the metal-containing systems, which is an interaction between the spin angular momentum and the orbital angular momentum, which permits mixing of the spin angular momentum with the orbital angular momentum of S_n and T_n states. Thus, these singlet and triplet states are no longer "pure" singlets and triplets, and the transition from one state to the other is less forbidden by multiplicity rules. With increasing of atomic number, a rate increase in intersystem crossing can also be achieved by the heavy atom effect, arising from an increased mixing of spin and orbital quantum number. This is accomplished through the introduction of heavy atoms into the molecule via chemical bonding or with the solvent, corresponding to internal heavy atom effect and external heavy atom effect, respectively. The spin-orbit interaction energy of atoms grows with the fourth power of the atomic number Z.

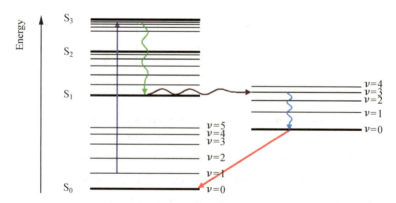

Figure 2.9　The Jablonski diagram of intersystem crossing (brown arrow)

Crossing between states of different multiplicity is in principle forbidden, but for some molecules spin-orbit coupling can be large enough to make it possible. The probability of intersystem crossing depends on the singlet and triplet states involved.

- If the electronic transition $S_0 \rightarrow S_1$ is $n \rightarrow \pi^*$ type, intersystem crossing is often efficient.
- The presence of heavy atoms increases spin-orbit coupling and thus favors intersystem crossing.

When a singlet state non-radiatively passes to a triplet state, or conversely a triplet transitions to a singlet, that process is known as intersystem crossing. In essence, the spin of the excited electron is reversed. This process is more likely to occur when the vibrational levels of the two excited states overlap, since little or no energy must be gained or lost in the transition. As the spin/orbital interactions in such molecules are substantial and a change in spin is thus more favorable, intersystem crossing is most common in heavy-atom molecules (e. g. those containing iodine or bromine). This process is called "spin-orbit coupling". Simply-stated, it involves coupling of the electron spin with the orbital angular momentum of non-circular orbits. In addition, the presence of paramagnetic species in solution enhances intersystem crossing.

Once a metal complex undergoes metal-to-ligand charge transfer, the system can undergo intersystem crossing, which, in conjunction with the tunability of metal-to-ligand charge transfer (MLCT) excitation energies, produces a long-lived intermediate. The energy of these intermediates can be adjusted by altering the ligands used in the complex. Another species can then react with the long-lived excited state via oxidation or reduction, thereby initiating a redox pathway via tunable photoexcitation. Complexes containing high atomic number d^6 metal centers, such as Ru(II) and Ir(III), are commonly used for such applications. The reason is that they favor intersystem crossing as a result of their more intense spin-orbit coupling.

Apart from the singlet and triplet states, complexes that have access to d orbitals are able to access spin multiplicities, as some complexes have orbitals of similar or degenerate energies so that it is energetically favorable for unpaired electrons. It is possible then for a single complex to undergo multiple intersystem crossings, which is the case in light-induced excited spin-state trapping (LIESST), where, at low temperatures, a low-spin complex can be irradiated and undergo two instances of intersystem crossing. For Fe(II) complexes, the first intersystem crossing occurs from the singlet to the triplet state, then followed by intersystem crossing between the triplet and the quintet state. At low temperatures, the low-spin state is favored. However, the quintet state is unable to relax back to the low-spin ground state due to their differences in zero-point energy and metal-ligand bond length. The reverse process is also possible for cases such as $[Fe(ptz)_6](BF_4)_2$, but the singlet state is not fully regenerated. The reason is that the energy needed to excite the quintet ground state to the necessary excited state to undergo intersystem crossing to the triplet state overlaps with multiple bands corresponding to excitations of the singlet state that lead back to the quintet state.

An example of intersystem crossing is the molecular transition in Fe(bpy)$_3^{2+}$ from low spin (LS) to high spin (HS) state on the Fe-N distance reaction coordinate (Figure 2.10). After optical excitation to the MLCT manifold, a decay towards the HS state occurs (120 fs) through the triplet state where a large fraction of energy is dissipated to other eigenmodes than breathing. The HS potential is reached around $\Delta r \approx 0.13$ Å from the equilibrium position and the Fe-N distance in the first Iron coordination shell expands and coherently oscillates (265 fs period) around the new equilibrium structure while losing energy. The wave packet disperses at 330 fs time constant and vibrationally cools inside the HS state potential within 1.6 ps.

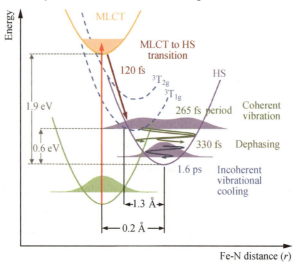

Figure 2.10 Schematic representation of the intersystem crossing dynamics during the molecular transition in Fe(bpy)$_3^{2+}$

Intersystem crossing is also observed in reactive collisions between atoms and organic molecules. The oxygen atom is one such reactive partner because its electronic ground state is a triplet (^{3}P). Therefore, the reactions can occur adiabatically on the triplet potential-energy surface (PES). A distinguishing characteristic of atomic oxygen reactions is the interaction between the triplet ground state PES of O(^{3}P) and the first singlet electronically excited PES of electronically excited O(^{1}D). Crossed-beam experiments have shown fingerprints of ISC in reactive collisions from the formation of strongly bound, long-lived addition complexes. When the oxygen atom is still approaching the organic molecule, these complexes increase the probability of non-radiative transition between the triplet and singlet PESs (ISC) occurring in the "entrance" channel of the system. A rebound mechanism, shown by the presence of a backward product angular distribution and a strongly repulsive product-energy release is expected for the reaction on the triplet PES. In contrast, observation of a backward-forward symmetric angular distribution along with high internal excitation of the products reveals a significant probability of non-adiabatic transition from the triplet to the singlet PESs. Then, the reactants experience a deep singlet well through the formation of a long-lived addition-complex intermediate. Ab initio calculations of stationary points along the reaction coordinates confirm these expectations, as illustrated in Figure 2.11 for the prototypical $O(^3P) + CH_3I$ reaction. The possibility of ICS occurring in the entrance is depicted in the diagram as a non-adiabatic crossing between the triplet ^3A and singlet ^1A PESs. Similar ISC mechanisms in the entrance channel have been found for reactive collisions between $O(^3P)$ atoms and unsaturated hydrocarbons.

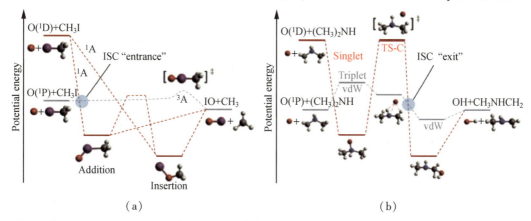

Figure 2.11 Stationary points along the reaction coordinate for the reactions $O(^3P) + CH_3I$ (a) and $O(^3P) + (CH_3)_2NH$ (b) [The singlet and triplet potential energy surfaces (PES) are depicted in red and blue, respectively. Molecular structures are depicted in the representative stationary points. The triplet pathway in both reactions highlights a "barrier-less" reaction coordinate. In contrast, the singlet pathways show the presence of deep wells and energy barriers. Intersystem crossing in the "entrance" (b) and "exit" (a) channels is shown with grey circles.]

2.3.2 Radiative Transitions

On irradiation with light of the frequency corresponding to the transition, an excited species may be induced to deactivate to the ground state with emission of a photon. Such a process, called stimulated emission, can be observed only under very specific conditions and constitutes the basis for laser operation.

Radiative transitions may be considered as vertical transitions and may therefore be explained in terms of the Franck-Condon principle. The intensity of any vibrational fine structure associated with such transitions will, consequently, be related to the overlap between the square of the wave functions of the vibronic levels of the excited state and the ground state. This overlap is maximized for the most probable electronic transition (the most intense band in the fluorescence spectrum). Figure 2.12 illustrates the quantum mechanical picture of the Franck-Condon principle applied to radiative transitions.

Fundamentals of Photophysics

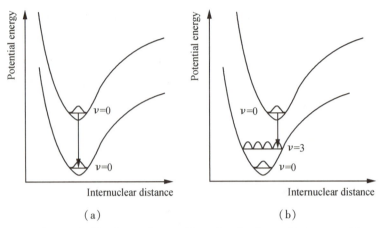

Figure 2.12 Most probable electronic transitions involved in radiative transitions where: (a) both electronic states have similar geometries; (b) the excited state and the ground state have very different geometries

According to the spin selection rule, radiative transitions could be classified as fluorescence, delayed fluorescence, and phosphorescence. As the spin selection rule plays an important role, it is customary to call: (1) fluorescence a spin-allowed emission (e.g. $S_1 \rightarrow S_0$); (2) phosphorescence a spin-forbidden emission (e.g. $T_1 \rightarrow S_0$).

i. Fluorescence

Fluorescence is the emission of light by a substance that has absorbed light or other electromagnetic radiation. Fluorescent dyes commonly seen in the laboratory are shown in Figure 2.13. Fluorescence is the property of some atoms and molecules to absorb light at a particular wavelength (the excitation: E_x) followed by a short-lived emission (E_m) of light at a longer wavelength. Fluorescence involves an external light source to excite the sample at a particular wavelength. When excited at the appropriate wavelength, the molecule is transformed from a ground to an excited state. When the molecule returns to the ground state, energy is released in the form of heat (loss of energy) and light at a different longer wavelength of lower energy (Figure 2.14).

Figure 2.13 Fluorescence luminescence

>>> **Chapter 2** Deactivation of Excited States

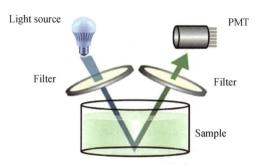

Figure 2.14 Schematic diagram of fluorescence emission process

During the 17th and 18th centuries, several investigators reported luminescence phenomena. It was British scientist Sir George G. Stokes who first described fluorescence in 1852 and was responsible for coining the term in honor of the blue-white fluorescent mineral fluorite (fluorspar). Stokes also discovered the wavelength shift to longer values in emission spectra that bears his name. Fluorescence was encountered in optical microscopy during the early part of the twentieth century for the first time. Several notable scientists (including August Köhler and Carl Reichert) initially reported that fluorescence was a nuisance in ultraviolet microscopy. The first fluorescence microscopes were developed between 1911 and 1913 by German physicists Otto Heimstädt and Heinrich Lehmann as a spin-off from the ultraviolet instrument. These microscopes were employed to observe autofluorescence in bacteria, animal, and plant tissues. Shortly thereafter, Stanislav von Provazek launched a new era when he used fluorescence microscopy to study dye binding in fixed tissues and living cells. However, it was not until the early 1940s that Albert Coons developed a technique for labeling antibodies with fluorescent dyes. Then the field of immunofluorescence was born. By the turn of the twenty-first century, the field of fluorescence microscopy was responsible for a revolution in cell biology, coupling the power of live cell imaging to highly specific multiple labeling of individual organelles and macromolecular complexes with synthetic and genetically encoded fluorescent probes.

Fluorescence involves a radiative transition (photon emission) between states of the same multiplicity (spin-allowed), usually from the lowest vibrational level of the lowest excited singlet state, $S_1(v=0)$.

It is indicated on a Jablonski diagram as a straight line going down on the energy axis between electronic states (Figure 2.15). Fluorescence is a slow process on the order of 10^{-9} to 10^{-6} seconds. Therefore, it is not a very likely path for an electron to dissipate energy especially at electronic energy states higher than the first excited state. While this transition is slow, it is an allowed transition with the electron staying in the same multiplicity manifold. In addition, fluorescence is most often observed between the first excited electron state and the ground state for any particular molecule because at higher energies it is more likely that energy will be dissipated through internal conversion and vibrational relaxation. At the first excited state, fluorescence can compete in regard to timescales with other non-radiative processes. The energy of the photon emitted in fluorescence is the same as the difference between the eigenstates and the transition. However, the energy of fluorescent photons is always less than that of the exciting photons. This difference is because of the energy loss in internal conversion and vibrational relaxation, where it is transferred away from the electron. The measured emission is usually distributed over a range of wavelengths because of the large number of vibrational levels that can be coupled into the transition between electronic states.

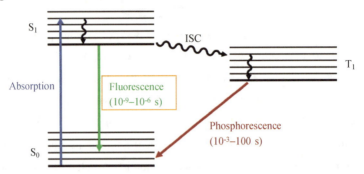

Figure 2.15 The Jablonski diagram of fluorescence and phosphorescence processes and their typical rate constants

As illustrated in Figure 2.16, fluorescence can be classified as spontaneous emission and stimulated emission. Spontaneous emission occurs when an excited atom or molecule emits a photon of energy equal to the energy difference between the two states without the influence of other atoms or molecules. The electrons in the excited state do not stay for a long period because the lifetime of electrons in the higher energy state or excited state is very small, of the order of 10^{-8} s. Hence, after a short period, they fall back to the ground state by releasing energy in the form of photons or light.

Stimulated emission occurs when a photon of energy equal to the energy difference between the two states interacts with an excited atom or molecule. In this process, the excited electron releases an additional photon of the same energy (the same frequency, the same phase, and in the same direction) while falling into the lower energy state. Such a process can be observed only under very specific conditions and constitutes the basis for laser operation.

More details about fluorescence will be introduced in Chapter 3.

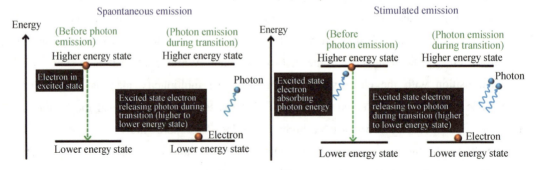

Figure 2.16 Spontaneous emission and stimulated emission

The use of fluorescent molecules in biological research is the standard in many applications. Their use is continually increasing due to their versatility, sensitivity and quantitative capabilities. Among their myriad of applications, fluorescent probes are employed to detect protein location and activation, identify protein complex formation and conformational changes and monitor biological processes in vivo. Immunofluorescence methods are used to provide this representative image (Figure 2.17).

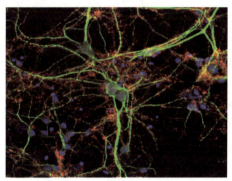

Figure 2.17 Detection of target proteins using fluorophore-conjugated antibodies

ii. Delayed Fluorescence

In certain compounds a weak emission has been observed with the same spectral characteristics (wavelengths and relative intensities) as fluorescence, but with a lifetime more characteristic of phosphorescence. This is called delayed fluorescence. Delayed fluorescence originates from the radiative transition of S_1 state regenerated from the first excited triplet (T_1). The lifetime of the singlet is generally 10^{-8} s, and the longest is 10^{-6} s, but it is sometimes observed that the singlet has a lifetime of 10^{-3} s. This long lifetime of fluorescence is called delayed fluorescence. These emissions of delayed fluorescence can arise by several mechanisms.

- Triplet-triplet annihilation (TTA, P-type delayed fluorescence)

P-type delayed fluorescence (Figure 2.18) is derived from the process of quenching two triplet states to generate a single state.

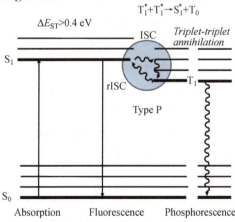

Figure 2.18 The Jablonski diagram of P-type delayed fluorescence

P-type delayed fluorescence is so called because it was first observed in pyrene. The fluorescence emission from a number of aromatic hydrocarbons shows two components with identical emission spectra. One component decay at the rate of normal fluorescence and the other has a lifetime approximately half that of phosphorescence. The implication of triplet species in the mechanism is given by the fact that the delayed emission can be induced by triplet sensitizers.

It is frequently observed for aromatic molecules in which the energy separation between the S_1 and T_1 excited states is too high to allow thermal activation. "P-type" delayed fluorescence depends indeed on the energy transfer between two triplet excited states T_1 to form an excited singlet state S_1 and a ground state S_0. The accepted mechanism is:

1. Absorption: $S_0 + h\nu \rightarrow S_1$
2. Intersystem crossing: $S_1 \rightarrow T_1$
3. Triplet-triplet annihilation: $T_1 + T_1 \rightarrow X \rightarrow S_1 + S_0$
4. Delayed fluorescence: $S_1 \rightarrow S_0 + h\nu$

It is the S_1 state produced by the triplet-triplet annihilation process that is responsible for the delayed fluorescence. Although it is emitted at the same rate as normal fluorescence, its decay is inhibited because it continues to be regenerated via Step 3.

Organic semiconductor materials have been widely used in various optoelectronic devices due to their rich optical or electrical properties, which are highly related to their excited states. Therefore, how to manage and utilize the excited states in organic semiconductors is essential for the realization of high-performance optoelectronic devices. TTA up-conversion is a unique process of converting two non-emissive triplet excitons to one singlet exciton with higher energy. Through harvesting sub-bandgap photons and TTA-based up-conversion, efficient optical-to-electrical devices can be realized.

For organic light emitting diodes (OLEDs) and organic light emitting transistors (OLETs) based on TTA materials, the internal conversion efficiency of up to 62.5% is possible. Although the maximum internal conversion efficiency is not as high as other materials (such as phosphorescent and TADF molecules), it also has some attractive advantages such as efficient blue emission, small efficiency roll-off, low operation voltage, and low cost. Efficiency roll-off in electroluminescence devices means the device efficiency decreases as the brightness increases, which may be caused by singlet-triplet annihilation, singlet-polar on annihilation, and loss of charge carrier balance, etc. Moreover, the unique structure of most TTA materials with relatively rigid conjugation enables their possibility for integrating high carrier mobility and efficient triplet utilization in one molecule, which are attractive for the fabrication of high-performance single component OLETs.

An overview summarized in Figure 2.19, which includes two working processes of optical-to-electrical conversion and electrical-to-optical conversion correlates with three typical optoelectronic devices, including photovoltaics, OLEDs and OLETs.

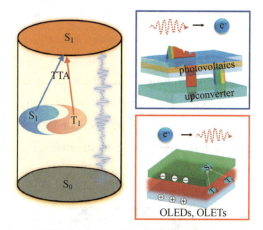

Figure 2.19 Illustration of TTA process and applications of TTA in optoelectronic devices, including photovoltaics, OLEDs, and OLETs

- Thermally activated delayed fluorescence (TADF)/E-type delayed fluorescence

E-type delayed fluorescence is so called because it was first observed in eosin. P-type delayed fluorescence is a collisional energy transfer process while E-type delayed fluorescence is not an energy transfer process but occurs due to smaller energy gap between the singlet and triplet states which results in reverse intersystem crossing of electron from the triplet state to the singlet state due to thermal activation.

E-type delayed fluorescence (Figure 2.20) refers to that when the energy of the triple excited state is close to that of the single excited state, the triple excited state can pass through the thermal activation reverse system to the single excited state, also known as thermal activation delayed fluorescence (TADF).

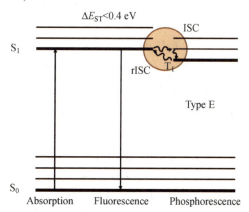

Figure 2.20 The Jablonski diagram of E-type delayed fluorescence

The intensity of the delayed fluorescence emission from eosin decreases as the temperature is lowered, which indicates that an energy barrier is involved. Since the delayed fluorescence is spectrally identical to normal fluorescence, emission must occur from the lowest vibrational level of S_1. However, the fact that the lifetime is characteristic of phosphorescence implies that the excitation originates from T_1. The explanation of this requires a small S_1-T_1 energy gap, where T_1 is initially populated by intersystem crossing from S_1. T_1 to S_1 intersystem crossing then occurs by thermal activation.

TADF is also a mechanism for enhancing the efficiency of OLEDs by harvesting triplet excitons. OLEDs are a type of light emitting diode where the emissive electroluminescent layer is composed of carbon based (organic) semiconductors, typically aromatic small molecules, which are in contrast to the inorganic crystalline semiconductors used in traditional LEDs. Furthermore, OLED displays are now often found in smartphones and high-end televisions due to their lower power consumption, high brightness, lightweight and higher contrast, compared to traditional LCD displays.

Figure 2.21 iPhone X with OLED display

In an OLED, a voltage is applied across an organic semiconductor layer, resulting in the injection of electrons and holes. These electrons and holes travel through the semiconductor, encountering each other and forming a strongly bound electron hole pair, which is called an exciton. Electrons and holes are fermions with half integer spin. The exciton can either have a total spin of zero ($S=0$) which is called a singlet exciton or a total spin of one ($S=1$) which is a triplet exciton depending on the relative orientations of the two spins. As shown in Figure 2.22, there are three combinations of half integer spins that form an $S=1$ exciton and only one combination which forms an $S=0$. This means that the excitons formed in an OLED will be 75% in the triplet state and only 25% in the singlet state. This is a major issue in the creation of an efficient OLED, as radiative decay from the triplet state (T_1) to the ground state singlet (S_0) is forbidden due to conservation of angular momentum. Therefore, 75% of the electrons and holes injected into the device are wasted and the maximum internal quantum efficiency of the OLED is limited to 25% (Figure 2.23, the 1st generation).

The first solution to this problem was to move away from the pure organic compounds used in the 1st generation OLEDs towards organic compounds that incorporate heavy metals such as iridium. The addition of heavy metals increases the spin orbit coupling (SOC) between the exciton spin angular momentum and the orbital angular momentum. The SOC results in the radiative transition from the T_1 to the S_0 being no longer strictly forbidden. So, the T_1 state therefore becomes emissive (the 2nd generation). Furthermore, the SOC promotes intersystem crossing (ISC) between the S_1 and the T_1 which further populates the T_1 state at the expense of the S_1. OLEDs using this mechanism are called the 2nd generation or PhOLEDs. They have internal quantum efficiencies approaching 100%.

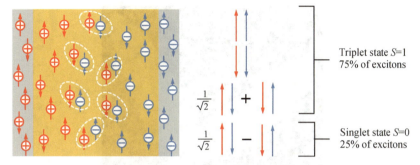

Figure 2.22   The triplet problem in OLEDs (Due to spin statistics 75% of the excitons formed inside the OLED will be in the triplet state.)

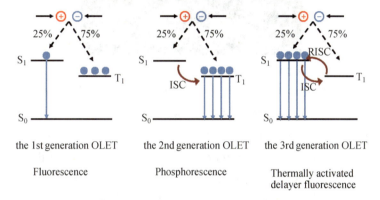

Figure 2.23   The operating principle of 1, 2 and 3 generations of OLEDs

Each pixel of an OLED display is actually made up of three separate OLEDs—a red, a green and a blue. The green and red emitters used in commercial OLED displays are 2nd generation phosphorescent materials, making them highly efficient. However, making a stable deep blue phosphorescent emitter remains a challenge. No suitable 2nd blue-ray emitter has yet been found. Therefore, commercial displays are forced to use the inefficient 1st generation fluorescent blue emitters and as a result, the blue OLED consumes much more power than the red and green. For portable electronics, where battery life is crucial, this is a major problem and there is a great demand for an efficient blue emitter.

iii. Phosphorescence

When molecules absorb light and go to the excited state, they have two options. They can either release energy and come back to the ground state immediately or undergo other non-radiative processes. If the excited molecule undergoes a non-radiative process, it emits some energy and comes to a triplet state where the energy is somewhat lesser than the energy of the exited state, but it is higher than the ground state energy. Molecules can stay a bit longer in this less energy triplet state. We call this state the metastable state. Then the metastable state (triplet state) can slowly decay by emitting photons, and come back to the ground state (singlet state). We call it phosphorescence when it happens (Figure 2.24).

Figure 2.24 Typical phosphorescent image

The study of phosphorescent materials dates back to at least 1602 when Italian Vincenzo Casciarolo described a "lapis solaris" (sun stone) or "lapis lunaris" (moon stone). The discovery was described in 1612 in philosophy professor Giulio Cesare la Galla's book *De Phenomenis in Orbe Lunae*. La Galla reported Casciarolo's stone emitted light on it after it had been calcified through heating. The stone received light from the sun and then (like the moon) gave out light in the darkness. The stone was impure barite, although other minerals also display phosphorescence including some diamonds (known to Indian king Bhoja as early as 1010 – 1055, rediscovered by Albertus Magnus and again rediscovered by Robert Boyle) and white topaz. The Chinese, in particular, valued a type of fluorite called chlorophane that would display luminescence from body heat, exposure to light, or being rubbed. Interest in the nature of phosphorescence and other types of luminescence eventually led to the discovery of radioactivity in 1896.

Besides a few natural minerals, phosphorescence is produced by chemical compounds. Probably the best-known of these is zinc sulfide, which has been used in products since the 1930s. Although phosphors may be added to change the color of light, zinc sulfide usually emits a green phosphorescence,

Phosphorescence (Figure 2.25) is a spin-forbidden radiative transition between states of different multiplicity, usually from the lowest vibrational level of the lowest excited triplet state, $T_1(v=0)$.

$$T_1 \rightarrow S_0 + h\nu$$

Typical timescales for photon emission by phosphorescence are of the order of 10^{-3} –

100 s.

More details about phosphorescence will be introduced in Chapter 3.

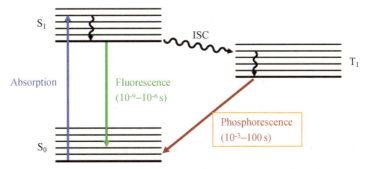

Figure 2.25 The Jablonski diagram of phosphorescence

At low temperatures or in a rigid medium, phosphorescence can be observed. The lifetime of the triplet state may, under these conditions, be long enough to observe phosphorescence on a timescale up to seconds, even minutes or more. The phosphorescence spectrum is located at wavelengths higher than the fluorescence spectrum because the energy of the lowest vibrational level of the triplet state T_1 is lower than that of the singlet state S_1.

A number of objects used in our daily life make use of the phosphorescence phenomenon to emit light energy into the environment. A few examples of phosphorescent light in real life are given below (Figure 2.26):

- Toys that glow in dark

Toys that glow in the dark are often one of the most prominent examples of objects that emit light radiations into the surroundings with the help of the phosphorescence phenomenon. Such toys are generally coated with a layer of phosphorescent material. After absorption of electromagnetic radiations they tend to radiate light energy in the surroundings.

- Stickers

Some stickers used for decoration purposes are able to exhibit a phosphorescence effect in real life. This is usually achieved by applying a thin layer of phosphorescent material to the surface of the stickers. The surface of the sticker releases the stored energy at a slow rate, thereby providing a glowing effect.

- Watches

Some watch dials glow when left in a dark place. This is mainly due to the phosphorescence of the element present in the structure of the watches and clocks. This helps improve visibility during the nighttime or in dark places.

- Paint

Some of the paints tend to provide a glowing effect due to the phosphorescent element present in them. Such paints are generally preferred for decoration purposes during festivals such as Halloween.

Fundamentals of Photophysics

• Safety signs

Safety signs are generally phosphorescent coated to improve the visibility of the caution during the nighttime. This helps prevent accidents and can be used to enhance the appearance of the signboards.

• Minerals

As a result of phosphorescence, various minerals such as celestite, colemanite, sphalerite, calcite, willemite, and fluorite, etc. tend to emit light radiations into the environment. Most of these minerals exist naturally; however, some of them can be created artificially. Various types of diamond are also known to exhibit phosphorescence in the environment. Other elements such as white topaz, chlorophane, etc. are also some of the prime minerals that display the phenomenon of phosphorescence in real life.

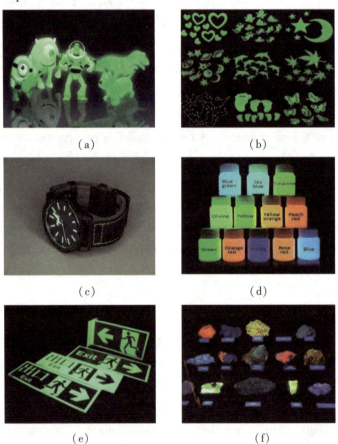

Figure 2.26　(a) Toys that glow in dark, (b) stickers, (c) watches, (d) paint, (e) safety signs, (f) minerals

There are some differences between fluorescence and phosphorescence (Table 2.3). Fluorescence is the emission of light by a substance that has absorbed light or other electromagnetic radiation while phosphorescence refers to the light emitted by a substance without combustion or perceptible heat. When we supply light to a sample of molecules, we immediately see the fluorescence. Fluorescence stops as soon as we take away the light source. But phosphorescence tends to stay little longer even after we remove the irradiating light source.

Table 2.3 Differences between fluorescence and phosphorescence

	Fluorescence	Phosphorescence
Definition	The emission of light by a substance that has absorbed light or other electromagnetic radiation.	The light emitted by a substance without combustion or perceptible heat.
Time	Stops as soon as we take away the light source.	Tends to stay little longer even after we remove the irradiating light source.
Mechanism	Takes place when excited energy is released and the molecule comes back to the ground state from the single excited state.	Takes place when a molecule is coming back to the ground state from the triplet excited state (metastable state).
Energy released	Higher than that in the phosphorescence.	Lower than that in the fluorescence.
Relationship between the absorbed and released energy	The absorbed amount of energy is released back.	The released energy is lower than what is absorbed.

2.4 Intermolecular Processes

Ground state intermolecular interactions are present in some systems and require measurements of the binding constants. These interactions are manifested by the spectral changes experienced in the absorption spectra. Therefore, these changes can be monitored as a function of the concentration of the substrates leading to the extraction of the binding constants. On the other hand, intermolecular and intramolecular excited state interactions refer to the energy and electron transfer operating in the excited states of different dyad or polyad systems. Moreover, these can also be excimers, dimers, or oligomers that are formed only in the excited states. Studies of photo-induced energy and electron transfers involve the measurement of their corresponding rates. The theory and methods used to characterize the different types of interactions are described next. Binding constant considerations are described elsewhere.

2.4.1 Vibrational Relaxation

Vibrational relaxation (VR), where molecules having excess vibrational energy, undergo rapid collision with one another and with solvent molecules to produce molecules in the lowest vibrational level of a particular electronic energy level.

Light absorption usually generates the excited state in a high vibrational level. Therefore, the newborn electronically excited molecules can be regarded as "hot" species with respect to the surrounding ground state molecules that have a Boltzmann equilibrium distribution largely centered on the zero vibrational level. The vibrationally excited molecules will tend to dissipate their excess vibrational energy (thermalize) by interaction (collisions) with surrounding molecules. This process is usually called vibrational relaxation (VR) (Figure 2.27). In solution, the collision rate is on the order of 10^{-13} s, and the total vibrational relaxation occurs in the picosecond time scale. In the gas phase, molecules cannot undergo vibrational relaxation by collision with other molecules at very low pressure. However, they can undergo intramolecular vibrational redistribution, that is, the energy originally localized in the mode populated by light absorption is rapidly distributed among the other vibrational modes. In other words, a molecule, in particular a large molecule, can act as its own heat bath. Vibrational relaxation and vibrational redistribution are the fastest processes occurring in the excited state. Thus, all the other processes, physical or chemical in nature, usually take place from thermally equilibrated excited states and compete with each other.

An electronically excited species is usually associated with an excess of vibrational energy in addition to its electronic energy (Figure 2.27), unless it is formed by a transition between the zero-point vibrational levels ($\nu=0$) of the ground state and the excited state ($0 \to 0$ transition). Therefore, vibrational relaxation involves transitions between a vibrationally-excited state ($\nu>0$) and the $\nu=0$ state within a given electronic state when excited molecules collide with other species such as solvent molecules, for example $S_2(\nu=3) \leadsto S_2(\nu=0)$. Typical timescales for the process are of the order of $10^{-12} - 10^{-10}$ s in condensed phases, and the excess vibrational energy is dissipated as heat.

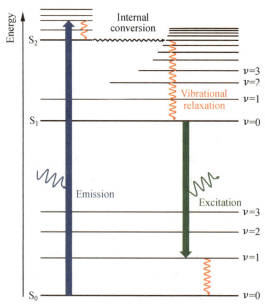

Figure 2.27　The Jablonski diagram of vibrational relaxation (orange arrows)

In vibrational relaxation process, the energy deposited by the photon into the electron is given away to other vibrational modes as kinetic energy. This kinetic energy may stay within the same molecule, or it may be transferred to other molecules around the excited molecule during collisions of the excited molecule with the surrounding molecules (solvent).

2.4.2 Energy Transfer

In presence of a molecule of a lower energy excited state (acceptor), the excited donor (D^*) can be deactivated by a process known as energy transfer which can be represented by the following sequence of equations.

$$D + h\nu \rightarrow D^* \quad (2\text{-}1)$$

$$D^* + A \rightarrow D + A^* \quad (2\text{-}2)$$

The prerequisite for energy transfer is that the energy level of the excited state of D^* has to be higher than that for A^* and the time scale of the energy transfer process must be faster than the lifetime of D^*. Two possible types of energy transfers are known, namely, radiative and non-radiative energy transfer.

Radiative transfer occurs when the extra energy of D^* is emitted in the form of luminescence and this radiation is absorbed by the acceptor (A).

$$D^* \rightarrow h\nu + D \quad (2\text{-}3)$$

$$h\nu + A \rightarrow A^* \quad (2\text{-}4)$$

For this to be effective, the wavelengths emitted by D^* need to overlap with those absorbed by A. This type of interaction operates although when the distance between the donor and acceptor is large (100 Å). However, this radiative process is inefficient because luminescence is a three-dimensional process. In this process, only a small fraction of the emitted light can be captured by the acceptor. The second type, non-radiative energy transfer, is more efficient. There are two different mechanisms used to describe this type of energy transfer: the Förster mechanism and the Dexter mechanism.

i. The Förster mechanism

The Förster mechanism is also known as the coulombic mechanism or dipole-induced dipole interaction. It was first observed by Förster. Here the emission band of one molecule (donor) overlaps with the absorption band of another molecule (acceptor). In this case, a rapid energy transfer may occur without a photon emission. In addition, this mechanism involves the resonant coupling of the electric dipole and the energy transfer from an excited molecule (donor) to an acceptor molecule. Based on the nature of interactions present between the donor and the acceptor, this process can occur over a long distance (30 ~ 100 Å). The mechanism of the energy transfer by this mechanism is illustrated in Figure 2.28.

Fundamentals of Photophysics

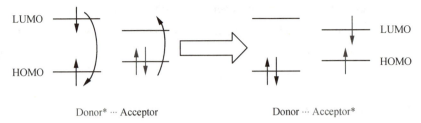

Figure 2.28 Mechanism of energy transfer action according to the Förster mechanism

In Figure 2.28, the excited donor electrons in the LUMO relax to the HOMO, and the released energy is transferred to the acceptor through coulombic interactions. As a result, an electron initially in the HOMO of the acceptor is promoted to the LUMO. This mechanism operates only in singlet states of the donor and the acceptor, which can be explained by the nature of the interactions (dipole-induced dipole), since only dipole-reserving transitions possess large dipole moments. This can be understood considering the nature of the excited state in both the singlet and the triplet states. Besides, the triplet state has a diradical structure, so it is less polar, making it difficult to interact over long distances (i.e. the Förster mechanism).

The rate of energy transfer (k_{ET}) according to this mechanism can be evaluated by the equation follows:

$$k_{ET} = k_D R_F^6 \left(\frac{1}{R}\right)^6 \tag{2-5}$$

where k_D is the emission rate constant for the donor, R is the interchromophore separation, and R_F is the Förster radius, which can be defined as the distance between the donor and the acceptor at which 50% of the excited state decays by energy transfer. That is, the distance where the energy transfer has the same rate constant as the excited state decay by the radiative and non-radiative channels ($k_{ET} = k_r + k_{nr}$). R_F is calculated by the overlap of the emission spectrum of the donor excited state (D*) and the absorption spectrum of the acceptor (A).

ii. The Dexter mechanism

The Dexter mechanism is a non-radiative energy transfer process that involves a double electron exchange between the donor and the acceptor (Figure 2.29). Although the double electron exchange is involved in this mechanism, no charge separated-state is formed.

The Dexter mechanism can be thought of as electron tunneling, by which one electron from the donor's LUMO moves to the acceptor's LUMO at the same time as an electron from the acceptor's HOMO moves to the donor's HOMO. In this mechanism, both singlet-singlet and triplet-triplet energy transfers are possible. This differs from the Förster mechanism, which operates only in singlet states. In addition, for this double electron exchange process to operate, there should be a molecular orbital overlap between the excited donor and the acceptor molecular orbital. For a bimolecular process, intermolecular collisions are also required. This mechanism involves short-range interactions (6 ~ 20 Å and shorter). Because it relies on tunneling, it decays exponentially with increasing molecular distance between donor and acceptor.

The rate constant can be represented by the following equation

$$k_{ET} = \frac{2\pi}{h} V_0^2 J_D \exp\left(-\frac{2R_{DA}}{L}\right) \tag{2-6}$$

where R_{DA} is the distance between the donor and the acceptor, J_D is the integral spectral overlap between the donor and the acceptor, L is the effective Bohr radius of the orbitals between which the electron is transferred, h is Plank's constant, and V_0 is the electronic coupling matrix element between the donor and the acceptor at the contact distance.

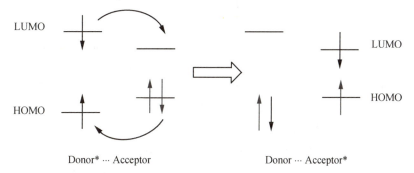

Figure 2.29 Mechanism of energy transfer action according to the Dexter mechanism

More details about energy transfer will be discussed in Chapter 5 specifically.

2.4.3 Electron Transfer

Electron transfer is considered as a photophysical process, involves a photoexcited donor molecule interacting with a ground state acceptor molecule. An ion pair is formed, which may undergo back electron transfer, resulting in quenching of the excited donor. The more details about electron transfer will be discussed in Chapter 5.

The energy contained in the excited state has an important effect on the ability of the excited state to donate or accept an electron. Treat excited substances as new chemical entities, and hence the use of thermodynamics in dealing with their redox behavior. It is reasonable in the condensed phase because the electronic relaxation time (over approximately 10^{-9} s) is usually several orders of magnitude longer than the time for thermal equilibration (on the order of 10^{-12} s) in all degrees of freedom of a polyatomic molecule.

Compared with the ground state species, electronically excited molecules are both better electron donors and better electron acceptors. To a first approximation, this can be easily understood by looking at Figure 2.30. Usually, light absorption promotes an electron from a lower to a higher orbital. The promoted electrons are therefore more easily removed, which means that the excited state has a smaller ionization potential than the ground state. At the same time, the promotion of an electron leaves behind a low-lying vacancy that can accept an electron, which means that the excited state has a higher electron affinity than the ground state.

Fundamentals of Photophysics

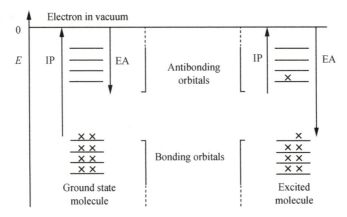

Figure 2.30 Orbital representation of changes in ionization potentials and electronic affinity of a molecule upon excitation

Photo-induced electron transfer involves an electron transfer within an electron donor-acceptor pair. The situation is represented in Figure 2.31. Photo-induced electron transfer represents one of the most basic photochemical reactions and at the same time it is the most attractive way to convert light energy or to store it for further applications. As shown in Figure 2.31, one can see a process taking place between a donor and a acceptor after excitation, resulting in the formation of a charge separated state, which relaxes to the ground state via an electron-hole recombination (back electron transfer).

Figure 2.31 Photo-induced electron transfer process

2.5 Dynamics of Excited State Deactivation

Generally speaking, the three unimolecular processes (radiative deactivation, non-radiative deactivation and chemical reaction) described earlier compete for deactivation of any excited state of a molecule. Therefore, their individual specific rates and the kinetics of their competition in each excited state are very important in determining the actual behavior of the excited molecule.

An excited state $^*A$ will decay according to overall first-order kinetics, with a lifetime, $(^*A)$, given by Equation.

$$\tau(^*A) = \frac{1}{k_p + k_r + k_{nr}} = \frac{1}{\sum_j k_j} \quad (2\text{-}7)$$

The probability of each deactivation process is related to its relative rate. For each process of the $^*A$ excited state, an efficiency $\eta_i(^*A)$ can be defined by the following equation.

$$\eta_i(^*A) = \frac{k_i}{\sum_j k_j} = k_i \tau(^*A) \qquad (2-8)$$

The quantum yield Φ_i of a given process originating from an excited state $^*A$ directly reached by light absorption is defined as the ratio between the number of molecules undergoing the process per unit time and the number of photons absorbed per unit time. In this case, the quantum yield is numerically equal to the efficiency and the rate constant of the process is given by the following equation:

$$k_i = \frac{\Phi_i}{\tau(^*A)} \qquad (2-9)$$

If the excited state $^*A$ (Figure 2.32) is populated after one or more non-radiative steps from other excited states, then the value of the quantum yield is given by this equation:

$$\Phi_i(^*A) = \eta_i(^*A) \Pi_n \eta_n \qquad (2-10)$$

where the η_n terms represent the efficiencies of the various steps involved in the population of $^*A$. In such a case,

$$k_i = \frac{\Phi_i(^*A)}{\tau(^*A) \Pi_n \eta_n} \qquad (2-11)$$

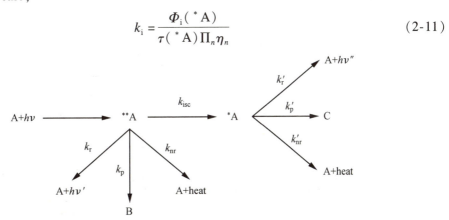

Figure 2.32 Schematic representation of the deactivation processes of two excited states, in which the $^*A$ excited state is populated by intersystem crossing from the upper lying $^{**}A$ excited state

Because non-radiative decay of upper excited states to the lowest one of the same multiplicity is very fast, generally it happens that the emitting level of a given multiplicity is the lowest excited level of that multiplicity (Kasha's rule). For compounds containing heavy atoms (including second- and third-row transition metal complexes), intersystem crossing has usually unitary efficiency because of the heavy-atom-induced spin-orbit coupling. Therefore, that emission (if any) occurs only from the lowest spin-forbidden excited state.

Although, in principle, the actual behavior of an excited molecule is the output of a complex system of consecutive/competitive processes, the exceedingly high rate of internal conversion between excited states is seen to considerably simplify the problem. In many occasions, a careful evaluation of the factors affecting the kinetics of unimolecular processes of the lowest excited states gives the possibility of rationalizing and, to some extent, predicting the photochemical behavior.

Taking transition-metal system as an example, investigation of its excited-state decay dynamics is a crucial step for the development of photo switchable molecular-based materials. Application fields include energy conversion, data storage, or molecular devices. Generally, after photoexcitation of the system into an electronic excited state with significant radiative coupling in the ground state, a sequence of interconnected processes take place during the excited state evolution that leads to a relatively stable occupation of the relevant final excited state. Several factors play a role in the dynamics of the excited state relaxation and control the deactivation to the proper electronic state. These factors include spin, spatial symmetry and characters of the electronic states involved in the process. Indeed, various electronic states of very different nature, occasionally lying in a narrow energy range, are involved. Those include states in which the excited electron is basically localized on the transition metal (metal-centered, MC), centered on the ligand (LC), and states involving charge separation, either charge transfer from the metal to the ligand (MLCT) or conversely, from the ligand to the metal.

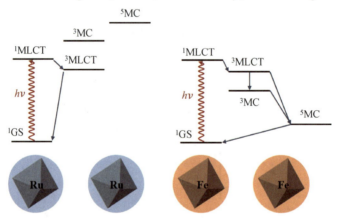

Figure 2.33 Schematic representation of the relative energies of the metal-centered (MC) and metal-to-ligand charge-transfer (MLCT) states in Ru^{II} (left) and Fe^{II} (right) polypyridyl complexes, illustrating how absorption leads to an expansion of the coordination sphere in the Fe complex, while the geometry of the Ru system remains practically unaffected

In addition, the photo-induced intramolecular electron transfer can be accompanied by important structural changes both in the molecular system and its nearest environment, which include cooperative effects, solvation effects or thermal distortions. The extent of the geometrical relaxation within the molecular complex as a result of the deactivation process relies on the particular system. To quote one typical example, Figure 2.33 shows how quasi-octahedral complexes with a $Fe^{II}N_6$ core, as found for example in the $[Fe^{II}(bpy)_3]$ complex, undergoing a spin crossover transition from a singlet to a quintet state, experience an enlargement of the Fe-N distance of around 0.2 Å, because of the occupation with two electrons of the anti-bonding orbitals in the quintet state. Conversely, the isoelectronic complex $[Ru^{II}(bpy)_3]$ only shows very small variations in the metal-ligand distance along the deactivation process, in which the triplet and quintet MC states lie higher in energy and the system remains in the ^{3}MLCT state.

A host of intramolecular non-radiative processes can potentially occur, such as intersystem crossing (ISC) between electronic states of different spin multiplicities, internal conversion (IC) between states of the same spin quantum number, and vibrational relaxation and/or intramolecular vibrational redistribution. Furthermore, in many cases all these processes can be extremely fast, and even take place in a shorter time scale than molecular vibrations.

One of the most important ingredients for an accurate account of the photochemistry of transition-metal complexes is the relative energy of the different electronic states involved in the deactivation cascade. There are several aspects that deserve a close inspection as graphically explained in Figure 2.34. First of all, it is of fundamental interest to have a good estimate for the adiabatic energy difference of the different electronic states relevant to the photophysical phenomenon under study. This often implies states with different spin moment and different equilibrium geometry. In spin-crossover complexes this is known as the high-spin (HS)/low-spin (LS) energy difference, ΔE^0_{HL}. Photo-induced processes are triggered by the absorption of photons, and hence, vertical excitation energies (DEFC, FC = Franck-Condon) are relevant to get information about the initial population of excited states. The deactivation of these excited states is largely determined by the potential energy surfaces on which the nuclei can move to adapt the nuclear configuration to the change in the electron distribution. In addition to minima on the excited state energy surface, other points are of special interest such as those that locate conical intersections or avoided crossings.

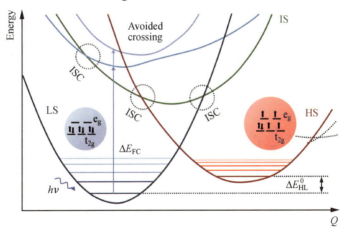

Figure 2.34 Schematic energy diagram as function of the nuclear displacement Q defining the high-spin (HS)/low-spin (LS) energy difference ΔE^0_{HL} [The vertical excitation energies DEFC and the location of avoided crossings and intersystem crossings (ISC) are indicated. IS = intermediate spin state]

References

[1] Brian W. *Principles and Applications of Photochemistry* [M]. Manchester: John Wiley & Sons, Ltd, 2009: 47 – 52.

[2] Aly, S. M., Carraher, J. C. E. & Harvey, P. D. *Macromolecules Containing Metal and Metal-like Elements: Photophysics and Photochemistry of Metal-containing Polymers* [M]. Manchester: John Wiley & Sons, Inc., 2010: 1 – 43.

[3] Lemke, H. T., Kjær, K. S. & Hartsock, R. Watching coherent molecular structural dynamics during photoreaction: beyond kinetic description [J]. *arXiv.*, 2015(1511): (1 – 16).

[4] Bañares, L. Unexpected intersystem crossing [J]. *Nat. Chem.*, 2019(11): 103 – 104.

[5] Gao, C., Wong, W. W. H. & Qin, Z. Application of triplet-triplet annihilation upconversion in organic optoelectronic devices: advances and perspectives [J]. *Adv. Mater.*, 2021(33): (1 – 25).

[6] Carmen, S., et al. Deactivation of excited states in transition-metal complexes: insight from computational chemistry [J]. *Chem. Eur. J.*, 2019(25): 52 – 64.

Chapter 3

Fluorescence and Phosphorescence

3.1 Introduction to Fluorescence

3.1.1 Luminescence in Nature

Luminescence is the emission of light from any substance, and occurs from electronically excited states. Luminescence is formally divided into two categories—fluorescence and phosphorescence—depending on the nature of the excited state.

Fluorescence involves a radiative transition (photon emission) between states of the same multiplicity (spin-allowed), usually from the lowest vibrational level of the lowest excited singlet state, $S_1(v=0)$.

$$S_1 \to S_0 + h\nu \tag{3-1}$$

Typical timescales for fluorescence emission are of the order of $10^{-9} - 10^{-6}$ s. Fluorescence occurs when short-wavelength electromagnetic radiation is absorbed and then reemitted at longer wavelength. This phenomenon is broadly distributed in marine and terrestrial environments and is found in distantly related organisms. Typical examples are fluorescence jellyfish and minerals showed in Figure 3.1 (a) and (b) respectively.

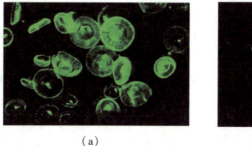

(a)　　　　　　　　　　　　(b)

Figure 3.1　(a) **Fluorescence jellyfish in marine environments**; (b) **fluorescence minerals in terrestrial environments**

Fundamentals of Photophysics

Phosphorescence is a spin-forbidden radiative transition between states of different multiplicity, usually from the lowest vibrational level of the lowest excited triplet state, $T_1(v=0)$.

$$T_1 \rightarrow S_0 + hv \tag{3-2}$$

Typical timescales for photon emission by phosphorescence are of the order of $10^{-3} - 10^2$ s, so that phosphorescence lifetimes are typically milliseconds to seconds. Even longer lifetimes are possible, as is seen from "glow-in-the-dark" toys. Following exposure to light, the phosphorescence substances glow for several minutes while the excited phosphors slowly return to the ground state. Typical examples are night pearl and night vision watch, just as show in Figure 3.2. This is because there exist many deactivation processes that compete with emission, such as non-radiative decay and quenching processes. Details of phosphorescence are described in Section 3.6 of this chapter.

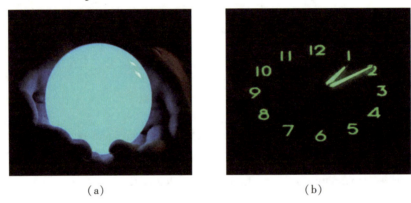

(a)   (b)

Figure 3.2   (a) **Night pearl**; (b) **night vision watch**

It should be noted that the distinction between fluorescence and phosphorescence is not always clear. Generally, fluorescent materials stop emitting light almost immediately when the UV light source stops, while phosphorescent materials continue emitting light for some time. In certain compounds a weak emission has been observed with the same spectral characteristics (wavelengths and relative intensities) as fluorescence, but with a lifetime more characteristic of phosphorescence.

3.1.2 History of Fluorescence

The discovery and study of fluorescence has a long history. Figure 3.3 reflects the history of fluorescence generally. An early observation of fluorescence was described in 1560 by Bernardino de Sahagún and in 1565 by Nicolás Monardes in the infusion known as lignum nephriticum (Latin for "kidney wood"). It was derived from the wood of two tree species, Pterocarpus indicus and Eysenhardtia polystachya. The chemical compound responsible for this fluorescence is matlaline, which is the oxidation product of one of the flavonoids found in this wood.

Edward D. Clarke in 1819 and René Just Haüy in 1822 described fluorescence in fluorites.

>>> **Chapter 3** Fluorescence and Phosphorescence

Sir David Brewster described the phenomenon for chlorophyll in 1833 and Sir John Herschel did the same for quinine in 1845. Fluorescence was further discovered in 1845 by Fredrick W. Herschel. He discovered that UV light can excite a quinine solution (e. g. tonic water) to emit blue light. In his 1852 paper on the "Refrangibility" (wavelength change) of light, George Gabriel Stokes described the ability of fluorspar and uranium glass to change invisible light beyond the violet end of the visible spectrum into blue light, and he named this phenomenon as fluorescence. The name of fluorescence was derived from the mineral fluorite (calcium difluoride), some examples of which contain traces of divalent europium, which serve as the fluorescent activator to emit blue light. In a key experiment he used a prism to isolate ultraviolet radiation from sunlight and observed blue light emitted by an ethanol solution of quinine exposed by it.

Many decades later, in the early 1900s, the first uses of fluorophores in biological investigations were performed to stain tissues, bacteria, and other pathogens. In the early 1940s, fluorescence labeling was achieved by Ellinger and Hirt for the first time. The cloning of green fluorescent protein (GFP) was achieved in the early 1990s and was easily applied to fluorescence microscopy.

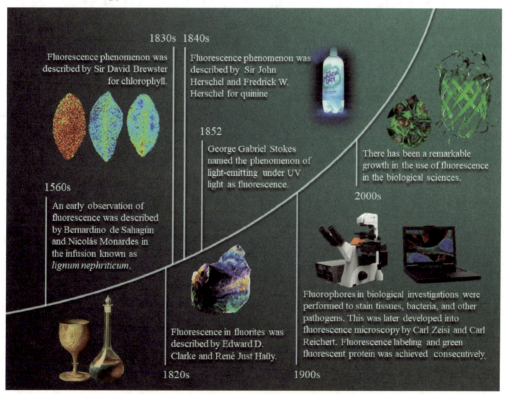

Figure 3.3 The history of fluorescence

During the past 20 years there has been a remarkable growth in the use of fluorescence in the biological sciences. Fluorescence spectroscopy and time-resolved fluorescence are considered to be primarily research tools in biochemistry and biophysics. In fact, over the years

Fundamentals of Photophysics

fluorescence has been widely used. Fluorescence as one of the most sensitive techniques available is now a dominant methodology used extensively in biotechnology, flow cytometry, medical diagnostics, DNA sequencing, forensics, and genetic analysis, to name a few. Fluorescence detection is highly sensitive, and there is no longer the need for the expense and difficulties of handling radioactive tracers for most biochemical measurements. There has been markedly growth in the use of fluorescence for cellular and molecular imaging. Fluorescence imaging can reveal the localization and measurements of intracellular molecules, sometimes at the level of single-molecule detection.

3.1.3 Fluorescence Processes

Fluorescence is the property of some atoms and molecules to absorb light at a particular wavelength (the excitation: E_x) followed by a short-lived emission (E_m) of light at a longer wavelength shown in Figure 3.4. The most striking example of fluorescence occurs when the absorbed radiation is in the ultraviolet region of the spectrum, and thus invisible to the human eye, while the emitted light is in the visible region, which gives the fluorescent substance a distinct color that can be seen only when exposed to UV light. In nature, fluorescence occurs frequently in certain minerals.

As one of the radiative transitions described in Chapter 2, fluorescence processes that occur between the absorption and emission of light are usually illustrated by Jablonski diagrams. Jablonski diagrams are often used as the starting point for discussing light absorption and emission. Jablonski diagrams are used in a variety of forms, to illustrate various molecular processes that can occur in excited states. These diagrams are named after Professor Alexander Jablonski, who is regarded as the father of fluorescence spectroscopy because of his many accomplishments. The specific fluorescence process in Jablonski diagrams can be described in Figure 3.4.

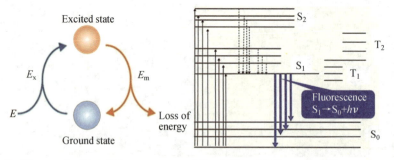

Figure 3.4 The production of fluorescence and the Jablonski diagram of fluorescence processes

De-excitation is a very important process in fluorescence, which can occur via a radiative decay, i.e. by spontaneous emission of a photon, shown in Figure 3.5. The radiative de-excitation process can be described as a monomolecular process:

$$\frac{dn_{exc}}{dt} = -k_F \cdot n_{exc} \qquad (3\text{-}3)$$

The reason is that the vibrational relaxation of any electronic state is always much faster than photon emission.

Most of the fluorescence spectrum is shifted to lower energies (longer wavelengths), compared with the absorption spectrum. In addition to this, the shape of the emission spectrum is approximately the mirror image of the absorption spectrum, providing that the ground and excited state have similar vibrational properties.

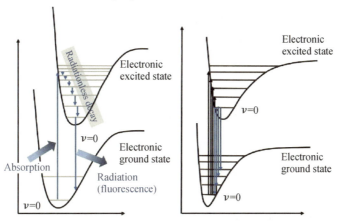

Figure 3.5 De-excitation processes of fluorescence

The rate constants for the various processes are denoted as follows:

- $k_f(k_r^S)$: fluorescence emission ($S_1 \rightarrow S_0$)
- $k_{ph}(k_r^T)$: phosphorescence emission ($T_1 \rightarrow S_0$)
- $k_i(k_{ic}^S)$: internal conversion ($S_1 \rightarrow S_0$)
- $k_x(k_{isc})$: intersystem crossing ($S_1 \rightarrow T_1$)
- $k_{nr}(k_{nr}^S)$: the overall non-radiative rate constant ($k_{nr}^S = k_{ic}^S + k_{isc}$)
- (k_{nr}^T): intersystem crossing ($T_1 \rightarrow S_0$)

Based on the timescale of various processes, the fluorescence can be observed if $k_f > k_i + k_x$.

3.1.4 Types of Fluorescence

According to the lifetime of the fluorescence, fluorescence can be divided into prompt fluorescence and delayed fluorescence.

- Prompt fluorescence: ~ 10^{-8} s

$S_1 \rightarrow S_0 + h\nu$

$S_1 + S_0 \rightarrow (S_1 S_0)^* \rightarrow 2S_0 + h\nu$

- Delayed fluorescence: the lifetime is similar to that of phosphorescence

Fundamentals of Photophysics

E-type delayed fluorescence:

T_1 + activation energy → S_1 → S_0 + $h\nu$

P-type delayed fluorescence:

$T_1 + T_1 \rightarrow S_1 + S_0$

$S_1 \rightarrow S_0 + h\nu$

According to the fluorescence wavelength (compared with excitation wavelength), fluorescence can be divided into stokes fluorescence, anti-stokes fluorescence and resonance fluorescence.

- Stokes fluorescence (in solutions): $\lambda_F > \lambda_E$
- Anti-stokes fluorescence (in gas): $\lambda_F < \lambda_E$
- Resonance fluorescence: $\lambda_F = \lambda_E$

3.2 Fluorescence Spectroscopy

Fluorescence spectroscopy analyzes fluorescence from a molecule based on its fluorescent properties. Fluorescence spectroscopy plays an important role in revealing the photochemical and photophysical properties of materials by analyzing the absorption, excitation and emission spectrum of samples (Figure 3.6). Fluorescence spectroscopy uses a beam of light that excites the electrons in molecules of certain compounds, and causes them to emit light. That light is directed towards a filter and onto a detector for measuring and identifying the molecule or changes in the molecule.

A fluorescence excitation spectrum is when the emission wavelength is fixed and the excitation monochromator wavelength is scanned. In this way, the spectrum gives information about the wavelengths at which a sample will absorb so as to emit at the single emission wavelength chosen for observation. It is similar to absorbance spectrum, but is a much more sensitive technique in terms of limits of detection and molecular specificity. In addition, excitation spectra are specific to a single emitting wavelength/species as opposed to an absorbance spectrum, which measures all absorbing species in a solution or sample. The emission and excitation spectra for a given fluorophore are mirror images of each other. Typically, compared to the excitation or absorbance spectrum, the emission spectrum occurs at higher wavelengths (lower energy).

These two spectral types (emission and excitation) are used to see how a sample is changing. The spectral intensity or peak wavelength may change with variants such as temperature, concentration, or interactions with other molecules around it. This includes quencher molecules and molecules or materials that involve energy transfer. Furthermore, some fluorophores are sensitive to solvent environment properties such as pH, polarity, and certain ion concentrations.

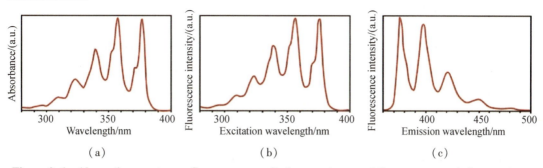

Figure 3.6 Absorption spectrum, fluorescence excitation spectrum and fluorescence emission spectrum of anthracene in cyclohexane measured using the FS5 spectrofluorometer

There are two points to note: (1) fluorescence emission occurs from S_1 and therefore its characteristics (except polarization) do not depend on the excitation wavelength; (2) the fluorescence spectrum is located at higher wavelengths (lower energy) than the absorption spectrum because of the energy loss in the excited state due to vibrational relaxation.

3.2.1 Absorption Spectrum

Absorption spectra (also known as UV-vis spectra, absorbance spectra and electronic spectra) show the change in absorbance of a sample as a function of the wavelength of incident light. A typical absorption spectrum is showed in Figure 3.6 (a). It is measured using a spectrophotometer. Absorption spectra are measured by varying the wavelength of the incident light using a monochromator and recording the intensity of transmitted light on a detector (Figure 3.7). The intensity of light transmitted through the sample, I_{Sample}, (such as an analyte dissolved in solvent) and the intensity of light through a blank, I_{Blank}, (solvent only) are recorded and the absorbance of the sample is calculated using:

$$A = \lg\left(\frac{I_{\text{Blank}}}{I_{\text{Sample}}}\right) \tag{3-4}$$

The absorbance is linearly proportional to the molar concentration of the sample, which enables the concentration of the sample to be calculated from the absorption spectrum using the Lambert-Beer Law.

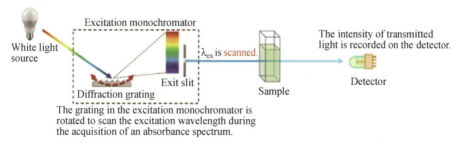

Figure 3.7 Schematic of the measurement of absorption spectra in a spectrophotometer

Fundamentals of Photophysics

3.2.2 Excitation Spectrum

Fluorescence excitation spectra show the change in fluorescence intensity as a function of the wavelength of the excitation light, and are measured using a spectrofluorometer. The measurement process of excitation spectra is showed in Figure 3.8. The wavelength of emission monochromator is set to a wavelength of known fluorescence emission by the sample, and the wavelength of the excitation monochromator is scanned across the desired excitation range and the intensity of fluorescence is recorded on the detector as a function of excitation wavelength. In addition, if the sample obeys Kasha's Rule, and Vavilov's Rule, then the excitation spectrum and absorption spectrum will be identical. Excitation spectra can therefore be thought of as fluorescence detected absorption spectra. A typical absorption spectrum is showed in Figure 3.6 (b).

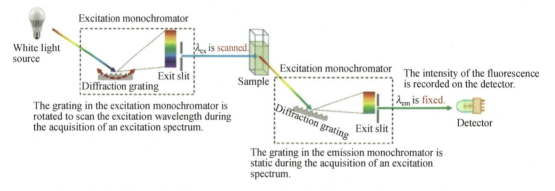

Figure 3.8 Schematic of the measurement of excitation spectra in a spectrofluorometer

3.2.3 Emission Spectrum

Fluorescence emission spectra show the change in fluorescence intensity as a function of the wavelength of the emission light, and are measured using a spectrofluorometer [Figure 3.6 (c)]. The wavelength of excitation monochromator is set to a wavelength of known absorption by the sample, and the wavelength of the emission monochromator is scanned across the desired emission range and the intensity of the fluorescence is recorded on the detector as a function of emission wavelength (Figure 3.9). Emission spectra vary widely and are dependent on the chemical structure of the fluorophore and the solvent in which it is dissolved. The spectra of some compounds, such as perylene, show significant structure due to the individual vibrational energy levels of the ground state and the excited state. Other compounds, such as quinine, show spectra devoid of vibrational structure.

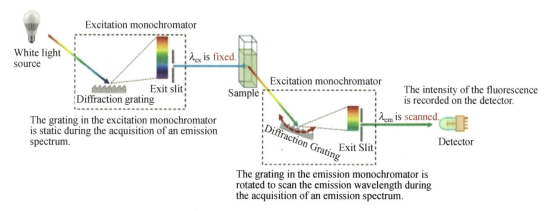

Figure 3.9 Schematic of the measurement of emission spectra in a spectrofluorometer

Figure 3.10 shows the absorption and fluorescence emission spectra of a solution of anthracene in benzene. Mirror-image symmetry exists between absorption and fluorescence spectra of a solution of anthracene in benzene. This mirroring only occurs when the geometries of the ground state (S_0) and the first excited state (S_1) are similar. The most noticeable features of the spectra, in addition to the mirror-image relationship, are that:

- The 0−0 bands for absorption and fluorescence occur at almost the same wavelength.
- Fluorescence emission occurs at longer wavelengths (lower energy) than the 0−0 band, while absorption occurs at shorter wavelengths (higher energy) than the 0−0 band.
- The absorption spectrum shows a vibrational structure characteristic of the S_1 state whereas the fluorescence spectrum shows a vibrational structure characteristic of the S_0 state of anthracene.

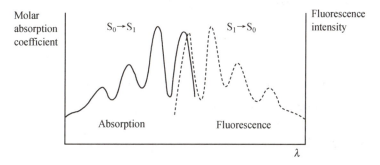

Figure 3.10 Absorption spectrum (solid line) and fluorescence spectrum (dashed line) of anthracene in benzene

Absorption occurs from $S_0(v=0)$ and, because of rapid vibrational relaxation, fluorescence occurs from $S_1(v=0)$. Fluorescence occurs at a lower energy (longer wavelength) than the exciting radiation because of the vibrational energy that is transferred from the excited anthracene to its surroundings prior to fluorescence emission from $S_1(v=0)$.

A simple energy-level diagram (Figure 3.11) shows that the 0−0 bands for fluorescence and absorption occur at the same wavelength, since the energy changes (represented by the lengths of the arrows) are equal.

Fundamentals of Photophysics

However, the 0 − 0 bands lie at slightly different wavelengths in absorption and emission. This separation results from energy loss to the solvent environment. For molecules in solution, the solvent cages surrounding the ground state molecules and the excited state molecules are different. Since electronic transitions occur at much faster rates than solvent-cage rearrangement, the energy change is different between absorption and emission.

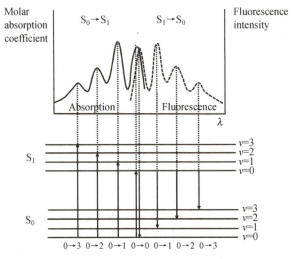

Figure 3.11 Energy-level diagram showing how the electronic and vibrational energy levels in the ground state (S_0) and excited state (S_1) of anthracene molecules are related to the absorption and fluorescence emission spectra

3.3 Several Rules about Fluorescence

There are several general rules that deal with fluorescence. Each of the following rules has exceptions but they are useful guidelines for understanding fluorescence (these rules do not necessarily apply to two-photon absorption).

3.3.1 Kasha's Rule

Kasha's Rule is named after the American molecular spectroscopist Michael Kasha. It is one of the main principles in fluorescence spectroscopy. Kasha published a seminal paper "Characterization of Electronic Transitions in Complex Molecules" in 1950. He established how to interpret and draw conclusions from the fluorescence emission spectra of complex molecules and contained the first statement of what is now known as Kasha's Rule. It is best stated exactly as it was written by Kasha in 1950: "The emitting level of a given multiplicity is the lowest excited level of that multiplicity." Because of the very rapid rate of deactivation to the lowest vibrational level of S_1 (or T_1), luminescence emission and chemical reaction by excited molecules will always originate from the lowest vibrational level of S_1 or T_1.

For a multistate molecule, the photon can only be emitted from the lowest excited state. The photons can be emitted as fluorescence or phosphorescence. Kasha's Rule indicates that the quantum yield of luminescence is independent of the wavelength of exciting radiation. This occurs because excited molecules usually decay to the lowest vibrational level of the excited state before fluorescence emission takes place. However, Kasha's Rule does not always apply and is violated severely in many simple molecules. So, we can put it another way, the fluorescence spectrum shows very little dependence on the wavelength of exciting radiation.

According to Kasha's Rule, fluorescence from organic compounds usually originates from the lowest vibrational level of the lowest excited singlet state (S_1). But as mentioned earlier, there are exceptions to Kasha's Rule. An exception to Kasha's Rule is the hydrocarbon azulene. It shows fluorescence from S_2.

This behavior may be explained by considering that the azulene molecule has a relatively large $S_2 \rightarrow S_1$ gap, which is responsible for slowing down the normally rapid S_2 to S_1 internal conversion such that the fluorescence of azulene is due to the $S_2 \rightarrow S_0$ transition. The fluorescence emission spectrum of azulene is an approximate mirror image of the $S_0 \rightarrow S_2$ absorption spectrum (Figure 3.12).

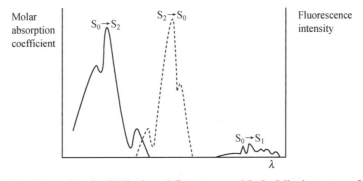

Figure 3.12 Absorption (solid line) and fluorescence (dashed line) spectra of azulene

Restricted by Kasha's Rule, almost all the organic light-emitting materials only feature one emission band, and it is difficult to achieve dual-or multi-emissive light-emitting materials. However, there is an urgent need of anti-Kasha emissive materials with dual- or multi-emissive bands for the development of high-efficient white organic light-emitting diodes (WOLEDs) and ratiometric probes as chemosensors or biosensors. Because of the different radiative decay pathways, the dual- or multi-emission bands of anti-Kasha emissive materials always display different responses under external stimuli. Therefore, the emission colors could be easily tuned using mechano-simulations or other kinds of simulations. These responsive properties enabled their broad applications in many fields including optical/chemical sensing, biological imaging, trademark anti-counterfeiting, and detections of mechano-signals. Therefore, in the scientific community and industrial arena, breaching Kasha's Rule and developing anti-Kasha emissive

Fundamentals of Photophysics

materials could have contribution.

The library of anti-Kasha emissive materials was summarized in three aspects according to their design strategies: (i) isomerization balancing; (ii) excited state balancing; and (iii) emissive building block combination.

i. Isomerization balancing

This class of compounds typically possess isomerization reactions under light excitations, electricity excitations, and protonation or deprotonation processes. Thus, the light-emitting molecules undergo obvious changes in structure, charge distribution, relaxation of excited states, and electron delocalization, which lead to obvious changes in the photophysical properties of the isomerized molecules. Therefore, emissive molecules with larger spectral widths and significantly different luminescence characteristics in original and isomerized forms can promote dual emission under appropriate conditions (light, electricity, and pH).

ii. Excited state balancing

In most light-emitting materials, the internal conversion rates are much higher than the radiative decay rates. Then photon emission generally occurs from the excited state with the lowest energy. On the contrary, under the influence of π-π stacking and intermolecular interactions, the overlay of energy level's overlap is small and the internal conversion processes might be prolonged. Therefore, the internal conversion processes and the radiative decay processes are competitive, resulting in dual/multi-emission bands. The strategy of balancing excited states to achieve anti-Kasha emissions is realized by the generation of new excited states, or the formation of a dimer composed of ground states. For example, the interactions between excimer or intermolecular could efficiently promote anti-Kasha emissions.

iii. Emissive building block combination

Combination of emissive building blocks to achieve anti-Kasha emission provides opportunities for fine-tuning of the photophysical properties. Anti-Kasha emissive materials obtained by the combination of emissive building blocks are the frequently found examples of copolymer, metal organic framework (MOF) materials and covalent organic framework (COF). In the emissive building block combination systems, anti-Kasha emissions are mostly achieved from the combined emission features of the different components and controlled by energy transfer processes. Moreover, anti-Kasha emissions also can be obtained through covalent linkages that connect red-, green- and blue-emitting molecular components, or through the interplay between interlayer and intralayer hydrogen bonding to form COF. This strategy provides new insights on how to obtain excellent anti-Kasha emissive materials, and can be tuned to achieve synergetic interactions for color-tuneable PL emission, thereby providing new strategies for designing color-tuneable anti-Kasha emissive materials.

3.3.2 Mirror-image Rule

In rigid molecules, where the geometries of the S_0 and S_1 states are similar, there is a mirror-image relationship between the absorption spectrum and the fluorescence spectrum. Emission spectrum is typically a mirror image of the absorption spectrum of the $S_0 \rightarrow S_1$ transition, but shift to longer wavelengths. This is due to the similarity of the energy spacing of the vibrational energy levels in the two states. Because the vibrational energy level spacing in the S_0 and S_1 levels are similar, the 0 – 1 emission band is at the same energy below the 0 – 0 band as the 0 – 1 absorption band is above it, and so on for the other vibrational bands (Figure 3.13).

The mirror-image rule is related to the Franck-Condon principle which states that electronic transitions are vertical, that is, energy changes without distance changing can be represented with a vertical line in the Jablonski diagram. This means the nucleus does not move and the vibration levels of the excited state resemble the vibration levels of the ground state. The same electronic transition is involved in both absorption and emission. In many molecules vibrational energy levels are not significantly altered by the different electronic distributions of S_0 and S_1.

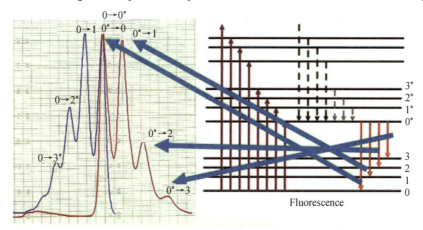

Figure 3.13 The absorption and emission spectra showing the mirror-image rule
(The numbers 0, 1, and 2 refer to vibrational energy levels.)

The above relation is a mirror relation rather than a translation relation. Take 1 – 4 for example (Figure 3.14): since $E = hc/\lambda$, E and λ have an inverse relationship. The energy of $S_1(1) \rightarrow S_0(4)$ of fluorescence is the lowest, so the wavelength is the longest. It is inferred that the display in the spectrum is a mirror-image relationship.

Fundamentals of Photophysics

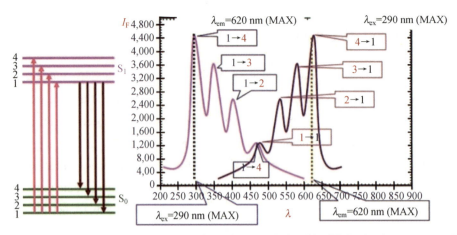

Figure 3.14 Schematic diagram of the mirror-image relationship (Take 1–4 as an example)

Although often true, there are many exceptions to the mirror-image rule. This is illustrated for the pH-sensitive fluorophore 1-hydroxypyrene-3,6,8-trisulfonate (HPTS) in Figure 3.15. The hydroxyl group is protonated at low pH. The absorption spectrum at low pH shows vibrational structure typical of an aromatic hydrocarbon. The emission spectrum shows a large Stokes shift and none of the vibrational structure is seen in the absorption spectrum. As a rule, the difference between the absorption emission spectra is due to the ionization of the hydroxyl group. The dissociation constant (pK_a) of the hydroxyl group decreases in the excited state, and this group becomes ionized. The emission usually occurs in a different molecular species, and this ionized species displays a broad spectrum. This form of HPTS with an ionized hydroxyl group can be formed at pH 13. The emission spectrum is a mirror image of the absorption of the high pH form of HPTS.

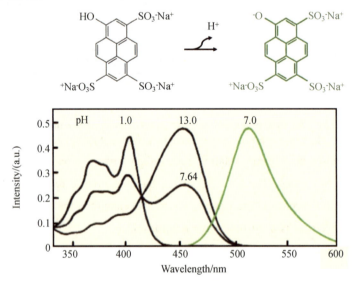

Figure 3.15 Absorption (pH 1, 7.64, and 13) and emission spectra (pH 7) of 1-hydroxypyrene-3,6,8-trisulfonate in water

Changes in pK_a in the excited state also occur for biochemical fluorophores. For example, phenol and tyrosine each shows two emissions, the long-wavelength emission being favored by a high concentration of proton acceptors. The pK_a of the phenolic hydroxyl group decreases from 11 in the ground state to 4 in the excited state. The phenolic proton is lost to proton acceptors in the solution after excitation. Depending upon the concentration of these acceptors, either the phenol or the phenolate emission may dominate the emission spectrum.

Other than proton dissociation, other excited state reactions can also result in deviations from the mirror symmetry rule. One example is shown in Figure 3.16, which shows the emission spectrum of anthracene in the presence of diethylaniline. The structured emission at shorter wavelengths is a mirror image of the absorption spectrum of anthracene. The unstructured emission at longer wavelengths is due to the formation of a charge-transfer complex between the excited state of anthracene and diethylaniline. The unstructured emission is from this complex. Furthermore, many polynuclear aromatic hydrocarbons, such as pyrene and perylene, also form charge-transfer complexes with amines. These excited state complexes are referred to as exciplexes.

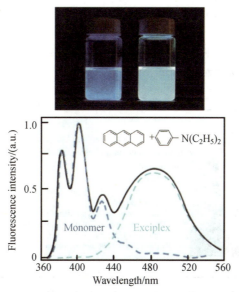

Figure 3.16 Emission spectrum of anthracene in toluene containing 0.2 mol · L^{-1} diethylaniline
(The dashed lines show the emission spectra of anthracene or its exciplex with diethylaniline.)

Some fluorophores can also form complexes with themselves. The most famous example is pyrene. At low concentrations pyrene displays a highly structured emission (Figure 3.17) while the previously invisible UV emission of pyrene becomes visible at 470 nm under higher concentrations. This long-wavelength emission is due to excimer formation. The term "excimer" is an abbreviation for an excited-state dimer.

Fundamentals of Photophysics

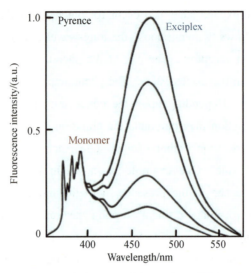

Figure 3.17 Emission spectra of pyrene and its excimer [The relative intensity of the excimer peak (470 nm) decreases as the total concentration of pyrene is decreased from 6×10^{-3} mol · L^{-1} (top) to 0.9×10^{-4} mol · L^{-1} (bottom).]

3.3.3 The Stokes Shift

Fluorescence typically occurs at lower energies or longer wavelengths. This phenomenon was first observed by Sir George Gabriel Stokes (Figure 3.18) in 1852 at the University of Cambridge.

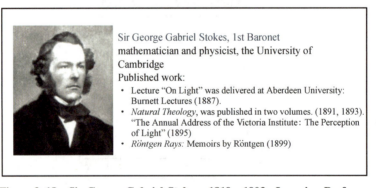

Sir George Gabriel Stokes, 1st Baronet
mathematician and physicist, the University of Cambridge
Published work:
- Lecture "On Light" was delivered at Aberdeen University: Burnett Lectures (1887).
- *Natural Theology*, was published in two volumes. (1891, 1893). "The Annual Address of the Victoria Institute: The Perception of Light" (1895)
- *Röntgen Rays:* Memoirs by Röntgen (1899)

Figure 3.18 Sir George Gabriel Stokes, 1819 – 1903, Lucasian Professor at the University of Cambridge

These early experiments used relatively simple instrumentation (Figure 3.19). The source of ultraviolet excitation was provided by sunlight and a blue glass filter, which was part of a stained-glass window. This filter selectively transmitted light below 400 nm, which was absorbed by quinine. The incident light was prevented from reaching the detector (eye) by a yellow glass (of wine) filter. Quinine fluorescence occurs near 450 nm and is therefore easily visible.

>>> **Chapter 3** Fluorescence and Phosphorescence

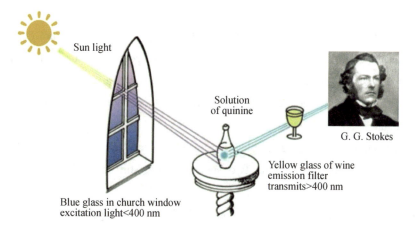

Figure 3.19 Experimental schematic for detection of the Stokes shift

It is clear from the Jablonski diagram that fluorescence always originates from the same level, irrespective of which electronic energy level is excited. The emitting state is the zeroth vibrational level of the first excited state. It is for this reason that the fluorescence spectrum is shifted to lower energy than the corresponding absorption spectrum. The Stokes shift is the distance (in wavenumbers) between the maximum of the first absorption band and the maximum of fluorescence (Figure 3.20). It is usually represented by $\Delta \nu$.

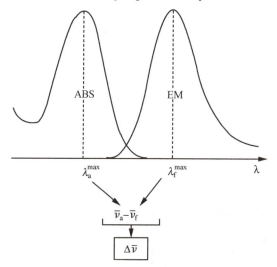

Figure 3.20 Schematic diagram of the Stokes shift

The Stokes shift has the following characteristics:

- A photon of energy $h\nu_{em}$ is emitted, returning the fluorophore to its ground state S_0. Owing to energy dissipation during the excited-state lifetime, the energy of this photon is lower, and therefore of longer wavelength, than the excitation photon $h\nu_{ex}$.

- The difference in energy or wavelength represented by $(h\nu_{ex} - h\nu_{em})$ is called Stokes shift. The Stokes shift is fundamental to the sensitivity of fluorescence techniques because it allows emission photons to be detected against a low background, isolated from excitation photons.

• The causes and magnitude of the Stokes shift can be complex and are dependent on the fluorophore and its environment. However, there are some common causes. It is frequently due to non-radiative decay to the lowest vibrational energy level of the excited state. Another factor is that the emission of fluorescence frequently leaves a fluorophore in a higher vibrational level of the ground state.

Taking Rhodamine B as an example, Figure 3.21 shows its absorption and luminescence spectra. The Stokes shift between the two spectra is the shift caused by the energy lost by vibration relaxation.

The detection of emitted fluorescence can be difficult to distinguish from the excitation light when using fluorophores with very small Stokes shifts, because the excitation and emission wavelengths greatly overlap. Therefore, fluorophores with large Stokes shifts are easy to distinguish because of the large separation between the excitation and emission wavelengths. Figure 3.22 shows two different Stokes shifts.

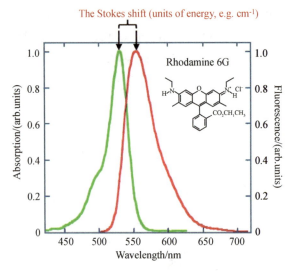

Figure 3.21 Absorption and luminescence spectra of Rhodamine B

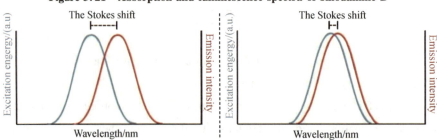

Figure 3.22 Two different Stokes shifts (left: large; right: small)

The Stokes shift is especially critical in multiplex fluorescence applications, because the emission wavelength of one fluorophore may overlap, and therefore excite another fluorophore in the same sample.

Using large Stokes-shifted fluorescent dyes in near-infrared fluorescence imaging has the advantages of low background interference, small light damage to biological samples, strong sample penetration, and high detection sensitivity. Therefore, near-infrared fluorescence imaging probes are used in chemical biology. And clinical laboratory diagnosis and other aspects show a good application prospect.

Fluorescent dyes with large Stokes shifts play a key role in avoiding self-quenching and scattered light of dyes in the process of biological imaging. For example, a novel fluorescent probe with large Stokes shifts could image mitochondria in different living cell lines for real-time. Introduction of pyridinium salt unit on the phenanthrenequinone imidazole-core can provide a new compound PI-C2. PI-C2 may exhibit large Stokes shifts and image mitochondria in different living cell lines.

Stokes shifts of PI-C2, MTR and MTG in H_2O, PBS, DMSO and MeOH have been list (Figure 3.23). The results demonstrate that probe PI-C2 has large Stokes shifts in various solutions, especially in DMSO and MeOH. The maximum value reaches to 219 nm in DMSO and MeOH. Thus, probe PI-C2 should play a key role in avoiding self-quenching in the process of biological imaging.

In some situations, choosing the proper pairs of fluorescent dyes (with large or small Stokes shifts) could be depleted using 660 nm light. In a report, the commercially available coumarin dye DY-485XL was used together with the small Stokes shift rhodamine dye NK51, which demonstrates the advantages and illustrates the principles of isoSTED (a combination of STED and 4Pi1 microscopies to improve resolution along Z-axis). The spatial distribution of two or more (biological) objects at the nanoscale could be imaged in 3D two-color isoSTED. In particular, the method of isoSTED strongly improves the resolution in z-direction: instead of being about 500 nm – 600 nm as in confocal microscopy, it becomes 30 nm – 50 nm (depending on the STED power, dye and sample properties). So put it another way, isoSTED enables uniform optical resolution in all 3D. The coumarin dye DY-485XL displays a similar emission spectrum as rhodamine dye NK51, whereas its excitation spectrum is blue-shifted by ~50 nm (Figure 3.23). This combination of fluorophores uses only one STED beam at 647 nm and two readily available excitation lasers with emission at 488 nm (for DY-485XL) and 532 nm (for NK51). In this study, mitochondrial outer membrane protein Tom20 was labeled by indirect immunostaining with coumarin DY-485XL (Dyomics GmbH), and the matrix protein mtHsp70 with rhodamine dye NK51. After that, the images were recorded by subsequent excitation of the sample at 488 nm (Channel 1) and 532 nm (Channel 2) and depletion of the fluorescence by stimulated emission triggered by 647 nm laser. As shown in Figure 3.23, rhodamine NK51 is excited also with 488 nm light, and the residual excitation cross-talk between two channels has to be removed by linear unmixing applied to the raw data. As a result, two proteins of an organelle inside the whole cell are imaged separately with nanoscale 3D resolution in the 50 nm

Fundamentals of Photophysics

range. These dual-color images demonstrate the possibilities and difficulties (such as cross-talk, uneven photobleaching) of two-color STED microscopy for the first time.

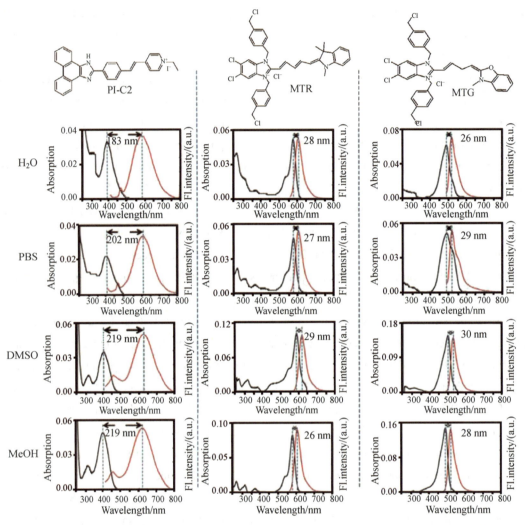

Figure 3.23 The absorption (black) and fluorescent spectra (red) in H_2O, PBS, DMSO, MeOH of the probe, MTR and MTG

3.4 Fluorescence Features

3.4.1 Fluorescence Lifetime

The fluorescence lifetime determines the time available for the fluorophore to interact with or diffuse in its environment, and hence the information available from its emission. Lifetime for single molecule is defined as the time that the molecule spends in the excited state prior to return to the ground state. In general, fluorescence lifetimes are near 10 ns. Similarly, the average lifetime is the average time that molecules spend in the excited states prior to return to the ground states.

Average lifetime (τ):

$$\tau = \frac{1}{\Gamma + k_{nr}} \quad (3\text{-}5)$$

where Γ is the emissive rate of the fluorophore, k_{nr} is the rate constant of non-radiative decay to S_0.

When non-radiative processes are not considered, the lifetime of the fluorophore is called the intrinsic or natural lifetime (τ_f):

$$\tau_f = \frac{1}{\Gamma} \quad (3\text{-}6)$$

Two methods of measuring time-resolved fluorescence are typically used, including the time-domain and frequency-domain methods. In time-domain or pulse fluorimeter measurements, the sample is excited with a pulse light (Figure 3.24). The width of the pulse needs to be short enough, and is preferably much shorter than the decay time (τ) of the sample. Following the excitation pulse, the time-dependent intensity is measured, and the decay time (τ) is calculated from the slope of a plot of ln $I(t)$ versus t, or from the time at which the intensity decreases to $1/e$ of the intensity at $t = 0$. In fact, the intensity decays are often measured through a polarizer oriented at 54.7° from the vertical z-axis. This condition is used to avoid the effects of rotational diffusion or anisotropy on the intensity decay.

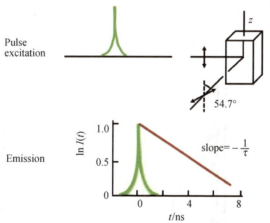

Figure 3.24 Pulse or time-domain lifetime measurements

Besides, the frequency-domain or phase-modulation method is also used to measure the decay time (τ). In this case the sample is excited with the intensity-modulated light, typically sine-wave modulation (Figure 3.25). It should be noted that the amplitude-modulated excitation should not be confused with the electrical component of an electromagnetic wave. What's more, the intensity of the incident light is varied at a high frequency typically near 100 MHz, so its reciprocal frequency is comparable to the reciprocal of the decay time (τ). When a fluorescent sample is excited in this method, the emission is forced to respond at the same modulation frequency. The lifetime of the fluorophore causes the emission to be delayed in time relative to the excitation, shown as the shift to the right in Figure 3.25. This delay is measured as a phase shift (φ). It can be used to calculate the decay time. Over here, magic-angle polarizer conditions are also used.

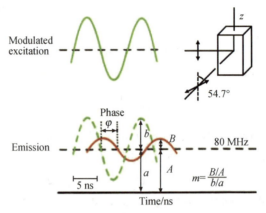

Figure 3.25 Phase-modulation or frequency-domain lifetime measurements
(The ratios B/A and b/a represent the modulation of the emission and excitation, respectively.)

Fluorescence decay is usually a single exponential decay process. Figure 3.26 shows the typical fluorescence decay curve. The lifetime of a population of fluorophores is the time measured for the number of excited molecules to decay exponentially to N/e (36.8%) of the original population via the loss of energy through fluorescence or non-radiative processes.

$$I_t = \alpha e^{-t/\tau} \tag{3-7}$$

where I_t is the intensity at time t, α is a normalization term (the pre-exponential factor).

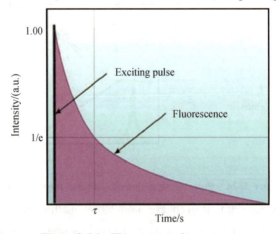

Figure 3.26 Fluorescence decay curve

The lifetime of fluorescence is an intrinsic property of fluorophore. It does not depend on fluorophore concentration, absorption, thickness of sample, measurement methods, fluorescence intensity, photo-bleaching and excitation intensity.

If the decay is single exponential and the lifetime is longer compared to that of the excitation light, then the lifetime can be determined directly from the slope of the curve. If the lifetime and the excitation piles width are comparable, certain type of deconvolution method must be used to extract the lifetime.

Fluorescence lifetime measurement has advantages in many aspects: (1) Lifetime measurements do not require wavelength-ratiometric probes to provide quantitative determination of many analytes. (2) The lifetime method expands the sensitivity of the analyte concentration range by the use of probes with spectral shifts. (3) Lifetime measurements may be used for analytes for which there are no direct probes. These analytes include glucose, antigens, or any affinity or immunoassays based on fluorescence energy transfer transduction mechanism.

An example of the use of lifetime data is given by a study of a rhodamine labeled peptide which can be cleaved by a protease, where lifetime data allow us to better understand the photophysics of this system. In the intact peptide the rhodamine molecules form a ground-state dimer with a low quantum yield (green curve in Figure 3.27). Upon cleavage of the peptide the rhodamine dimer breaks apart and the fluorescence is greatly enhanced (blue curve in Figure 3.27).

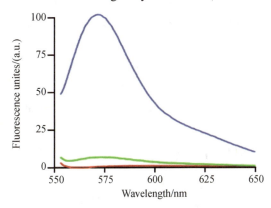

Figure 3.27 Fluorescence spectra of the rhodamine molecules

As the lifetime data in Table 3.1 indicate, before protease treatment the rhodamine lifetime was biexponential with 95% of the intensity due to a long component and 5% due to a short component. Hence one can argue that the intact peptide exists in an equilibrium between open (unquenched) and closed (quenched) forms.

Table 3.1 Fluorescence lifetime parameters of intact and cleaved labeled peptides

Sample	τ_1/ns	f_1	α_1	τ_2/ns	f_2	α_2
pepF1-5R	2.44	0.95	0.52	0.14	0.05	0.48
pepF1-5R + Pronase	2.43	1.00				
pepF1-6R	2.50	0.95	0.52	0.076	0.05	0.63
pepF1-6R + Pronase	2.50	1.00				

3.4.2 Quantum Yield

The preparation of integrated circuit chips depends on the process involving photochemistry, which has the fairly large impact scale. Therefore, it is significant to comprehend the intrinsic mechanism of photochemistry, which is beneficial to obtain the maximum reaction efficiency and reduce the sizable energy consumption of the light source. The quantum yield is a measure of how efficiently the absorbed photons are utilized. Quantum yields can provide information about the electronic excited state relaxation processes, such as the rates of radiative and non-radiative processes. Moreover, they can also find applications in the determination of chemical structures and sample purity. The emission quantum yield can be defined as the fraction of molecules that emits a photon after direct excitation by a light source. Therefore, emission quantum yield is also a measure of the relative probability for radiative relaxation of the electronically excited molecules.

During the process of fluorescence production, the excited molecules emit photons when they return to the ground state S_0. The fluorescence quantum yield (QY_f) is defined as the ratio of the number of emitted photons (over the whole duration of the decay) to the number of absorbed photons.

$$QY_f = \frac{\#\text{photon emitted}}{\#\text{photon absorbed}} = \frac{k_f}{k_f + k_{nr}} \quad (3\text{-}8)$$

where k_f is the rate constant of fluorescence emission, k_{nr} is the rate constant of non-radiative decay to S_0.

Fluorescence quantum yield is proportional to fluorescence lifetime:

$$QY = \frac{k_f}{k_f + k_{nr}} = \frac{k_f}{k} = \frac{\tau}{\tau_r} \approx \tau \quad (3\text{-}9)$$

The radiation lifetime τ_r is practically constant for a given molecule. The fluorescence lifetime τ depends on the environment of the molecule through k_{nr}.

The measurement of fluorescence lifetime is more robust than the measurement of fluorescence intensity (from which the QY is determined) because it depends on the intensity of excitation instead of the concentration of the fluorophores.

In general, the presence of heavy atoms as substituents of aromatic molecules (e.g. Br, I) results in fluorescence quenching (internal heavy atom effect) because of the increased probability of intersystem crossing. In fact, intersystem crossing is favored by spin-orbit coupling.

As shown in Table 3.2, the heavy atom has an effect on emissive properties of naphthalene.

Table 3.2   The heavy atom effect

	Φ_F	$K_{isc}/(s^{-1})$	Φ_p	τ_T/s
Naphthalene	0.55	1.6×10^6	0.051	2.3
1-Fluoronaphthalene	0.84	5.7×10^5	0.056	1.5
1-Chloronaphthalene	0.58	4.9×10^7	0.30	0.29
1-Bromonaphthalene	0.016	1.9×10^9	0.27	0.02
1-Iodonaphthalene	<0.0005	$>6.0 \times 10^9$	0.38	0.002

The heavy atom effect can be weak for some aromatic hydrocarbons if the fluorescence quantum yield is large so that de-excitation by fluorescence emission dominates all other de-excitation processes, or there is no triplet state energetically close to the singlet state (e.g. perylene).

The shortest approach to estimating the quantum yield of a fluorophore is to make comparisons with the standards of known quantum yield. Some of the most used standards are listed in Table 3.3. The excitation wavelength ordinarily has little effect on the fluorescence quantum yields, so the standards can be used wherever they display useful absorption.

Table 3.3   Standard materials and their corresponding quantum yield values in literature

Compound	Solvent	Quantum yield	Emission range / nm
Cresyl violet	Methanol	0.54	600–650
Rhodamine 101	Ethanol + 0.01% HCl	1.00	600–650
Quinine sulfate	$0.1\ mol \cdot L^{-1}\ H_2SO_4$	0.54	400–600
Fluorescein	$0.1\ mol \cdot L^{-1}\ NaOH$	0.79	500–600
Norharmane	$0.1\ mol \cdot L^{-1}\ H_2SO_4$	0.58	400–550
Harmane	$0.1\ mol \cdot L^{-1}\ H_2SO_4$	0.83	400–550
Harmine	$0.1\ mol \cdot L^{-1}\ H_2SO_4$	0.45	400–550
2-methylharmane	$0.1\ mol \cdot L^{-1}\ H_2SO_4$	0.45	400–550
Chlorophy II A	Ether	0.32	600–750
Zinc phthalocyanine	1% pyridine in toluene	0.30	660–750
Benzene	Cyclohexane	0.05	270–300
Tryptophan	Water, pH 7.2, 25℃	0.14	300–380
2-Aminopyridine	$0.1\ mol \cdot L^{-1}\ H_2SO_4$	0.60	315–480
Anthracene	Ethanol	0.27	360–480
9,10-diphenyl anthracene	Cyclohexane	0.90	400–500

The quantum yield of the unknown is universally acquired by comparing the wavelength integrated intensity of the unknown with that of the standard. During the testing, it is required to keep the optical density (absorbance) below 0.05 in order to avoid inner filter effects, unless the excitation of the optical densities of the sample matches that of the reference. At this

Fundamentals of Photophysics

condition, the quantum yield of the unknown can be calculated by the equation:

$$QY = QY_{ref} \frac{n^2}{n_{ref}^2} \frac{I}{A} \frac{A_{ref}}{I_{ref}} \qquad (3\text{-}10)$$

where QY_{ref} is the quantum yield of the reference compound, n is the refractive index of the solvent, I is the integrated fluorescence intensity (area under the plot), and A is the absorbance at the excitation wavelength. In this expression it is assumed that the sample and the reference are excited at the same wavelength, so that it is not necessary to correct for the different excitation intensities of different wavelengths. The reference standards for quantum yield measurements are as follows: (1) The lowest energy absorption band of the reference and the sample should overlap as much as possible. (2) The QY of the reference should match the expected QY of the sample. The details of the quantum yield measurement by comparing the fluorescence spectra (Figure 3.28) are shown as follows:

- Choose a standard with a quantum yield Φ_s.
- At λ_1 determine the absorption of the standard, and integrate the absorption spectrum of standard, get A_s.
- At $\lambda_{ex} = \lambda_1$ excite the standard, and integrate the emission spectrum of standard, get F.
- Repeat the 3rd step with the blank solvent. Minus the emission from blank, get F_s.
- Repeat the 2nd to the 4th steps with the sample, and get A_x and F_x.

$$\Phi_x = \Phi_s \frac{A_s}{A_x} \times \frac{F_x}{F_s}$$

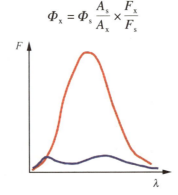

Figure 3.28 Illustration of comparing the fluorescence spectra

In order to obtain reliable fluorescence quantum yields (the relative uncertainties is less than 5%), instrument calibration and selection of fluorescence standards are important. In particular, the instrument needs to be checked for linearity over the dynamic range of spectral radiance/fluorescence intensity of the samples, which are required to be measured, spectrally characterized and corrected. In order to reduce errors caused by imperfect calibration of nonlinearity and spectral characteristics, the selection standards of fluorescence quantum yield should be carried out, according to the following criteria: ① Absorption and excitation spectrum are similar in shape and wavelength range to the samples under test; ② Emission spectrum is similar in shape and wavelength range to the samples under test; ③ Because of the overlap of excitation and emission spectra, a large Stokes shift is needed to minimize errors; ④ The quantum yield is slightly higher than that of the samples under test; ⑤ Good photostability is important to acquire a quantum yield that is insensitive to environmental conditions (temperature, oxygen levels, time of exposure, impurities); ⑥ The quantum yield is relatively insensitive to sampling conditions (e.g. excitation wavelength, pH, instrument polarization).

3.4.3 Fluorescence Intensity

The fluorescence intensity (I_f) is proportional to the concentration c. For high concentrations, linear relationship no longer holds due to self-quenching. If the quantum yield is higher, the absorbed light is mainly found in fluorescent emission. If the quantum yield is small, most absorbed energy is lost by thermal effect.

The fluorescence intensity is proportional to the luminous I_a absorbed and the fluorescence quantum efficiency φ:

$$I_f = \varphi I_a \quad (3-11)$$

According to the Lambert-Beer law:

$$I_a = I_0(1 - 10^{-t/c}) \quad (3-12)$$

$$I_f = \varphi I_0(1 - 10^{-t/c}) = \varphi I_0(1 - e^{-2.3t/c}) \quad (3-13)$$

where I_0 is the intensity of the incident light, t is the time, c is the concentration.

In general, the absorbance A of the solution does not exceed 0.05.

It is important to recognize if the concentration is in the limited range of optical densities where fluorescence intensities are proportional to the concentration (Figure 3.29). Therefore, two methods are used to estimate the fluorescence intensity of an unknown sample.

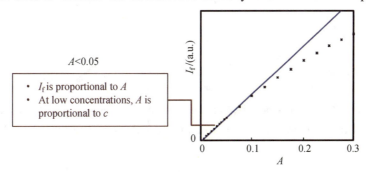

Figure 3.29 Relationship between fluorescence and optical absorbance

- The standard curve method describes the utilization of a series of standard concentration patterns to draw the standard curve. Then the fluorescence intensity of the unknown samples is measured under the same conditions, and the concentration can be calculated on the standard curve.
- The comparison method is that the fluorescence intensities of the standard samples and the unknown samples are measured in a linear range and then compared.

Fundamentals of Photophysics

The brightness is defined as the number of photons emitted per second by a single fluorophore observed by a given set of optical conditions. Different from the quantum yield, the brightness is not a molecular property of the fluorophore, but a fundamental property depends on the precise optical conditions including the light intensity, the light collection efficiency of the instrument, and the counting efficiency of the detector. It is proportional to the product of the extinction coefficient (ε) and the quantum yield (Φ):

$$\text{Brightness} \propto \varepsilon \Phi \tag{3-14}$$

The brightness of a fluorophore labelled molecule is proportional to the extinction coefficient (ε), the quantum yield (Φ) and the number of dyes per molecule (n):

$$\text{Brightness} \propto n\varepsilon\Phi \tag{3-15}$$

3.4.4 Fluorescence Polarization

Polarization of fluorescence is determined by the orientation of the fluorophores' transition moment at the instant of fluorescence emission. This gives an opportunity to determine rotational diffusion of fluorophores by detecting the anisotropy of their fluorescence.

In 1926, Francis Perrin described the theoretical basis of fluorescence polarization based on the observation that the emission from a small fluorescent molecule excited by plane-polarized light is largely depolarized due to rotational diffusion during the lifetime of the fluorescence. If linearly polarized light is used to excite an ensemble of fluorophores, only those fluorophores aligned with the plane of polarization will be excited (Figure 3.30).

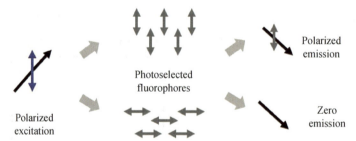

Figure 3.30 Linearly polarized light used to excite an ensemble of fluorophores

Reversely, if those fluorophores are random oriented, it will result in unpolarized emission (Figure 3.31).

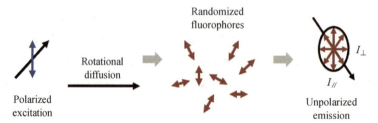

Figure 3.31 Fluorophores random oriented

As shown in Figure 3.32, small molecules rotate fast in solution and exhibit depolarized light. Large molecules rotate slower and exhibit a higher degree of polarization. Thus, changes in fluorescence polarization can reflect the association or dissociation between molecules.

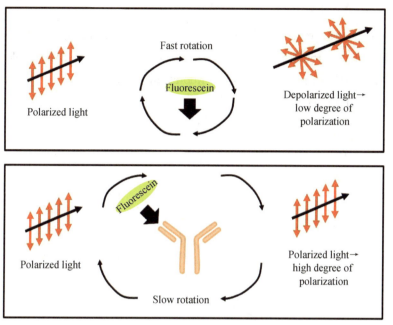

Figure 3.32 Schematic diagram of the change of fluorescence polarization

In the 1950s, Weber extended Perrin's theoretical work to ellipsoids of revolution, and invented the first instrument to measure fluorescence polarization. To measure fluorescence polarization in the laboratory, a fluorescent sample is excited by polarized light and emission intensities are collected from channels that are parallel ($I_{//}$) and perpendicular ($I_\perp$) to the electric vector of the excitation light.

Numerically, the fluorescence polarization can be expressed as Figure 3.33.

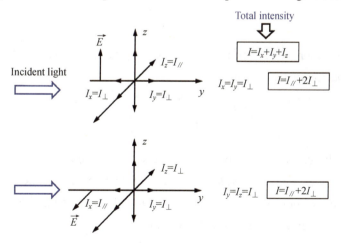

Figure 3.33 Illustration of the fluorescence polarization

Fundamentals of Photophysics

I_x, I_y, and I_z are the intensities of these sources, and the total intensity is $I = I_x + I_y + I_z$. If the emission polarizer is positioned parallel ($//$) to the direction of the polarized excitation, the observed intensity is called $I_{//}$. If the polarizer is perpendicular ($\perp$) to the excitation, the intensity is called $I_\perp$.

The measurement approach is illustrated in Figure 3.34. A beam of light emits from the light source through a vertical polarizer as vertically polarized light. The sample is excited by vertically polarized light to produce polarized fluorescence and the intensity of the emission can be measured through a polarizer.

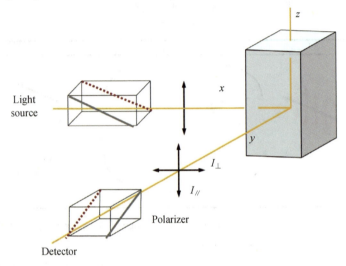

Figure 3.34 Schematic diagram for measurement of fluorescence anisotropies

The polarization state of fluorescence is characterized either by the polarization ratio (P) or the emission anisotropy (r):

$$P = \frac{I_{//} - I_\perp}{I_{//} + I_\perp} \quad (3\text{-}16)$$

$$r = \frac{I_{//} - I_\perp}{I_{//} + 2I_\perp} \quad (3\text{-}17)$$

Anisotropy is preferred because it is normalized by the total intensity $I = I_{//} + 2I_\perp$, which results in simplification of the equations.

The fluorescence polarization is influenced by many factors:

$$\frac{1}{P} - \frac{1}{3} = \left(\frac{1}{P_0} - \frac{1}{3}\right)\left(1 + \frac{RT}{V_\eta}\tau\right) \quad (3\text{-}18)$$

Where P_0 is the limiting intensity of fluorescence polarization. R is gas constant. T is the temperature. V is molar molecular volume. τ is the fluorescence lifetime of fluorophore. η is the viscosity of the medium. When the temperature and viscosity of the solution are fixed, the P value is mainly determined by the molecular volume.

The principles of fluorescence polarization have been extensively applied in assays to enable high-throughput screening of small molecule libraries for the purposes of probe development and drug discovery. A search of the PubChem BioAssay database for "fluorescence polarization" and "fluorescence anisotropy" on October 4, 2015 identified 1,000 unique fluorescence polarization and fluorescence anisotropy assays. As shown in Figure 3.35, fluorescence polarization assays have been developed to interrogate an impressive diversity of biological target classes, including receptors, ion channels, epigenetic regulators, transcription factors, kinases, proteases and isomerases. This emphasizes the utility of fluorescence polarization assays to report on a range of critical biological activities for HTS applications.

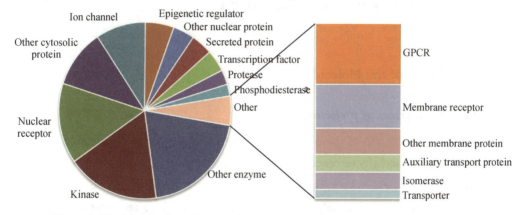

Figure 3.35 Distribution of targets interrogated by fluorescence polarization assays in drug discovery

Another application of fluorescence polarization is the determination of crystal axis. As depicted in Figure 3.36, the left panels are microscopic images of the emission from GFP crystals while the right panels show color images of the crystals when transilluminated with polarized white light.

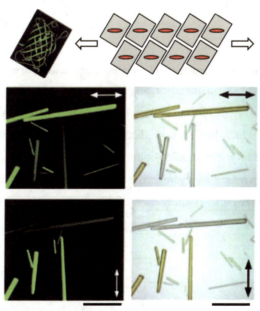

Figure 3.36 Optical images of the GFP crystals

Fundamentals of Photophysics

The crystals aligned with the incident polarization are yellow. However, the crystals appear very pale blue clearly when rotated 90° from the incident polarization. The results indicate that the crystals only absorb the polarized light with which the transition moment is aligned. Therefore, it is easy to deduce that the chromophore in GFP is aligned along the long axis of the crystals.

3.5 Factors Affecting Fluorescence

3.5.1 Effect of the Molecular Structure

i. Effect of the extent of p-electron system

An increase in the extent of the p-electron system (i.e. degree of conjugation) leads to a shift of the absorption and fluorescence spectra to longer wavelengths and to an increase in the fluorescence quantum yield. This simple rule can be illustrated by a series of aromatic hydrocarbons (Figure 3.37): naphthalene, anthracene, naphthalene and pentacene fluorescein ultraviolet, blue, green and red, respectively.

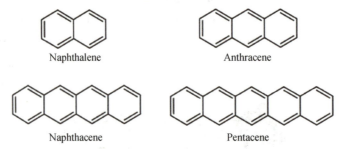

Figure 3.37 Structures of a series of aromatic hydrocarbons

As shown in Figure 3.38, a metal-organic framework (MOF) is designed and prepared from luminescent Tb (III), adenosine diphosphate (ADP) and bipyridyl (Bipy). Its green fluorescence at 545 nm, which has been marked, enables the fluorometric detection of cyanide ion based on the principle of π-conjugation-induced fluorescence enhancement. The fluorescence of the probe is strongly increased by cyanide due to extended π-conjugation between probe MOF and cyanide which sensitizes the fluorescence of Tb (III). This effect has been applied to quantify cyanide at levels as low as 30 nM in aqueous solution.

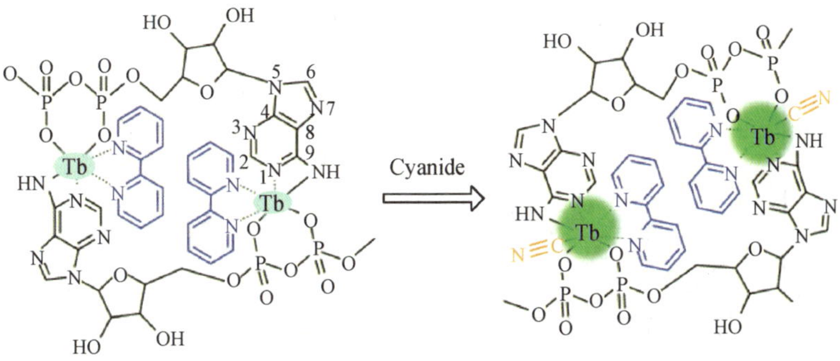

Figure 3.38 Fluorescence enhancement mechanism of Tb-ADP-Bipy

ii. Effect of substituent groups

Substituent groups have a noticeable effect on the fluorescence quantum yield of many compounds. Electron-donating groups such as -OH, $-NH_2$ and $-NR_2$ intensify the fluorescence efficiency, while electron-withdrawing groups such as -CHO, $-CO_2H$ and $-NO_2$ reduce it, as shown by naphthalene and its derivatives in Table 3.4.

Table 3.4 The effect of substituent groups on fluorescence efficiency of naphthalene and its derivatives, and fluorescence quantum yields measured in fluid solution at room temperature

Compound	Φ_f
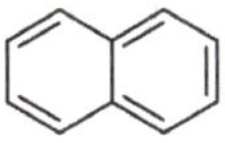 naphthalene	0.19
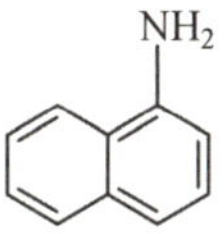 1-naphthylamine (NH_2)	0.38
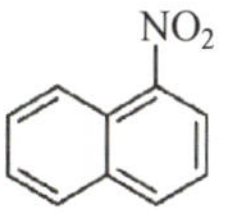 1-nitronaphthalene (NO_2)	0.001

Previous studies have reported that substituent affects the properties of pH fluorescence probes containing pyridine group. A series of 4-styrylpyridine (SYP) pH fluorescence probes with different substituents ($-NH_2$, -OH, -H, -Br, $-NO_2$) on para-benzene ring are developed. As shown in Figure 3.39, the designed SYP derivatives serve as pH ratiometric fluorescence probes in acidic region and almost the entire acid range can be covered by pH response range of SYP-A, SYP-B, SYP-C. The relationship between pK_a and Hammett constant of substituent appears good linearity. These compounds exhibit strong pH-dependent performance so that their spectra red-shift upon protonation of pyridine N-atom and blue-shift upon deprotonation. These sensors employ the substituent on the para-benzene ring as electron donor and are functioned by intramolecular charge transfer (ICT) under acidic pH conditions.

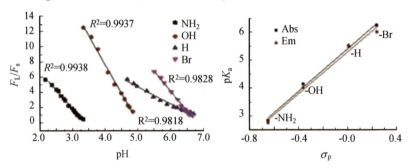

Figure 3.39 (a) The linearity between fluorescence intensity and pH of SYP-A ~ D;
(b) Correlation of pK_a with Hammett constant (σ_p) for various substituents

iii. Effect of heavy atom

In general, the presence of heavy atoms as substituents of aromatic molecules (e.g. Br, I) results in fluorescence quenching (internal heavy atom effect). This is because the probability of intersystem crossover increases. The heavy atom effect of halogens reduces fluorescence, and in molecules containing heavy atoms, the probability of band crossing increases, which weakens fluorescence and enhances phosphorescence. The heavy atom effect is illustrated in Table 3.5 and 3.6.

Table 3.5 The effect of the internal heavy atom on the fluorescence efficiency of naphthalene and its derivatives, and fluorescence quantum yields determined in solid solution at 77 K

Compound	Φ_f
naphthalene	0.55
1-bromonaphthalene	0.0016
1-iodonaphthalene	0.0005

Table 3.6 The effect of the external heavy atom on the fluorescence efficiency of naphthalene, and fluorescence quantum yields determined in solid solution at 77 K

Solvent	Φ_f
ethanol/methanol	0.55
1-bromopropane	0.13
1-iodopropane	0.03

iv. Effect of rigid structure

The rigidity of the molecule prevents the loss of energy through rational and vibrational energetic level changes. Any subsistent on a luminescent molecule that can cause increased vibration or rotation can quench the fluorescence. The planer structure of fluorescent compounds allows delocalization of the π-electrons in the molecule. That increases the chance in turn that luminescence can occur because the electrons can move to the proper location to relax into a lower energy localized orbital.

Molecular rigidity can be improved either by increasing the structural rigidity of the molecule (e. g. by preventing rotation or bending of bonds) or by increasing the rigidity of the medium (e. g. use a rigid glass made by the frozen fluid solution instead of a fluid solution at room temperature). Molecular rigidity favors efficient fluorescence emission, as illustrated in Table 3.7.

The fluorescence quantum yield of trans-stilbene is 0.75 when measured in a rigid glass at 77 K, showing that the rigid medium results in more efficient fluorescence.

Table 3.7 The effect of molecular rigidity on fluorescence quantum yield measured in solution at room temperature

Compound	Structure	Φ_f
trans-stilbene		0.05
5,10-dihydroindeno[2,1-a] indene		1.00
biphenyl		0.15
fluorene		0.66

3.5.2 Effect of the Environment

i. Effect of the temperature

The temperature of the environment has a large effect on fluorescence, and a greater effect on phosphorescence. The radiative transition has almost nothing to do with temperature and the non-radiative transition increases significantly with temperature.

Generally, the temperature increase results in the decrease of the fluorescence quantum yield and the lifetime because the non-radiative processes related to thermal agitation (collisions with solvent molecules, intramolecular vibrations and rotations, etc.) are more efficient at higher temperatures. As mentioned above, phosphorescence is observed only under certain conditions because the triplet states are very efficiently deactivated by collisions with solvent molecules (or oxygen and impurities). These effects can be reduced and may even disappear when the molecules are in a frozen solvent, or in a rigid matrix (e. g. polymer) at room temperature.

At low temperatures, the solvent becomes more viscous, so the process of solvent reorientation takes more time. Figure 3.40 shows a simplified description of solvent relaxation. Upon excitation, it is assumed that the fluorophore is initially in the Franck-Condon state (F) and solvent relaxation proceeds with a rate k_S. If the rate of solvent relaxation is much slower than the decay rate ($\gamma = 1/\tau$), the emission spectrum of the unrelaxed F state will be detected with large probability. If the rate is much faster than the emission rate ($k_S \gg \gamma$), then emission from the relaxed state (R) will be observed. At intermediate temperatures, where $k_S \approx \gamma$, emission and relaxation will occur simultaneously. Under these conditions an intermediate emission spectrum will appear. Generally, this intermediate spectrum is broader on the wavelength scale due to the contributions from both the F and R states.

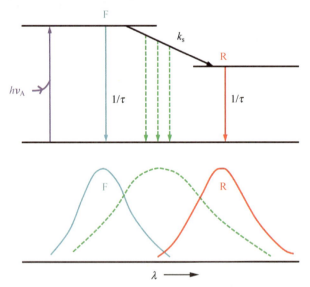

Figure 3.40 The Jablonski diagram for solvent relaxation

ii. Solvent polarity

The fluorophore is typically excited to the first singlet state (S_1), usually to an excited vibrational level within S_1 and the excess vibrational energy is rapidly lost to the solvent. If the fluorophore is excited to the second singlet state (S_2), it will rapidly decay to the S_1 state in 10^{12} s due to the internal conversion. Solvent effects shift the emission to still lower energy due to the stabilization of the excited state by the polar solvent molecules. Commonly, the dipole moment of the fluorophore in the excited state (E) is larger than that in the ground state (G). Following excitation, the solvent dipoles can reorient or relax around E, which decreases the energy of the excited state. As the solvent polarity is increased, this effect enhances, which causes energy emission at lower energy states or longer wavelengths. Normally, only fluorophores that are themselves polar, display a large sensitivity to solvent polarity, while nonpolar molecules, such as unsubstituted aromatic hydrocarbons, are much less sensitive to solvent polarity. For example, the solvent polarity effects of CCl_4, $CHCl_3$ and CH_3CN are illustrated in Table 3.8.

Table 3.8 The solvent polarity effect on fluorescence wavelength and fluorescence quantum yield

Solvent	Dielectric constant	λ_f	Φ_f
CCl_4	2.24	390	0.002
$CHCl_3$	5.2	398	0.041
CH_3CN	38.8	410	0.064

Solvent polarity can play an important role in emission spectra. Figure 3.41 shows a photograph of the emission from 4-dimethylamino-4'-nitrostilbene (DNS) in solvents with increasing polarity and the emission spectra are shown in the lower panel. The color shifts from deep blue ($\lambda_{max} = 450$ nm) in hexane to orange in ethyl acetate ($\lambda_{max} = 600$ nm), and red in n-butanol ($\lambda_{max} = 700$ nm).

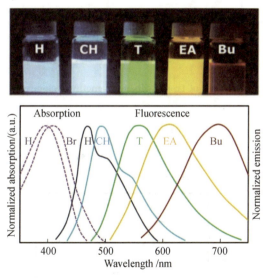

Figure 3.41 Photograph and emission spectra of DNS in solvents of increasing polarity (H, hexane; CH, cyclohexane; T, toluene; EA, ethyl acetate; Bu, n-butanol)

iii. Solvent viscosity

A decrease in solvent viscosity also increases the probability of external conversion and leads to the decrease in quantum efficiency. The fluorescence of a molecule is decreased by solvents containing heavy atoms or other solutes with such atoms in their structure; carbon tetrabromide and ethyl iodide are examples. The effect is similar to what occurs when heavy atoms are substituted into fluorescing compounds; orbital spin interactions result in an increase in the rate of triplet formation and a corresponding decrease in fluorescence. Compounds containing heavy atoms are frequently incorporated into solvents when enhanced phosphorescence is desired.

Viscosity can have a dramatic effect on the emission intensity of fluorophores. For example, the intensity decays of stilbene (Figure 3.42) rely on temperature strongly. The rotation about the central ethylene double bond in the excited state is considered as the inducement of the effect commonly (Figure 3.43). In the ground state there is a large energy barrier to rotation about this bond, whereas the energy barrier is much smaller in the excited state. Rotation about this bond occurs in about 70 ps, providing a return path to the ground state. Rotation of cis-stilbene is even more rapid, and it also leads to very short fluorescence lifetime. Such rotations in the excited state are regarded as the significant factor to affect the emission of many other fluorophores.

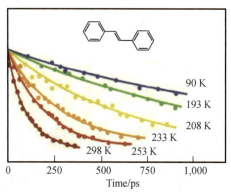

Figure 3.42 Intensity decays of trans-stilbene in methylcyclohexane: isohexane (3:2)

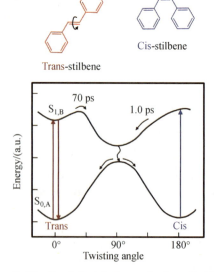

Figure 3.43 Excited-state isomerization of stilbene

iv. pH effect

The fluorescence of aromatic compound with basic or acid substituent rings are usually pH-dependent. For example, 1-naphthol-4-sulfonic acid, a compound that is not detectable with the eye because it occurs in the ultraviolet region, is converted to phenol ions with the addition of a base, and the emission band shifts to the visible wavelength where it can be visually seen (Figure 3.44).

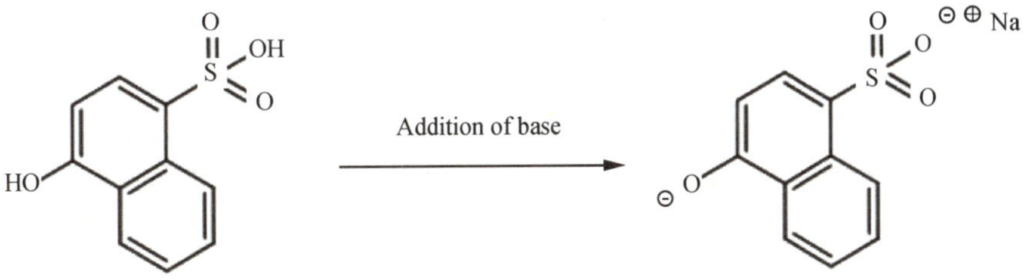

1-naphtohol-4-sulfonic acid (Phenolic form) → Addition of base → 1-naphtohol-4-sulfonic acid (Phenolate ion)

Figure 3.44 1-naphthol-4-sulfonic acid affected by adding base

Coumarin (2H-1-benzopyrone-2-one), a colorless crystalline solid, has a sweet Vanilla-like smell. (Figure 3.45) It is naturally found in many plants and contains huge category of phenolic material present in plants, behaving as a chemical protecting agent against chemicals. Its derivatives are a type of organic heterocycle that has a wide range of biological activities and can be used in a variety of natural and artificial medicine molecules.

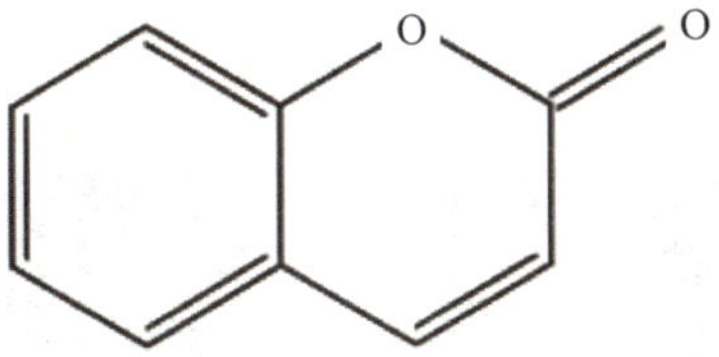

Coumarin

Figure 3.45 Structural formula of coumarin

Figure 3.46 shows the pH dependence of fluorescent probe of Coumarin (7-diethylaminocoumarin-3-aldehyde) in the presence of bisulfite ion. The studies have shown the pH dependence of fluorescent probe of Coumarin (7-diethylaminocoumarin-3-aldehyde) in the presence of bisulfite ion. The pH range selected is from 2.0 to 8.0 and it is prepared by mixing $0.2 \text{ mol} \cdot \text{L}^{-1}$ Na_2HPO_4 and $0.1 \text{ mol} \cdot \text{L}^{-1}$ citric acid in specific volume ratios. Initially, the change in fluorescence is very slight in free probe in Na_2HPO_4 citric acid buffer. However, fluorescence peak increases sharply when pH increases from 3.0 to 5.0 for the probe solution in the presence of bisulfite anion.

Fundamentals of Photophysics

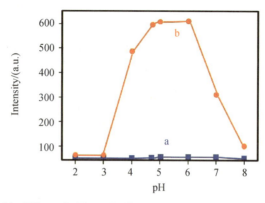

Figure 3.46 Effect of pH on the fluorescence intensity of the coumarin (7-diethylaminocoumarin-3-aldehyde), in the presence and absence of bisulfite ion probe

v. Dissolved oxygen effect

The presence of dissolved oxygen often reduces the intensity of fluorescence in a solution. This effect may originate from a photochemically induced oxidation of the fluorescing species. More commonly, however, the quenching takes place as a consequence of the paramagnetic properties of molecular oxygen, which promotes intersystem crossing and conversion of excited molecules to the triplet state. Other paramagnetic species also tend to quench fluorescence.

For instance, the typical oxygen-sensitive composite coating could quench the fluorescence. The O_2 quenching effect on a silica-Ni-P composite coating can be visualized and acquired by an inverse fluorescence microscope, as shown in Figure 3.47. It is facile to observe the change in the red color intensity of the silica-Ni-P composite coating, which proves that the O_2-sensitive film has been effectively quenched by O_2.

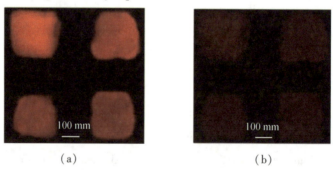

Figure 3.47 Fluorescence microscopy images of the silica-Ni-P composite coating: (a) in the absence of O_2 and (b) in the presence of air (21%, V/V, O_2)

3.6　Phosphorescence

3.6.1　Definition of Phosphorescence

The phosphorescence emission takes place from a triplet excited state. Phosphorescence is in principle a forbidden transition but it is weakly allowed through spin-orbit coupling. As shown in Figure 3.48, phosphorescence arises as the result of a radiative transition between states of different multiplicity, $T_1 \rightarrow S_0$.

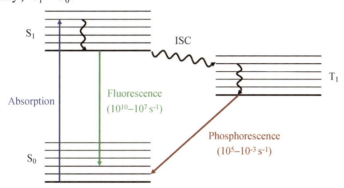

Figure 3.48　Schematic diagram of phosphorescence generation mechanism

Since the phosphorescence emission process is spin-forbidden, it has a much smaller rate constant, k_p, than that of fluorescence, k_f.

$$k_f(10^{-6} - 10^{-9} \text{ s}) > k_p(10^{-3} - 100 \text{ s})$$

Population of the triplet manifold by direct singlet-triplet absorption is a very inefficient process, being spin-forbidden. Instead, the triplet manifold is populated indirectly by excitation into the singlet manifold followed by intersystem crossing.

When the excited triplet state is populated, the process of vibrational relaxation and possibly internal conversion may occur (if intersystem crossing takes place to an excited triplet of higher energy than T_1). Thus, the excited molecule will relax to the lowest vibrational level of the T_1 state, and then phosphorescence emission will occur in compliance with Kasha's Rule.

Phosphorescence is spin-forbidden and thus phosphorescence emission is weaker and slower than fluorescence. Notice that in Figure 3.49 the intensity of absorption (molar absorption coefficient) and emission is plotted against wavenumber, which is proportional to energy. Because the energy of T_1 is lower than that of S_1, the phosphorescence spectrum is always found at lower wavenumbers (longer wavelengths) than the fluorescence spectrum.

Fundamentals of Photophysics

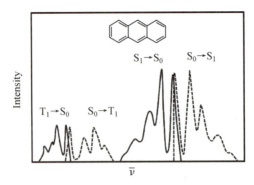

Figure 3.49 Absorption spectra (dashed line) and emission spectra (solid line) of anthracene in solution in cyclohexane

3.6.2 Phosphorescence Features and Measurement

Phosphorescence process can be described using the same parameters: quantum yield, lifetime and intensity. Compared with fluorescence, phosphorescence has a longer wavelength and lifetime. Phosphorescence lifetime and intensity are sensitive to heavy atoms and oxygen (spin-orbit coupling increases the deactivation rate constant).

i. Quantum yield

It has been considered that the triplet state is produced by intersystem crossing from S_1, and the phosphorescence emission from T_1 state is imperfect.

Thus, the triplet quantum yield is given by:

$$\Phi_T = \frac{k_{isc(ST)}}{k_{isc(ST)} + k_{ic} + k_f} = k_{isc(ST)} \cdot \frac{1}{\tau} \tag{3-19}$$

k_f, k_{isc} and k_{ic} are the first order rate constants of fluorescence, intersystem crossing and internal conversion, respectively.

The fraction of triplet states that phosphoresce is given by the phosphorescence quantum efficiency (θ_P):

$$\theta_P = \frac{k_p}{k_p + \Sigma k_{nr(ST)}} \tag{3-20}$$

The phosphorescence quantum yield (Φ_P) (the fraction of photons emitted from T_1 when S_1 is excited) is given by:

$$\Phi_P = \Phi_T \theta_P \tag{3-21}$$

The value of the phosphorescence quantum yield can be determined by measuring the total luminescence spectrum under steady irradiation. If the fluorescence quantum yield is known, the phosphorescence quantum yield can be calculated by comparing the relative areas under the two corrected spectra.

ii. Lifetime

The concentration of molecules in the triplet state decays exponentially with a time constant

τ_{ph} representing the lifetime of the triplet state (phosphorescence lifetime):

$$\tau_{ph} = \frac{1}{k_r^T + k_{nr}^T} \qquad (3\text{-}22)$$

For organic molecules, the lifetime of the singlet state ranges from tens of picoseconds to hundreds of nanoseconds ($10^{-12} - 10^{-7}$ s), whereas the triplet lifetime is much longer ($10^{-6} - 10^2$ s). As shown in Figure 3.50, the phosphorescence spectrum is located at wavelengths longer than the fluorescence spectrum because the energy of the lowest vibrational level of the triplet state (T_1) is lower than that of the singlet state (S_1).

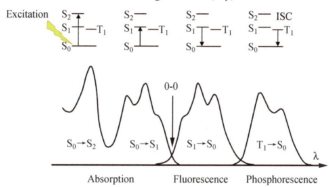

Figure 3.50 Comparison between phosphorescence and fluorescence spectrum

For all that, such a difference cannot be used to make a distinction between fluorescence and phosphorescence because some inorganic compounds (for instance, uranyl ion) or organometallic compounds may have a long lifetime.

3.6.3 Low-temperature and Room-temperature Phosphorescence

Because $S_1 \rightarrow T_1$ absorption has a very small molar absorption coefficient (ε), it is expected that the T_1 state has a much longer luminescent lifetime (τ_0) than that of the S_1 state (because of the inverse relation between ε and τ_0). As a result of this longer lifetime, the T_1 state is particularly susceptible to quenching. This leads to not readily observe the phosphorescence in fluid solution as the T_1 state is quenched before emission can occur. This quenching in solution involves the diffusion together of either two T_1 molecules or the T_1 molecule and a dissolved oxygen molecule or some impurity molecules. In order to observe phosphorescence, it is necessary to reduce or prevent the above-mentioned diffusion processes, which are mainly influenced by temperature. Phosphorescence is usually not observed in fluid solution near room temperature. One reason of this phenomenon is the long lifetime of phosphorescence and the presence of dissolved oxygen and other quenchers. For instance, the phosphorescent lifetime of tryptophan in water has been reported to be 1.2 ms in the absence of oxygen. In order to observe phosphorescence, the low-temperature and room-temperature phosphorescence techniques are most often used.

Fundamentals of Photophysics

In low-temperature techniques, the liquid nitrogen (77 K) is used to freeze the solution. The used solvent needs to meet two conditions. One of the conditions is that it is easy to purify and has no strong absorption and emission in the analysis wavelength region, and the other condition is that the transparent rigid vitreous body with sufficient viscosity can be formed at low temperature. The fluorescence is determined by using a rotating-can phosphoroscope (Figure 3.51). In addition to phosphorescence, fluorescence also normally exists. The two forms of luminescence can be differentiated by exploiting the lifetimes of phosphorescence and fluorescence, because T_1 states are much longer-lived than S_1 states. The rotation of the rotating can is set so that the path to the detector is blocked when the exciting light reaches the sample and open when the exciting light is blocked and the fluorescence has decayed.

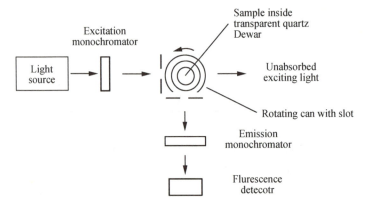

Figure 3.51 Schematic diagram of a rotating-can phosphoroscope

In room-temperature techniques, it is achieved by rigidizing the sample point on a solid substrate or by using a protective medium to reduce non-radiative impact inactivation and quenching effects such as oxygen. The solid matrix (such as cellulose, etc.) is used as the adsorbed phosphor at room temperature to increase the molecular rigidity and reduce the triplet quenching and other non-radiative transitions, so as to improve the phosphor quantum efficiency. Besides, the use of surfactants in the critical concentration of the formation of a heterogeneous micelle, change the phosphorescent microenvironment, increase the directional binding, so as to reduce the internal conversion and collision, such as the probability of deactivation, and improve the triplet stability.

References

[1] Wang, H., Wang, J. & Zhang, T. Breaching Kasha's rule for dual emission: mechanisms, materials and applications [J]. *J. Mater. Chem. C.*, 2021(9): 154 – 172.

[2] Birks, J. B. *Photophysics of Aromatic Molecules* [M]. New York: John Wiley &

Sons, 1970.

[3] Meng, F., et al. A novel fluorescent probe with a large Stokes shift for real-time imaging mitochondria in different living cell lines[J]. *Tetrahedron. Lett.*, 2017(58): 3287 – 3293.

[4] Zwinkels, J. C., DeRose, P. C. & Leland, J. E. Spectral fluorescence measurements [J]. *Experimental Methods in the Physical Sciences*, 2014(46): 221 – 290.

[5] Hall, M. D., et al. Fluorescence polarization assays in high-throughput screening and drug discovery: a review [J]. *Methods. Appl. Fluores.*, 2016(4): 022001(1 – 20).

[6] Xiong, X., Xiao, D. & Choi, M. M. F. Dissolved oxygen sensor based on fluorescence quenching of oxygen-sensitive ruthenium complex immobilized on silica-Ni-P composite coating[J]. *Sensor. Actuat. B:Chem.*, 2006(117): 172 – 176.

Fundamentals of Photophysics

Chapter 4

Supramolecules in Photophysics

4.1 Introduction to Supramolecular Chemistry

4.1.1 Definition of Supramolecule

Supramolecule refers to a system in which two or more molecular entities hold together and organize through intermolecular (non-covalent) binding interactions (Figure 4.1). A supramolecule is a well-defined distinct system generated through interactions between a molecule (receptor or host) with convergent binding sites such as donor atoms, sites for formation of hydrogen bonds and sizable cavity, and another molecule (analyte or guest) with different binding sites such as hydrogen bond acceptor atoms.

The term "supermolecule" (or "supramolecule") was introduced by Karl Lothar Wolf et al. (übermoleküle) describing hydrogen-bonded acetic acid dimers in 1937. The study of non-covalent association of molecular complexes has been developed into the field of supramolecular chemistry. The term supermolecule is sometimes used to describe supramolecular assemblies, which are complexes of two or more non-covalently bonded molecules (often macromolecules). The term supermolecule is also used in biochemistry to describe complexes of biomolecules, such as peptides and oligonucleotides composed of multiple chains.

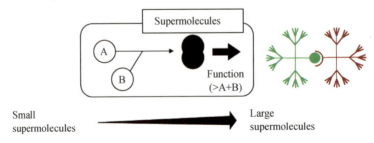

Figure 4.1 Schematic illustration of a supramolecule

Molecular assemblies are produced by spontaneous self-assembly of multiple components under a given set of conditions. The guiding force of the self-assembling process is derived from molecular binding events. In short, for the generation of ordered molecular assemblies, the process is supramolecular, but the method is molecular.

The mechanism of self-assembly of molecules in solution can be broadly classified into two categories based on the dependence of the association constant K on the size of oligomers between species. These are isodesmic and cooperative mechanisms. As the name suggests, in the isodesmic mechanism (or equal-K model), the association constant is independent of the size of the oligomer. The degree of polymerization is found to be broad in isodesmic processes, even under favorable thermodynamic conditions. On the other hand, the cooperative mechanism of self-assembly, also known as the nucleation-elongation process, has two or more equilibrium constants of different oligomer sizes. The formation of the nuclei is related to a nucleation equilibrium constant (K_n), while the elongation process is governed by another equilibrium constant (K_e). Supramolecular polymers formed by such mechanisms can be topologically 1-dimensional, 2-dimensional or 3-dimensional structures.

Organic molecules that self-assemble to form 1-dimensional supramolecular polymers, typically, have three components [see schematic in Figure 4.2 (a)]: an aromatic planar core, a self-assembling moiety and a linker connecting them which can be used to provide functionality to the molecule. The self-assembling moiety will usually contain one or more groups larger than an alkyl group (such as, say, cholesterol). The optimal inter-alkyl distance between adjacent molecules in a 1-dimensional supramolecular polymer is around 4.2 Å, while the corresponding aromatic cores prefer to be located at distances of around 3.5 Å. In order to meet these two needs, the polymer exhibits a helical twist as shown in Figure 4.2 (b).

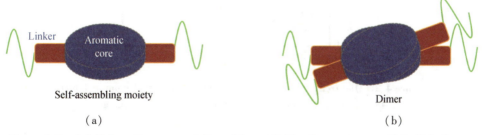

Figure 4.2 (a) Schematic representation of key moieties of a monomer and its (b) dimer, exhibiting a helical twist

Chirality is often a central theme in supramolecular polymers because the preferred helical handedness of the polymer can be used as an experimental probe by circular dichroism spectroscopy. While achiral molecules will form 1-dimensional stacks of either handedness in equal proportions, stacks formed by molecules with a chiral center show a preference for one particular helical orientation. The position of the chiral center in the self-assembling moiety is not only seen to determine the helical orientation of the supramolecular stack, but also affects its stability. The solvent also plays an important role in the conformation of the molecule in a stack. They can be embedded into the stacks, thereby influencing its conformation. The orientation of functional groups, mainly at the linker moiety, can give rise to several low-lying geometries of the monomer, the existence of which can be obtained from crystallographic studies.

Superstructures formed by 1-dimensional supramolecular polymers are largely fibers. These fibers exhibit dynamic self-assembly and disassembly in response to environmental changes, which is the advantages of supramolecular polymers over the traditional ones. Therefore, in this review, we focus on modelling studies of 1-dimensional polymers, a topic recently reviewed by Bochicchio and Pavan. The experimental interest in these materials is from the perspective of optoelectronic applications. UV-vis absorption and fluorescence spectroscopy are the main experimental tools used. These spectra, in turn, can be calculated for single molecules as well as for very short oligomers in gas phase using quantum chemical calculations. The oligomerization process can be studied using all-atom molecular dynamics simulations, while nano- and meso-scale structures and organizations can be studied using coarse grain methods. Formation of fibers and their morphology from polymers can be studied using continuum methods. The length and time scales associated with these processes and techniques are shown in Figure 4.3.

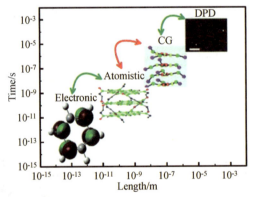

Figure 4.3 Hierarchy of multiscale modeling of supramolecular polymers
(DPD = dissipative particle dynamics, CG = coarse-grained)

4.1.2 Difference Between Molecular Chemistry and Supramolecular Chemistry

i. What is supramolecular chemistry?

Supramolecular chemistry is the branch of chemistry that deals with chemical systems that contain a discrete number of molecules. Different interactions can exist between these molecules, including intermolecular forces, electrostatic forces, hydrogen bonds, strong covalent bonds, etc. Supramolecular chemistry deals mainly with weak, reversible non-covalent bonds. Such bonds include metal coordination, hydrophobic forces, van der Waals forces, π-π interactions, and electrostatic interactions.

Supramolecular chemistry has been defined as the "chemistry of molecular assemblies and of the intermolecular bond" by one of its main proponents, Jean-Marie Lehn, who won the Nobel Prize in 1987 for his work in this field. More informally, this may be expressed as "chemistry beyond the molecule". Other definitions include such phrases as "the chemistry of the non-covalent bond" and "non-molecular chemistry".

Supramolecular chemistry can be defined as "chemistry beyond the molecule", bearing on the organized entities of higher complexity that results from two or more chemical species bound together by intermolecular forces. Its development requires the utilization of all resources of molecular chemistry, combined with the designed manipulation of non-covalent interactions, to form supramolecular entities, that is, supermolecules possessing features as well defined as those of molecules themselves. One might say that supermolecules-to-molecular and the intermolecular bond are like molecules to atoms and the covalent bond.

In addition, some advanced chemical concepts are discussed under supramolecular chemistry. These include molecular self-assembly (Figure 4.4), molecular folding, molecular recognition, dynamic covalent chemistry, etc. Since this subject area reveals the behavior of non-covalent interactions, it is very important for understanding many biological processes that rely on these chemical interactions.

When considering the concepts in supramolecular chemistry, the most important branch is the molecular assembly, where we discuss building systems without guidance from an external source. In other words, this concept explains how molecules can be guided to assemble through non-covalent interactions.

Molecular complexation and recognition is another important area of supramolecular chemistry, including the specific binding of a guest molecule to a complementary host molecule to form a host-guest complex.

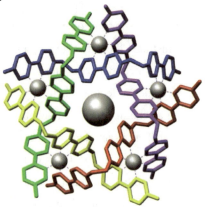

Figure 4.4　An example of molecular assembly

Fundamentals of Photophysics

Supramolecular chemistry is usually accompanied by the formation of non-covalent bonds. It can also be divided into two categories: one is the reformation of intermolecular chemical bonds after breaking, and the other is the formation of supramolecular molecules through non-covalent interactions. Figure 4.5 illustrates both of these processes.

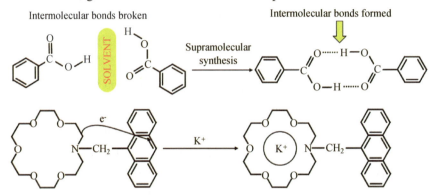

Figure 4.5 Two main types of reactions in supramolecular chemistry

ii. What is molecular chemistry?

Molecular chemistry, the chemistry of the covalent bond, is concerned with uncovering and mastering the rules governing the structures, properties and transformations of molecular species. Molecular chemistry is the branch of chemistry that studies the formation and breaking of chemical bonds between molecules. This subject area falls under molecular science; there are two subject areas under molecular science, i.e. molecular chemistry and molecular physics (where we discuss laws that govern the structures and properties of molecules).

According to molecular chemistry, the molecule is a stable system (we call it a bound state) composed of two or more atoms (polyatomic). Polyatomic ions are considered as charged molecules, while the term labile molecule is used for very reactive species. For example, short-lived nuclei, radicals, molecular ions, etc.

Molecular chemistry is often accompanied by the formation of covalent bonds. There are two main types of reactions in molecular chemistry: one is the breaking and reformation of covalent bonds, and the other is the formation of new molecules based on atoms through covalent interactions. Figure 4.6 illustrates these two processes.

Figure 4.7 summarizes the difference between supramolecular chemistry and molecular chemistry. Supramolecular chemistry and molecular chemistry are two fields of chemistry. The key difference between supramolecular chemistry and molecular chemistry is that supramolecular chemistry deals with weak, reversible non-covalent interactions between molecules whereas molecular chemistry deals with the regularity of the formation and breakage of chemical bonds between molecules. Originally, supramolecular chemistry was defined in terms of the non-covalent interaction between a "host" and a "guest" molecule, as highlighted in Figure 4.8, which illustrates the structural and functional differences between molecular and supramolecular chemistry.

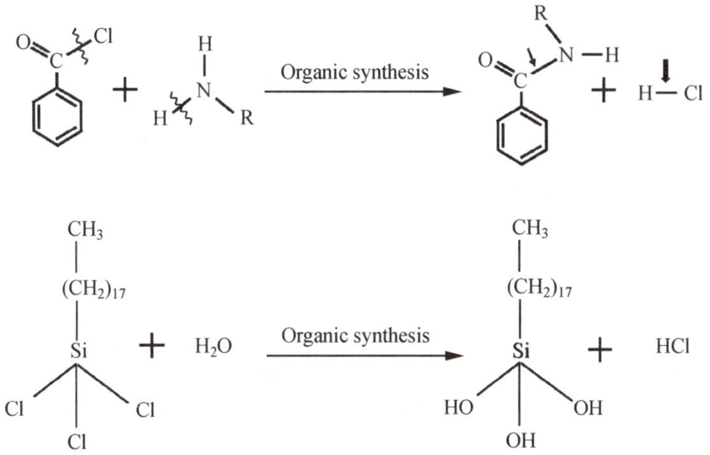

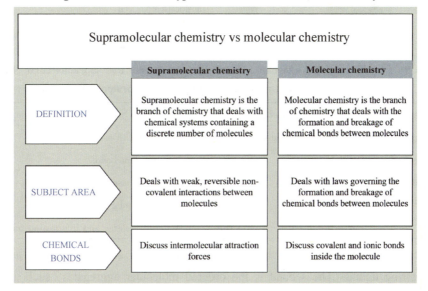

Figure 4.6 Two main types of reactions in molecular chemistry

Supramolecular chemistry vs molecular chemistry		
	Supramolecular chemistry	Molecular chemistry
DEFINITION	Supramolecular chemistry is the branch of chemistry that deals with chemical systems containing a discrete number of molecules	Molecular chemistry is the branch of chemistry that deals with the formation and breakage of chemical bonds between molecules
SUBJECT AREA	Deals with weak, reversible non-covalent interactions between molecules	Deals with laws governing the formation and breakage of chemical bonds between molecules
CHEMICAL BONDS	Discuss intermolecular attraction forces	Discuss covalent and ionic bonds inside the molecule

Figure 4.7 Difference between supramolecular chemistry and molecular chemistry

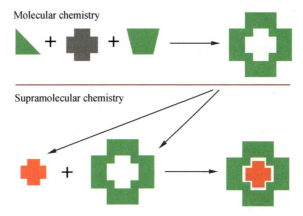

Figure 4.8 Schematic illustration of molecular chemistry and supramolecular chemistry

The classification and systematic knowledge of molecular chemistry and supramolecular chemistry are shown in Figure 4.9. Molecular chemistry and supramolecular chemistry are also compared in Table 4.1.

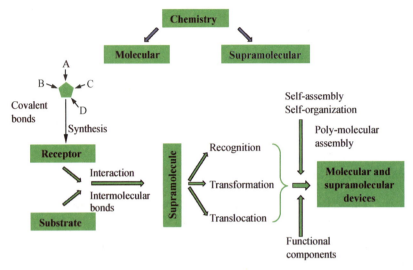

Figure 4.9　Classification and systematic knowledge of molecular and supramolecular chemistry

Table 4.1　Comparation of molecular chemistry and supramolecular chemistry

Items	Molecular chemistry	Supramolecular chemistry
Structural units	Atom or atomic group	Molecule with assembly ability
Binding force	Covalent bonds	Non-covalent bonds
Structure realization	Synthetic chemistry	Molecular assembly
Structure	Molecular structure	Supramolecular structure
Property	Physical and chemical properties	Material, energy and information transmission

The scope of supramolecular chemistry is as follows. The first is host-guest system consisting of ring ligand. The second is the ordered molecular aggregates, such as enzymes, L-B membranes, cell membranes, etc. The third is supramolecular compounds, such as aggregates of polymers, consisting of two or more groups linked by flexible or rigid chains.

4.1.3　The Importance of Supramolecular Chemistry and Its Applications in Nature

Supramolecular chemistry is about intermolecular interactions and chemistry of molecular aggregates. Supramolecular chemistry involves intermolecular interactions, including self-assembly, molecular recognition, supermolecules, molecular assemblies, supramolecular chirality, molecular and supramolecular devices, and supramolecular polymers. At the supramolecular level, molecular assembly is just as important as synthesis in molecular chemistry (Figure 4.10).

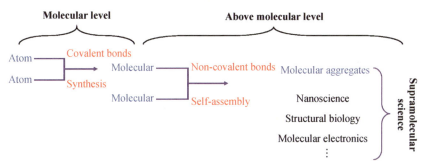

Figure 4.10 Diagram of supramolecular science

If the 20th century is the century of covalent bonds, the 21st century will be the century of supramolecular chemistry that studies non-covalent interactions between molecules (Figure 4.11). Supramolecular science is an important source of new concepts and high-tech in the 21st century. The object of study in supramolecular science is molecular aggregates. The essence of the formation of molecular aggregates is intermolecular interaction. Molecular self-assembly is an effective means to build various molecular aggregates.

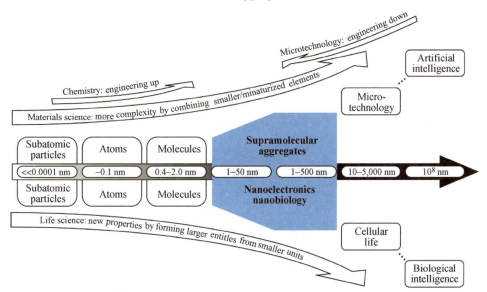

Figure 4.11 Schematic illustration of the significance of supramolecular chemistry

Molecular self-assembly is the spontaneous organization of molecules into structurally well-defined and stable arrangements through non-covalent interactions under conditions close to thermodynamic equilibrium. Such interactions typically include hydrogen bonding, electrostatic attraction, and van der Walls interactions. Generally, molecular self-assembly relies on chemical complementarity and structural compatibility. These fundamental principles are critical for designing the molecular building blocks required to fabricate functional macrostructures. Molecular fabrication involves not only the understanding of molecular self-assembly, but also the knowledge of surface science. Materials communicate with their environment through their interfaces. Such communications are determined by the interfacial properties of the materials. In particular, understanding the molecular mechanisms and signaling cascades between living cells and their environment is crucial for the fabrication of novel biomaterials; biomineralization requires functional templates to direct and regulate the mineral deposition and the crystal growth. The molecular interactions in molecular self-assembly, guest-host molecular recognition, and molecular biomimetic are governed by the interfacial communications, eventually controlling the material structures and their mechanical, electronic, magnetic, and solution properties.

The principle of self-organization that creates functional units is not an invention of modern natural sciences. It was already a fundamental idea of the ancient philosophies in Asia and Europe: only the mutuality of the parts creates the whole and its ability to function. Translated into the chemical language, this means: The self-organization of molecules leads to supramolecular systems and is responsible for their functions. Thermotropic and lyotropic liquid crystals are such functional units, which form through self-organization. As highly oriented systems, they exhibit novel properties. The importance of lyotropic liquid crystals to the life sciences has been known for a long time. They are a prerequisite for the development of life and for the functioning of cells. In materials science, this concept of organization for function has led to the development of novel liquid crystal materials. Basic requirements for working in this frontier area among organic chemistry, membrane biology, life science, and materials science will be the delight of scientific adventures as well as the courage to go ahead (Figure 4.12).

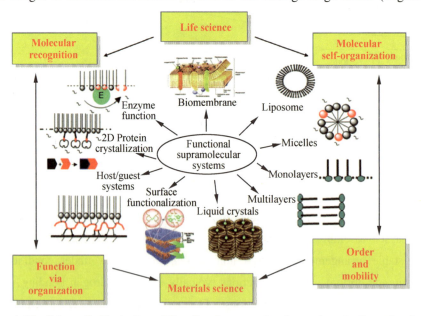

Figure 4.12 Schematic illustration of functional supramolecular systems in the natural world

i. Chemical recognition

Semiochemistry is the chemistry of signalling and sensing and, like the binding of neurotransmitters and hormones, a great deal of sensing and signalling in the nature utilizes chemical recognition events, usually between a chemical messenger molecule such as a hormone or pheromone and a protein-based receptor. Hormones are responsible for intercellular signalling in the body, and they are involved in the regulation of cellular function. Insects, reptiles, and mammals can use pheromones to induce particular kinds of behavior or to signal information such as the whereabouts of food or sexual availability. There are many classes of pheromone, including aggregation, alarm and epideictic (signal the presence of insect eggs) pheromones; releaser, primer, territorial, trail and sex pheromones. The sense of smell is also based on molecular recognition, with odorants dissolved in nasal mucus, where they bind to and activate odorant-binding proteins, or in insects, chemosensory proteins (isolated for the first time as recently as generating a nerve signal to the brain).

A good example of a pheromone is bombykol, (10E,12Z)-hexadeca-10,12-dien-1-ol, a pheromone used by the female silkworm moth to attract a mate [Figure 4.13 (a)]. Its predominantly hydrocarbon structure makes it insoluble in water, so it remains localized near the source. Bombykol, one of the first pheromones to be isolated (in 1959), was used in small amounts to confuse male moths as to the location of females and thus control numbers.

(a) (b)

Figure 4.13 (a) **Paired male and female Bombyx Mori silkworm moths;** (b) **jasmine blossom**

Another example of a particularly sweet small-molecule messenger is methyl jasmonate, the principal scent component (ca. 2.5 per cent) of jasmine oil. Methyl jasmonate can be detected in very small quantities by scent. In fact, it takes some 10,000 flowers to make just 1 g of the oil [Figure 4.13 (b)], and the oil is used as a perfume in detergents and is also a flavour ingredient in black tea. Methyl jasmonate has two chiral centers and therefore exists in four stereoisomeric forms. The fact that only one, methyl (+)-epijasmonate, (3R,7S)-(+)-methyl 3-oxo-2-[2-(Z)-pentenyl] cyclopentane-1-acetate has any distinct odour, highlighting the remarkable specificity of odour proteins (which are themselves chiral, of course) for the molecular shape recognition. When a plant is attacked by an insect, it produces higher levels of methyl jasmonate, which build up in the damaged parts of the plant. The injury is thought to activate a lipase that results in the release of linolenic acid (the precursor of jasmonic acid) and activation of proteinase inhibitor genes. The proteinase produced is ingested by the insect and interferes with its digestive system, preventing further feeding. Methyl jasmonate is more volatile than the strongly hydrogen bonding jasmonic acid, and hence can signal the occurrence of insect attack to nearby undamaged plants, stimulating the production of defensive chemicals. Methyl jasmonate may also act as an attractant for insect predators.

ii. Biological nanostructures

Mosquitoes have two eyes, but their eyes are compounded into hundreds of small lenses (Figure 4.14). This makes a mosquito's vision highly fragmented and less focused, but mosquitoes do have the ability to detect motion very effectively this way. Their compound eyes also have a slightly weaker texture-defining abilities. Compound eyes, found in most arthropods, consist of many microscopic lenses arranged in a curved array. Each tiny lens captures a separate image, and a mosquito's brain integrates all of the images to achieve peripheral vision without head or eye movement. The simplicity and versatility of compound eyes make them ideal candidates for miniaturized vision systems, which could be used by drones or robots to rapidly image their surroundings.

Some of the lenses also have adopted specific roles: Some have evolved into what are known as ocelli, which are simple light-sensitive eyes located on the top of their heads. These unique lenses can detect some fundamental changes in light, helping to differentiate between objects and locations.

Figure 4.14 Compound eyes of a mosquito

A butterfly's wings [Figure 4.15 (a)] do more than just facilitate flight. Their patterns and colors can serve as a form of camouflage or mimicry, and the fine scales on the wings keep the insect insulated during the cold months. The special shape and scales of the wings give butterflies their characteristic sublime flight patterns and amazing aerial maneuverability. A butterfly's wings also act as a kind of heat sink, which the insect uses to capture light and warm its body.

A butterfly's wings are mainly composed of thin layers of a protein called chitin, the same protein in the exoskeletons and hard shells of insects and arthropods. Each of the two wings is divided into two parts, the upper forewing and the lower hindwing. Both sides of the wings are flanked by overlapping layers of tiny scales, also made of chitin. The scales are responsible for much of the wings' interesting properties: their color, iridescence, and air flow around the wing. Scales usually contain pigments, substances that reflect a specific wavelength of light. The concentration of the paint determines the oranges, browns, blacks, and whites on a butterfly's wings. Greens, blues, and reds are often not produced by pigments, but by the scattering of light by the microstructure of the scale itself. The tiny scales have microscopic ridges that scatter incoming light, producing the shiny colors and iridescence associated with a butterfly's wings [Figure 4.15 (b)].

A butterfly's wings perform a number of functions. Most obviously, their colors and patterns can be used as camouflage, a warning sign, or a form of mimicry. Butterflies can use their wings to blend into the background foliage to avoid predators, and the bright colors of wings can disorient or signal that the insect is not tasty. In some cases, butterflies evolve to mimic the wing color and patterns of harmful species, tricking predators into ignoring them. Another function of the wings is to capture heat and insulate the insect. Small scales on the wings absorb heat from the light, thereby regulating the insect's body temperature. Even a tiny difference in the thickness of the scales can drastically affect how well they absorb heat. In this sense, a butterfly's scales can be seen as similar to mammalian hair.

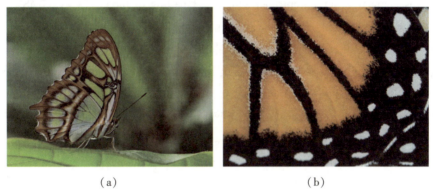

(a)　　　　　　　　　　　　(b)

Figure 4.15　(a) Wings of a butterfly; (b) a close-up of the scales on the wings of a monarch butterfly

iii. Detection of melamine

The identification and detection of melamine are of great significance in food industries. Since 2008 there have been scandals involving the illegal addition of melamine to dairy products to falsely increase protein content (Figure 4.16). This has resulted in serious kidney damages in tens of thousands of infants and even several cases of death. The damaging effect is thought to be caused by the aggregates of melamine with associated impurities or metabolites existing in body fluids such as cyanuric acid and uric acid bearing imide moieties. It is clear that recognition and detection of melamine have become more important than ever.

Fundamentals of Photophysics

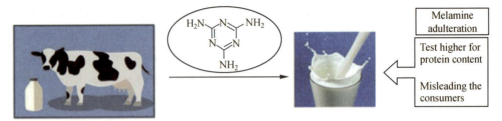

Figure 4.16 Melamine in milk

With increasing public health and nutrition awareness, there has been a lot of significant research in the development of methods to detect melamine. There are many methods for detection of melamine, such as liquid chromatography-tandem mass spectrophotometry (LC-MS/MS), gas chromatography-mass spectrometry (GC-MS), high performance liquid chromatography (HPLC), matrix-assisted laser desorption/ionization mass spectrophotometry (MALDI-TOF) and nuclear magnetic resonance (NMR). Each method has its own strengths and weaknesses.

Mass spectrometry is a highly sensitive method for confirming melamine in various food samples and is being rigorously improved with new labelling methods. LC-MS/MS and GC-MS detection limits for melamine in milk and milk products are 10 and 0.002 mg · kg^{-1}, respectively. The detection limit of urine melamine by MALDI-MS is 12.5 mg · mL^{-1}. However, lengthy sample preparation and high establishment cost are a major obstacle to this method. HPLC offers an alternative with a detection limit of 0.035~0.110 mg · kg^{-1} in milk and dairy products, but sometimes fails to confirm the target analyte. Although these techniques are highly sensitive, they are also expensive, which limits their availability in general clinical laboratories, and additionally demands high expertise for operation.

In order to develop a simple and user-friendly testing tool for melamine detection, a preliminary study was performed to analyze the intrinsic fluorescent capabilities of melamine alone and also in the presence.

Figure 4.17 shows the excitation of some complex forming agents (cyanuric acid and barbituric acid) at different wavelengths. The fluorescence emission of melamine at various excitation wavelengths at acidic, neutral and basic pH was analyzed. The fluorescence properties of melamine-cyanuric acid complex and melamine-barbituric acid complex were further analyzed. No intrinsic fluorescence was observed for melamine itself, but a milky-colored complex was formed on reaction of melamine with barbituric acid and cyanuric acid. In fact, melamine-cyanuric acid complex at neutral pH emits fluorescence at 700 nm upon excitation at 350 nm in a dose-dependent manner. The observed fluorescence emission property of melamine-cyanuric acid complex can be further explored for point-of-care detection of melamine in biological samples. POCT rapid detection of melamine will be helpful in alarming society beforehand if melamine tainted food items are consumed. In areas where sporadic incidences of melamine adulteration have been found to occur, POCT testing can be effective.

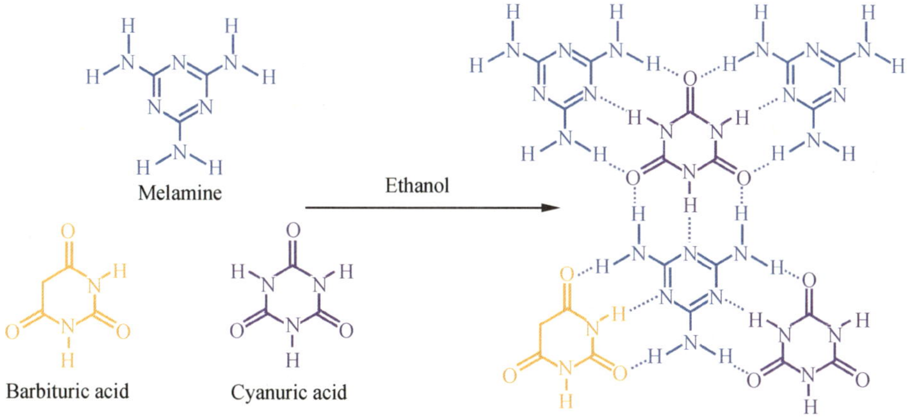

Figure 4.17　Supramolecular complex formed by melamine

4.1.4　The Development of Supramolecular Chemistry

Supramolecular chemistry, as it is now defined, is a young discipline that dates back to the late 1960s and early 1970s. However, its concepts and roots, along with many simple (and not-so-simple) supramolecular chemical systems, may be traced back almost to the beginnings of modern chemistry itself. Table 4.2 gives an illustrative (although necessarily subjective and non-comprehensive) chronology. Much of supramolecular chemistry has sprung from the developments of macrocyclic chemistry in the mid-to-late 1960s, especially the development of macrocyclic ligands for metal cations. Four systems of fundamental importance may be identified, prepared by the groups of Curtis, Busch, Jäger and Pedersen, three of which used the Schiff base condensation reaction of an aldehyde with an amine to form an imine. Conceptually, these systems may be seen as a development of naturally occurring macrocycles (ionophores, hemes, porphyrins etc.). To these may be added the work of Donald Cram on macrocyclic cyclophanes (which dates back to the early 1950s) and, subsequently, on spherands and carcerands, and the great contribution of Jean Marie Lehn, who prepared the cryptands in the late 1960s, and has since continued to shape many of the recent developments in the field.

Table 4.2　Timeline of supramolecular chemistry

Year	Event
1810	Sir Humphry Davy: discovery of chlorine hydrate
1823	Michael Faraday: formula of chlorine hydrate
1841	C. Schafhäutl: study of graphite intercalates
1849	F. Wöhler: β-quinol H_2S clathrate
1891	Villiers and Hebd: cyclodextrin inclusion compounds

Fundamentals of Photophysics

(continued)

Year	Event
1893	Alfred Werner: coordination chemistry
1894	Emil Fischer: lock and key concept
1906	Paul Ehrlich: introduction of the concept of a receptor
1937	K. L. Wolf: the term Übermoleküle is coined to describe organized entities arising from the association of coordinatively saturated species (e. g. the acetic acid dimer)
1939	Linus Pauling: hydrogen bonds are included in the groundbreaking book *The Nature of the Chemical Bond*
1940	M. F. Bengen: urea channel inclusion compounds
1945	H. M. Powell: X-ray crystal structures of β-quinol inclusion compounds; the term "clathrate" is introduced to describe compounds where one component is enclosed within the framework of another
1949	Brown and Farthing: synthesis of [2.2]paracyclophane
1953	Watson and Crick: structure of DNA
1956	Dorothy Crowfoot Hodgkin: X-ray crystal structure of vitamin B12
1959	Donald Cram: attempted synthesis of cyclophane charge transfer complexes with $(NC)_2 C = C(CN)_2$
1961	N. F. Curtis: first Schiff's base macrocycle from acetone and ethylene diamine
1964	Busch and Jäger: Schiff's base macrocycles
1967	Charles Pedersen: crown ethers
1968	Park and Simmons: Katapinand anion hosts
1969	Jean-Marie Lehn: synthesis of the first cryptands
1969	Jerry Atwood: liquid clathrates from alkyl aluminium salts
1969	Ron Breslow: catalysis by cyclodextrins
1973	Donald Cram: spherand hosts produced to test the importance of preorganization
1978	Jean-Marie Lehn: introduction of the term "supramolecular chemistry", defined as the "chemistry of molecular assemblies and of the intermolecular bond"
1979	Gokel and Okahara: development of the lariat ethers as a subclass of host
1981	Vögtle and Weber: podand hosts and development of nomenclature
1986	A. P. de Silva: fluorescent sensing of alkali metal ions by crown ether derivatives
1987	Award of the Nobel Prize for Chemistry to Donald J. Cram, Jean-Marie Lehn and Charles J. Pedersen for their work in supramolecular chemistry
1996	Atwood, Davies, MacNicol & Vägtle: publication of *Comprehensive Supramolecular Chemistry* containing contributions from many key groups and summarizing the development and state of the art
1996	Award of the Nobel Prize for Chemistry to Kroto, Smalley and Curl for their work on the chemistry of the fullerenes
2003	Award of the Nobel Prize for Chemistry to Peter Agre and Roderick MacKinnon for their discovery of water channels and the characterization of cation and anion channels, respectively
2004	J. Fraser Stoddart: the first discrete Borromean-linked molecule, a landmark in topological synthesis

i. Charles J. Pedersen

The crown ethers are among the simplest and most attractive macrocyclic (large ring) ligands, and are ubiquitous in supramolecular chemistry as hosts for both metallic and organic cations. They consist solely of a cyclic array of ether oxygen atoms linked by organic spacers, typically -CH_2CH_2- groups. Although the metal binding ability of unidentate ethers such as the common solvent diethyl ether is very poor, the crown ethers are much more effective due to the chelating effect and the partial preorganization arising from their macrocyclic structure.

The discovery of the crown ethers in 1967 gained a share of the 1987 Nobel Prize for Chemistry for Charles J. Pedersen, a chemist working at the American E. I. du Pont de Nemours company (Figure 4.18). Charles J. Pedersen was born in Pusan, Korea in 1904. He was the first DuPont scientist to win the Nobel Prize, and he was one of just a few scientists to get the honor while not having earned a doctorate. Pedersen started his career as a scientist at the University of Dayton in Ohio, where he studied chemical engineering. He subsequently went on to the Massachusetts Institute of Technology for his master's degree. Although being a talented student, Pedersen did not want to be supported by his father and thus abandoned his studies to work. He went to work for the DuPont Company, where he stayed until his retirement. Pedersen's magnum work came after the age of 60, unlike most scientists who peak in their careers in their mid-30s or early 40s. He was investigating the effects of bi- and multi-dentate phenolic ligands on the catalytic properties of the vanadyl group, VO, when he discovered unknown crystals of a by-product by mistake. Fascinated by it, he continued his research, unaware that it would lead to a new chapter in chemistry. Crown ethers are molecules that include hydrogen, carbon, and oxygen atoms, which he discovered. For this discovery, he won the Nobel Prize.

Figure 4.18 Charles J. Pedersen (1904 – 1989)

Curiously, however, Pedersen's initial synthesis of the first crown ether, dibenzo[18]crown-6 (4.4) was accidental. In attempting to carry out the synthesis of the linear di-ol (4.3), which he hoped would serve as a ligand for the catalytic vanadyl ion, Pedersen performed the reaction shown in Figure 4.19. The starting material was the catechol (1,2-dihydroxybenzene) derivative (4.1), where one of the hydroxyl groups is protected by a tetrahydropyran ring to prevent its reaction. Unknown to Pedersen, his beginning material was slightly contaminated with some free catechol (4.2). The resulting product was a mixture of the desired compound (4.3) along with a small amount of dibenzo[18]crown-6, formed in only 0.4% yield. It is a tribute to Pedersen's skills that he was able to isolate and characterize this small amount of by-product. Pedersen's interest was motivated by the interesting solubility properties of (4.4), and by its high degree of crystallinity (indicating a molecular compound rather than a polymer). The compound dissolved sparingly in methanol, but its solubility was significantly enhanced upon addition of alkali metal salts. Pedersen soon synthesized the compound in much better yield. He found that it could dissolve inorganic salts like $KMnO_4$ in organic solvents like benzene, giving it a purple coloration (this was termed "purple benzene"). Pedersen also realized the crown ether's ability to dissolve the alkali metals themselves to give interesting blue solutions of what are now known to be alkalide and electride salts. He eventually concluded that "the potassium ion had fallen into the hole at the center of the molecule", a bold and highly imaginative statement at the time. Events soon proved him to be entirely correct. This initial result quickly led to the synthesis of a related family of species, which Pedersen named "crown ethers" because of the crown-like shape of the capsular complex between **4.4** and K^+ (Figure 4.20).

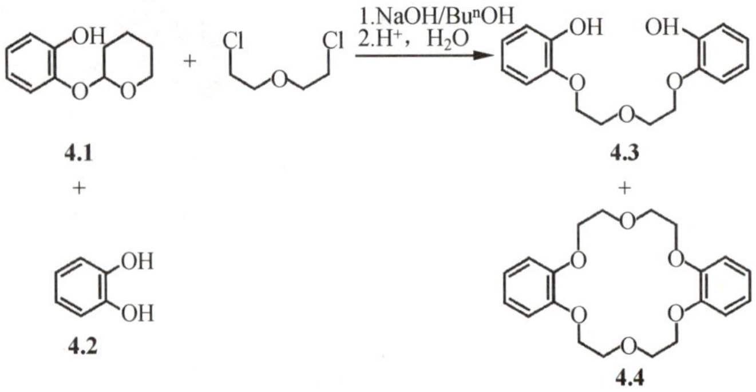

Figure 4.19 Accidental synthesis of the first crown ether, dibenzo[18]crown-6

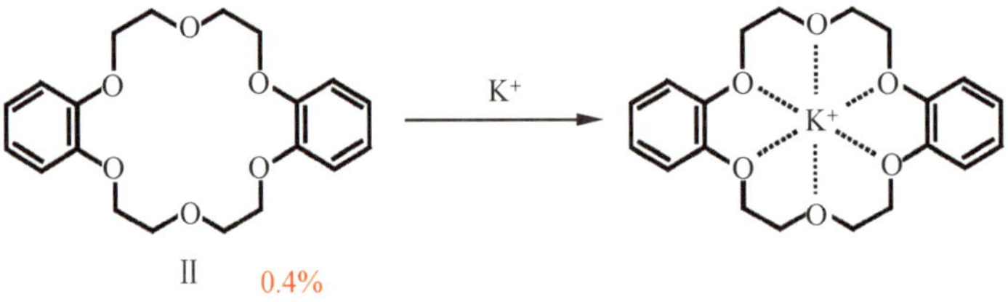

Figure 4.20 Synthesis of crown ether complex according to Pedersen

ii. Donald J. Cram

In addition to Pedersen and Lehn, the 1987 Nobel Prize for Chemistry was shared by a third supramolecular chemist, Donald J. Cram (Figure 4.21), for his development of a further type of macrocyclic cation host: the spherands. Cram was born in Chester, Vermont, U.S., 1919. He amplified and expanded upon Pedersen's ground-breaking synthesis of the crown ethers—essentially two-dimensional organic compounds that are able to recognize and selectively combine with the ions of certain metal elements. Cram synthesized molecules that took this chemistry into three dimensions, creating a series of differently shaped molecules

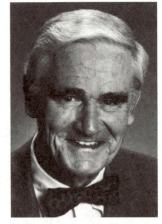

Figure 4.21 Donald J. Cram (1919-2001)

that could selectively interact with other chemicals because of their complementary three-dimensional structures. His work represents a major step toward the synthesis of functional laboratory-made mimics of enzymes and other natural molecules whose peculiar chemical behavior is due to their characteristic structure.

Driven by the crystal structure of enzyme and nucleic acid as well as the immune system specificity, D. J. Cram always wanted to design some organic compounds to imitate the functions of some natural materials. It was not until Pederson published his paper that he realized they were very similar. D. J. Cram proposed "Host-guest chemistry" in *Science* in 1974.

While the crown ethers and even the cryptands are relatively flexible in solution, Cram realized that if a rigid host could be engineered whose donor sites were forced to converge on a central binding pocket even before the addition of a metal cation, then strong binding and excellent selectivities between cations should be observed. Using space-filling molecular models (termed Corey-Pauling-Koltun or CPK models, Figure 4.22), Cram and co-workers designed the rigid, three-dimensional spherands **4.5** and **4.6**, whose cation-binding oxygen atoms are pre-organized in an octahedral array, ready to receive a metal ion (Figure 4.23).

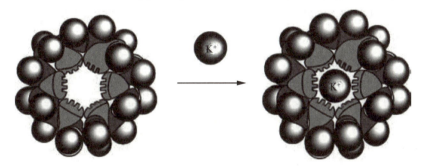

Figure 4.22 CPK models of crown ether and spherand complexes used in early supramolecular design (Modern computational techniques, such as molecular mechanics have, to some extent, replaced such tactile representations.)

Figure 4.23 Three-dimensional spherands designed by Cram and co-workers

In the case of compound **4.5**, three of the aryl rings are pointing upwards (out of the page) and three downwards. This results in the anisyl oxygen atoms being fixed in an almost perfect octahedral array, while the p-methyl and anisyl methyl groups present a lipophilic surface to the solvent. This host selectively binds small cations, such as Li^+ and, to a lesser extent,

Na$^+$, in its cavity. In fact, **4.5** is one of the strongest complexing agents for Li$^+$ known. All other cations are excluded because they are simply too large to fit into the binding pocket. Spherand **4.6** has a similarly sized binding pocket, formed from the tethering of the rings in pairs with diethylene glycol linkages, resulting in four rings being down and two up. The analogous fluoro compound **4.7**, as well as an octameric analogue, has also been synthesized. Although X-ray crystallography confirms that the hosts possess very similar cavities to **4.5**, the fluoro species display no metal ion binding properties. Apparently, the intrinsic affinity of fluoro substituents for alkali metal cations is so low that even incorporation of multiple binding sites into a highly organized host fails to bring the binding on to a measurable scale. Interestingly, however, for cryptand, a definite interaction of cation guest with a fluoro substituent is observed, both in the solid state by X-ray crystallography and in solution by ^{19}F NMR spectroscopy.

In 1986, Cram and his colleagues published a paper regarding preorganization and complementarity of supramolecular host-guest complexes. Complexes consist of two or more molecules or ions that are held together in unique structural relationships by electrostatic forces rather than being fully covalently bonded. The host component is defined as an organic molecule or ion whose binding sites converge in the complex. The guest component is defined as any molecule or ion that differs in its binding site in the complex.

In supramolecular chemistry, we generally consider a molecule (a host) needs to bind another molecule (a guest) to create a host-guest complex or supramolecule (Figure 4.24). However, not all subjects and objects can be recombined. The recombination between subjects and objects is identifiable, which needs to meet the following two principles.

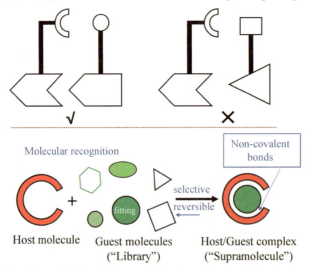

Figure 4.24   Host-guest complex or supramolecule

(1) Complementarity principle: In order to bind, a host must have binding sites of the correct electronic character to be complementary to the binding site of the guest. The binding

sites must be separated on the host in such a way to make it possible for them to interact with the guest.

(2) The lock-and-key principle: The size, shape and location of the binding sites within the active site of a host (e.g. an enzyme) are ideal for specific recognition of the guest (e.g. substrate).

iii. Jean-Marie Lehn

Jean-Marie Lehn, born in Rosheim, France, is a French chemist (Figure 4.25). In 1987, he received the Nobel Prize for Chemistry together with Donald J. Cram and Charles Pedersen for synthesizing cryptands. Lehn was an early innovator in the field of supramolecular chemistry, i.e. the chemistry of host-guest molecular assemblies resulting from intermolecular interactions, and he continues to innovate in this field.

The early research of Jean-Marie Lehn laid the chemical foundation of "molecular recognition", that is, understanding the nature and process of molecular recognition and its application in chemical and biological sciences based on chemistry and biology. By introducing biological science

Figure 4.25 Jean-Marie Lehn (1939 –)

concepts into materials science, Jemari Lane's work has also led to new research areas, such as supramolecular polymer chemistry, supramolecular liquid crystals, and the exploration of how inorganic arrays can be self-organized to construct molecular electronics devices and nanomaterials. He expanded the study of supramolecular chemistry into a new and broader area at the intersection of chemistry and biology: the study of self-organizing processes.

Shortly after Pedersen's work, Jean-Marie Lehn, then a young researcher at the University of Strasbourg, decided to design 3D analogues of the crown ethers. In this way it was expected that metal ions could be completely encapsulated within a crown-like host with consequent gains in cation selectivity and enhancements in ionophore-like transport properties. Therefore, the bicyclic cryptands (named for their ability to spherically surround or "entomb" a metal cation, as in a crypt; from the Greek kruptos meaning "hidden") and a growing number of related compounds were preorganized using the high-dilution technique, the majority in remarkably high yields (Figure 4.26). The first and foremost member of the series is [2.2.2]cryptand (**4.9**), which is sold commercially under the trade name Kryptofix®. Because it is based on rings of a similar size to [18]crown-6, this host also exhibits selectivity for K^+ over the other alkali metals. However, the binding of K^+ by [2.2.2]cryptand in methanol is approximately 104 times stronger than its crown analogue. Similarly, [2.2.1]cryptand (**4.10**) is selective for

Na$^+$. The key to the dramatically enhanced metal cation binding capacity of cryptands to crown ethers is the preorganized, three-dimensional nature of their cavity, which enables spherical recognition of the M$^+$ ion to take place.

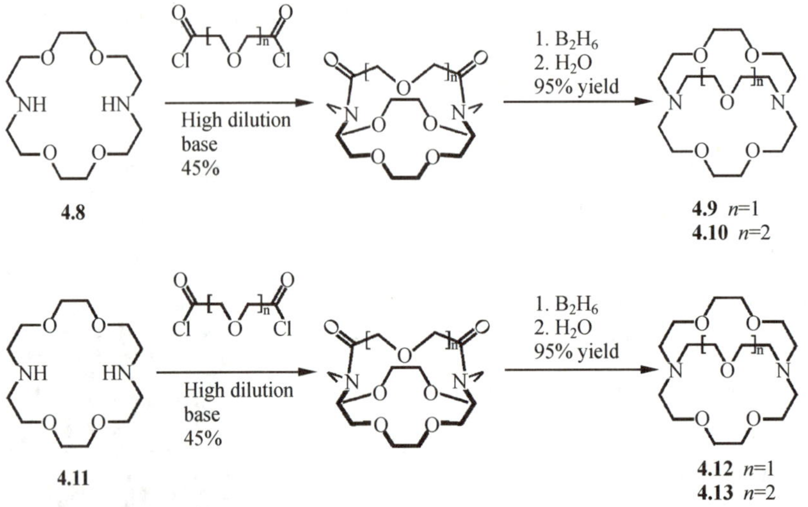

Figure 4.26 High dilution synthesis of [2.2.1] cryptand ($n=1$) and [2.2.2] cryptand ($n=2$)

High-dilution syntheses, although highly versatile, do not readily prepare large quantities of material and typically involve many steps, especially the final reductive decarbonylation with diborane. Since the early work of Lehn, a wide variety of synthetic procedures have been developed to prepare a large number of cryptands of varying degrees of sophistication, including chiral species and those containing three different bridges. Many of these follow a step-by-step approach, which may be summarized as follows:

(1) Build two linear chains with suitable reactive groups at each chain end;
(2) Cyclization reaction of these two chains produces a corand (crown ether-like macrocycle);
(3) Add a third chain to the corand to obtain a macrobicyclic compound.

Due to the length and tedium of such approaches, much simpler routes, often exploiting template effect, have been devised, which, in many simple cases, can be remarkably effective. Such alternative syntheses for **4.9** and related compounds are shown in Figure 4.27.

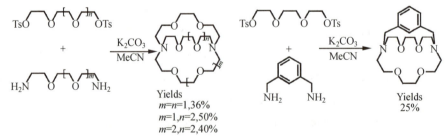

Figure 4.27 Simple template synthesis of selected cryptands with yields

4.2 Supramolecules Formation

4.2.1 Gibbs Free Energy

Gibbs free energy, also known as the Gibbs function, Gibbs energy, or free enthalpy, is a quantity that is used to measure the maximum amount of work done in a thermodynamic system when the temperature and pressure are held constant. Gibbs free energy is represented by the symbol "G". Its value is usually expressed in Joules or Kilojoules. Gibbs free energy can be defined as the maximum amount of work that can be extracted from a closed system.

This property was identified in 1876 by the American scientist Josiah Willard Gibbs while he was conducting experiments to predict the behavior of systems when combined together or whether a process could occur simultaneously and spontaneously. Gibbs free energy was also previously known as "available energy". It can be visualized as the amount of useful energy present in a thermodynamic system that can be used to perform some work.

Gibbs free energy equals the enthalpy of the system minus the product of the temperature and entropy. The formula is given as:

$$G = H - TS \tag{4-1}$$

where G is Gibbs free energy, H is enthalpy, T is temperature, and S is entropy.

Gibbs free energy is a state function, so it doesn't depend on the path. Therefore, the change in Gibbs free energy is equal to the change in enthalpy minus the product of the change in temperature and entropy of the system.

$$\Delta G = \Delta H - \Delta(TS) \tag{4-2}$$

If the reaction is carried out under constant temperature ($\Delta T = 0$):

$$\Delta G = \Delta H - T\Delta S \tag{4-3}$$

This equation is known as the Gibbs Helmholtz equation.

$\Delta G > 0$: the reaction is non-spontaneous and endergonic.

$\Delta G < 0$: the reaction is spontaneous and exergonic.

$\Delta G = 0$: the reaction is at equilibrium.

According to the equation, several aspects should be addressed:

(1) According to the second law of thermodynamics, the entropy of the universe always increases with spontaneous process.

(2) ΔG determines the direction and degree of chemical change.

(3) ΔG is only meaningful for reactions where the temperature and pressure are kept constant. The system is normally open to the atmosphere (constant pressure) and we start and

end the process at room temperature (after any heat we have added or which is liberated by the reaction has dissipated).

(4) ΔG is used as the single master variable that determines whether a given chemical change is thermodynamically possible. Thus, if the free energy of the reactants is greater than that of the products, then when the reaction occurs, the entropy of the world will increase, so the reaction will tend to take place spontaneously. ΔS universe = ΔS system + ΔS surroundings.

(5) If ΔG is negative, the process will occur spontaneously and is called exergonic.

(6) Therefore, spontaneity depends on the temperature of the system.

Supramolecular formation is a spontaneous process, that is to say, the Gibbs free energy (ΔG) is negative (Figure 4.28). In order for the Gibbs free energy to be less than 0, the change in enthalpy (ΔH) needs to be negative and the change in entropy (ΔS) needs to be positive. In other words, a formation process of supramolecules is a process of enthalpy reduction and entropy increase.

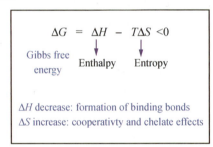

Figure 4.28 Formation conditions of supramolecules in thermodynamics

4.2.2 Enthalpy Reduction Factor

i. What is enthalpy?

Enthalpy is the measurement of energy in a thermodynamic system. The quantity of enthalpy is equal to the total heat of a system, which is equal to the internal energy of the system plus the product of volume and pressure. When a process is carried out under constant pressure, the heat absorbed or released is equal to the enthalpy change. Enthalpy is sometimes called "heat content", but "enthalpy" is an interesting and unusual word, so most people like to use it. Etymologically, the word "entropy" is also derived from the Greek, meaning "turning" and "enthalpy" is also derived from the Greek meaning "warming".

When a process starts at constant pressure, the heat released (either absorbed or released) equals the change in enthalpy. The enthalpy change is the sum of internal energy expressed in U and the product of volume and pressure, denoted by pV, expressed in the following way:

$$H = U + pV \tag{4-4}$$

Enthalpy is also described as a state function based entirely on state functions p, T and U. It is usually represented by the change in enthalpy (ΔH) of a process between the beginning and the final states.

$$\Delta H = \Delta U + \Delta pV \qquad (4-5)$$

If the pressure and temperature do not change throughout the process and the task is limited to the pressure and volume, the change in enthalpy is given by:

$$\Delta H = \Delta U + p\Delta V \qquad (4-6)$$

The heat flow (q) at constant pressure in a process equals the change in enthalpy, based on the following equation:

$$\Delta H = q \qquad (4-7)$$

Knowing whether q is endothermic or exothermic, the relationship between q and ΔH can be defined. An endothermic reaction is the one that absorbs heat and shows that heat is consumed in the reaction from the surroundings, hence $q > 0$ (positive). If q is positive, then ΔH is also positive, at constant pressure and temperature for the above equation. Similarly, if the heat is released through an exothermic reaction, the heat is transferred to the surroundings. Therefore, ΔH will be negative if q is negative.

Generally, supramolecular chemistry concerns non-covalent bonding interactions. The term "non-covalent" encompasses a large number of attractive and repulsive effects. The most important, along with an indication of their approximate energies, is explained below. When considering a supramolecular system, it is vital to consider the interactions of all these interactions and effects related to the host and guest as well as their surroundings (e.g. solvation, ion pairing, crystal lattice, gas phase etc.). Table 4.3 shows some common interactions in supramolecular chemistry that will cause enthalpy reduction, and these interactions will be discussed in later sections.

Table 4.3 Common interactions in supramolecular chemistry and corresponding strength range

Supramolecular interactions	Strength range of interactions
Electrostatic	100 – 350 kJ · mol^{-1} for ion-ion interactions 50 – 200 kJ · mol^{-1} for ion-dipole interactions 5 – 50 kJ · mol^{-1} for dipole-dipole interactions
Hydrogen bonding	4 – 120 kJ · mol^{-1}
π-π stacking interactions	0 – 50 kJ · mol^{-1}
Cation-π interactions	5 – 80 kJ · mol^{-1}
van der Waals forces	< 5 kJ · mol^{-1}
Hydrophobic effects	—

ii. Electrostatic (e.g. ion-ion, ion-dipole and dipole-dipole)

The intermolecular forces of attraction where fully or partially ionic species attract each other are called electrostatic interactions. These attraction forces do not include any electron

Fundamentals of Photophysics

sharing between atoms. Therefore, they are also named as non-covalent bonds. The term electrostatic interactions include both attractive and repulsive forces ionic species, which means ions with opposite charges attract each other while the similar charges repel from each other. In liquids, molecules held together by intermolecular interactions are comparably weaker than the intramolecular interactions within polyatomic ions and atoms of molecules. These are very important for describing the formation of different molecules. A transition occurs between the solid and liquid or the liquid and gas phase due to changes in intermolecular interactions. Generally speaking, there are three types of electrostatic interactions: ion-ion, ion-dipole and dipole-dipole electrostatic, as shown in Figure 4.29.

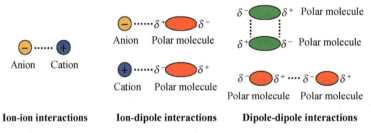

Figure 4.29　Three types of electrostatic interaction involving recognize

An ion is defined as an atom or molecule that gains or loses one or more of its valence electrons, giving it a net positive electrical charge (cation) or negative electrical charge (anion). In other words, there is an imbalance in the number of protons (positively charged particles) and electrons (negatively charged particles) in a chemical species.

A dipole refers to a separation of opposite electrical charges. A dipole is quantified by its dipole moment (μ). A dipole moment is the distance between charges times the charge. The unit of the dipole moment is the Debye, of which 1 Debye is 3.34×10^{-30} C·m. The dipole moment is a vector quantity that has both magnitude and direction. The direction of an electric dipole moment is from the negative charge toward the positive charge. The greater the difference in electronegativity, the greater the dipole moment. The distance separating opposite electrical charges also affects the magnitude of the dipole moment. In chemistry, a dipole usually refers to the separation of charges between two covalently bonded atoms or atoms that share an ionic bond within a molecule. For example, a water molecule (H_2O) is a dipole. The oxygen side of the molecule has a net negative charge, while the side with the two hydrogen atoms has a net positive electrical charge. The charges of a molecule, like water, are partial charges, which means they don't add up to the "1" for a proton or electron. All polar molecules are dipoles.

Electrostatic interactions are of great significance in the formation of various self-assembled nanostructures. They provide stable and robust interaction among various enzymes, peptides, and proteins.

- **Ion-ion interaction**

Ion-ion interactions are an attractive force between ions with opposite charges. Also known as ionic bonds, they are the forces that hold ionic compounds together.

Like charges repel each other and opposite charges attract. These Coulombic forces act over relatively long distances in the gas phase. The force depends on the product of the charges (Z_1, Z_2) divided by the square of the separation distance (d^2):

$$F = -\frac{Z_1 Z_2}{d^2} \tag{4-8}$$

If a vacuum contains two oppositely-charged particles [such as a sodium cation and a chloride anion, Figure 4.30 (a)], they will be attracted to each other, and the force will increase as the two objects come closer to each other. Eventually, they will stick together, requiring considerable energy to separate them again [Figure 4.30 (b)]. They form an ion-pair, a new particle which has a positively- and negatively-charged region [Figure 4.30 (c)]. There are fairly strong interactions between these ion pairs and free ions, so the clusters grow. For example, a saturated sodium chloride solution contains more ions than the solution can support, and sodium cations will start attracting chloride anions to form solid salt crystals. Ion-ion interactions can be seen in precipitation reactions, where a solid form upon the mixing of two aqueous solutions.

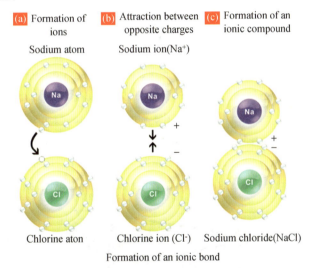

Figure 4.30　Ion-ion interactions forming the strong bonds that hold ionic solids together

- **Ion-dipole interaction**

An ion-dipole interaction is the result of an electrostatic interaction between a charged ion and a molecule with a dipole. It is an attractive force that is commonly found in solutions, particularly ionic compounds dissolved in polar liquids. A cation can attract parts of the negative end of a neutral polar molecule, while an anion attracts the positive end of a polar molecule. As the charge on the ion increases or the dipole size of the polar molecule increases, the attractive force of the ionic dipole becomes stronger. These interactions can be very significant factors in many chemical situations, so it is important to learn how to use them.

The ions are aligned with a polar molecule in such a way that the positive ions are close to the negative part of the dipole and vice versa. For example, when sodium chloride (NaCl) is dissolved in water, the sodium cation and the partial negative charge on the oxygen atom in a water molecule is the ion-dipole interaction (Figure 4.31). Another similar force of attraction is the ion-induced dipole force, which occurs when the charges on the ions induce a temporary charge in a nonpolar molecule by distorting its electronic cloud.

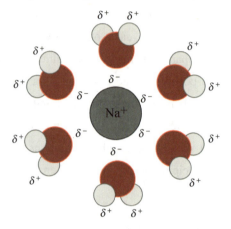

Figure 4.31 Ion-dipole interactions (The oxygen end of the dipole is closer to the sodium than the hydrogen end, so the net interaction is attractive.)

For ion-dipole interactions, the interaction occurs between a dipole moment (μ), which is actually a vector with a magnitude of $\delta \cdot r$, where δ is the value of the partial charge in the dipole moment (δ^+ or δ^-, noting they have the same magnitude, just opposite in sign), with an ion of charge q. (We are using δ to indicate the partial charge in the dipole, just to prevent it from being confused with q, which is the charge of the ion). Therefore, there are two ion-dipole interactions, with one being attractive and the other repulsive, as shown in Figure 4.32.

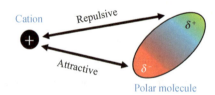

Figure 4.32 Interactions between a positive cation and a polar molecule

Ion-dipole interactions are electrostatic attraction forces between an ion and a polar molecule. Ion-dipole interactions also include coordinative bonds, which are predominantly electrostatic in nature in the case of the interactions of nonpolarizable metal cations and hard bases. Ion-dipole interactions are stronger than dipole-dipole interactions because an ion has a much stronger charge than a dipole when compared. Ion-dipole interactions are even stronger than hydrogen bonding.

The key differences between ion-ion interactions and ion-dipole interactions are as follows:

(1) Ions have integer charges (+1, +2, +3, ... for cations and -1, -2, -3, ... for anions), while dipoles have partial charges, and the partial charges (δ^+ or δ^-) can be very small fractions.

(2) Ion-ion interactions decay more slowly than ion-dipole interactions. Tripling the distance between two ions reduces the energy by 1/3, while tripling the distance between an ion and a dipole reduces it by 1/9. That is, one is inversely proportional to the distance between them, and the other is proportional to the inverse square of the distance.

- **Dipole-dipole interaction**

Dipole-dipole interactions are a type of intermolecular force between two molecules with net dipole moments (asymmetrical charge distributions, where polar molecules create partial positive and partial negative charges). Molecules tend to align themselves so that the positive end of one dipole is close to the negative end of another, and vice versa. When a positive dipole and a negative dipole approach each other, an attractive intermolecular interaction will be created, and two positive dipoles or two negative dipoles will create a repulsive intermolecular interaction [Figure 4.34 (a)]. Alignment of one dipole with another can lead to significant attractive interactions through the matching of either a single pair of poles on adjacent molecules or opposing alignment of one dipole with the other. Dipole-dipole interactions and London dispersion forces are collectively known as van der Waals force. Dipole-dipole interactions are an electrostatic interaction, as are ion-ion interactions and ion-dipole interactions.

Dipole-dipole forces are attractive forces between the positive end of one polar molecule and the negative end of another polar molecule. The strength of the dipole-dipole forces ranges from 5 kJ to 20 kJ per mole. They are much weaker than ionic or covalent bonds and only have a significant effect when the molecules involved are close together (touching or almost touching). Figure 4.33 shows two arrangements of polar iodine monochloride (ICl) molecules that create dipole-dipole attractions.

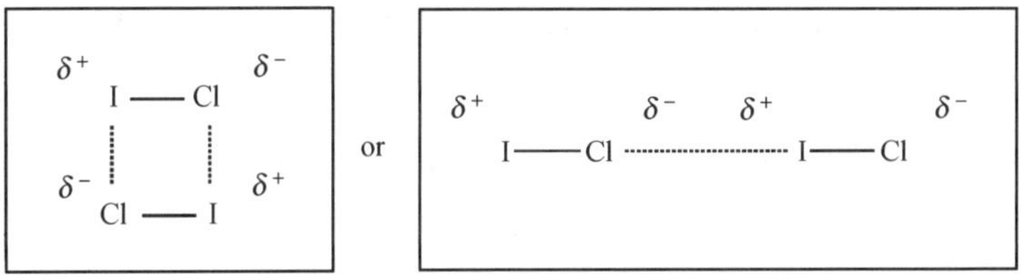

Figure 4.33 Two arrangements of polar iodine monochloride (ICl) molecules with dipole-dipole interactions

When considering a series of compounds with similar molar masses (which have dipole-dipole interaction forces between molecules), the strength of dipole-dipole forces increases with the increasing polarity. That happens because when the polarity is high, it means that the charge separation is high [Figure 4.34 (b)]. When the molecule has a high charge separation (high charged positive and negative terminals in the same molecule), it tends to strongly attract opposite charges. This also causes an increase in the boiling point of compounds. Greater the dipole-dipole forces, greater the boiling point.

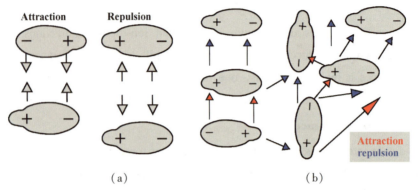

Figure 4.34 Dipole-dipole interactions (a) between two polar molecules and (b) in a system existing abundant dipole molecules

The potential energy of dipole interactions is important for living organisms. The greatest impact of dipole interactions on living organisms is protein folding. Every process of protein formation, from the binding of individual amino acids to the formation of secondary to tertiary and even quaternary structures, depends on dipole-dipole interactions. A prime example of quaternary dipole interaction that is critical to human health is the formation of erythrocytes. Erythrocytes, commonly called red blood cells, are the type of cells responsible for the gas exchange (i.e. respiration). Inside the erythrocytes, the molecule involved in this key process, is "hemoglobin", which consists of four protein subunits and a heme group. For a heme to form properly, multiple steps must occur, all of which involve dipole interactions. The four protein subunits—two alpha chains, two beta chains—and the heme group, interact through a series of dipole-dipole interactions that allow the erythrocyte to take its final shape. Any mutation that disrupts these dipole-dipole interactions prevents the erythrocyte from forming properly, and impairs their ability to transport oxygen to the tissues of the body. Therefore, we can see that without the dipole-dipole interactions, proteins would not be able to fold properly and all life would cease to exist.

- **Similarities and differences between ion-dipole and dipole-dipole forces**

Ion-dipole forces and dipole-dipole forces are intermolecular forces that exist between different chemical species, such as cations, anions, and polar molecules. Polar molecules are covalent compounds with dipoles (electrical charge separations). A polar molecule has a positively charged end and a negatively charged end in the same molecule. Therefore, these terminals can have electrostatically attractive forces with opposite charges. The difference between ion-dipole and dipole-dipole forces is that ion-dipole forces exist between ionic substances and polar molecules whereas dipole-dipole forces exist between polar molecules.

Table 4.4 Similarities and differences between ion-dipole and dipole-dipole forces

Similarities between ion-dipole and dipole-dipole forces	
• Both ion-dipole and dipole-dipole forces are types of intermolecular interactions. • Both ion-dipole and dipole-dipole forces are electrostatic forces.	
Differences between ion-dipole and dipole-dipole forces	
Ion-dipole forces are attractive forces between ionic species and polar molecules.	Dipole-dipole forces are intermolecular forces that occur between polar molecules.
Strength	
Ion-dipole forces are stronger than hydrogen bonds and dipole-dipole forces.	Dipole-dipole forces are weaker than hydrogen bonds and ion-dipole forces.
Components	
Ion-dipole forces arise between ions (cations or anions) and polar molecules.	Dipole-dipole forces arise between polar molecules.

Electrostatic interactions become important for gases at high pressure because they are responsible for their observed deviations from the ideal gas law at high pressure. Such interactions are important for maintaining the 3D structure of larger molecules, for example, nucleic acids and proteins. Such interactions are involved in some biological processes, where larger molecules bind specifically but transiently to one molecule. Furthermore, these interactions also profoundly affect crystallinity and the design of materials, in general, for the synthesis of many organic molecules and particularly for self-assembly.

iii. Hydrogen bonding

Hydrogen bonding refers to the formation of hydrogen bonds, which are a special type of intermolecular attraction due to the dipole-dipole interaction between a hydrogen atom bonded to a highly electronegative atom and another highly electronegative atom located nearby the hydrogen atoms produced by the interaction. Here are some common examples in Figure 4.35, take water molecules (H_2O) for an explanation, hydrogen is covalently bonded to the more electronegative oxygen atom. Thus, hydrogen bonding is created in water molecules due to the dipole-dipole interaction between the hydrogen atom of one water molecule and the oxygen atom of another H_2O molecule [Figure 4.35 (b)]. An excellent example of hydrogen bonding in supramolecular chemistry is the formation of carboxylic acid dimers [Figure 4.35 (d)], which leads to the shift of the $\nu(OH)$ infrared stretching frequency from about 3,400 cm^{-1} to about 2,500 cm^{-1}, accompanied by a significant broadening and intensifying of the absorption.

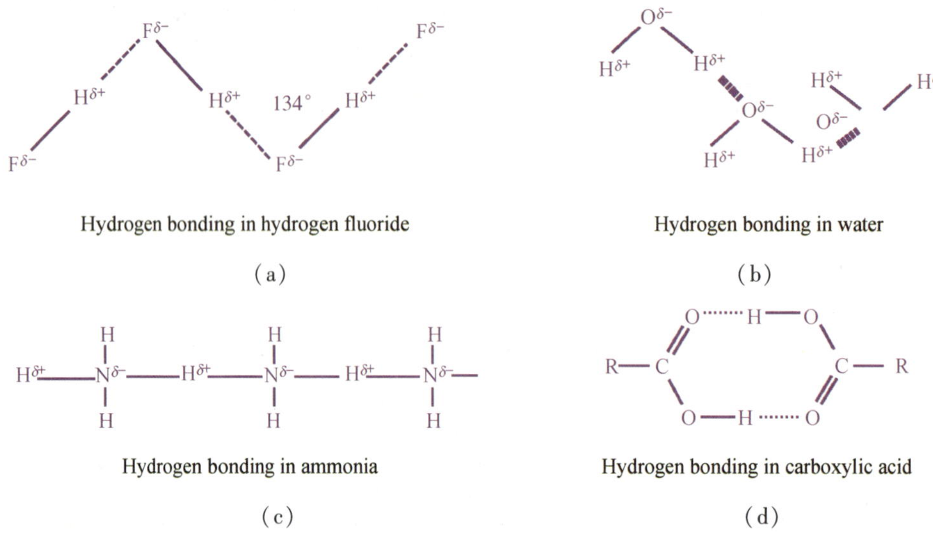

Fundamentals of Photophysics

Figure 4.35 Examples of hydrogen bonding in (a) hydrogen fluoride, (b) water, (c) ammonia and (d) carboxylic acid

Consider two water molecules coming close together. The δ^+ hydrogen is strongly attracted to the lone pair electrons, almost as if you are beginning to form a coordinate (dative covalent) bond. It does not go that far, but the attraction is much stronger than an ordinary dipole-dipole interaction. Hydrogen bonds are about a tenth of the strength of an average covalent bond, and are constantly broken and reformed in water.

Water is an ideal example of hydrogen bonding. Note that each water molecule may potentially form four hydrogen bonds with surrounding water molecules: two with the hydrogen atoms and two with the oxygen atoms. The correct numbers of δ^+ hydrogens and lone pairs of electrons are involved in hydrogen bonding (Figure 4.36).

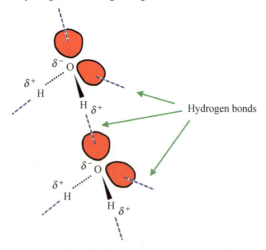

Figure 4.36 Hydrogen bonds formation in water

Although hydrogen bonding is well known as a type of intermolecular force, these bonds can also occur within a single molecule, between two identical molecules, or between two different molecules.

Intramolecular hydrogen bonding occurs within one single molecule. This happens when two functional groups of molecule can form hydrogen bonds with each other. In order for this to happen, both a hydrogen donor and a hydrogen acceptor must be present within one molecule, and they must be within close proximity to each other in the molecule. For example, intramolecular hydrogen bonding takes place in ethylene glycol $[C_2H_4(OH)_2]$ between its two hydroxyl groups due to the molecular geometry (Figure 4.37).

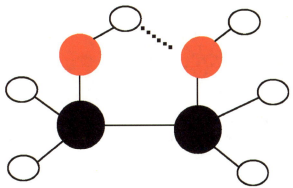

Figure 4.37 Intramolecular hydrogen bonds in Ethylene Glycol molecule

Intermolecular hydrogen bonding occurs between different molecules of a substance. Hydrogen donors and acceptors can occur between any number of similar or dissimilar molecules, as long as they are present in positions where they can interact with one another. For example, intermolecular hydrogen bonding can occur between individual NH_3 molecules, between individual H_2O molecules, or between NH_3 and H_2O molecules (Figure 4.38).

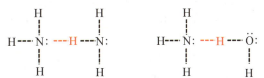

Figure 4.38 Intermolecular hydrogen bonds between two NH_3 molecules (left) and between NH_3 and H_2O molecules (right)

The types of geometries that can be used for hydrogen-bonding complexes are summarized in Figure 4.39. These geometries are called primary hydrogen bond interactions, which imply a direct interaction between the donor group and the acceptor group. Secondary interactions between adjacent groups must be considered. The partial charges on adjacent atoms can either increase the binding strength through the attraction between opposite charges, or decrease the affinity due to the repulsion between similar charges. Figure 4.40 shows two cases where the arrays of hydrogen bond donors and acceptors are in close proximity. An array of three donors (DDD) facing three acceptors (AAA) [Figure 4.40 (a)] has attractive interactions only between adjacent groups, therefore the binding is enhanced in such a situation. Mixed donor/acceptor arrays (ADA, DAD) suffer from repulsions due to the primary interactions that bring partial charges of the same sign close together [Figure 4.40 (b)].

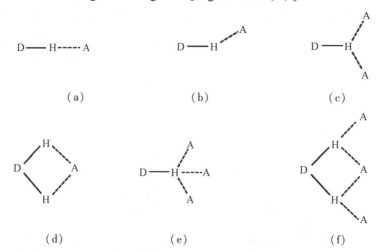

Figure 4.39　Various types of hydrogen bonding geometries: (a) linear, (b) bent, (c) donating bifurcated, (d) accepting bifurcated, (e) trifurcated, (f) three center bifurcated

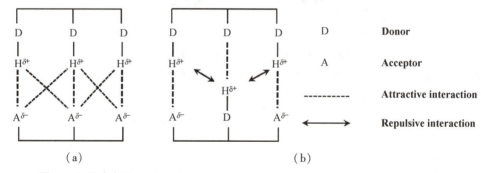

Figure 4.40 (a) Secondary interactions providing attractions between neighboring groups between DDD and AAA arrays (primary interactions in bold);
(b) repulsions from mixed donor/acceptor arrays (ADA and DAD)

Hydrogen bonds come in a variety of lengths, strengths and geometries and can be grouped into three broad categories, the properties of which are listed in Table 4.5. A strong interaction is somewhat similar in nature to a covalent bond, in which the hydrogen atom is near the center point of the donor and acceptor atoms. Strong hydrogen bonds are formed between a strong acid and a good hydrogen bond acceptor, such as in complexes of H_5O_2 ion or "proton sponge", which are practically linear with the hydrogen atom between the two electronegative atoms. Moderately strong hydrogen bonds are formed between neutral donor and neutral acceptor groups via lone electron pairs, such as the self-association of carboxylic acids, or amide interactions in proteins. Moderate hydrogen bonding interactions do not have a linear geometry, but are slightly curved. Weak hydrogen bonds play a role in structural stabilization and can be important when large numbers of hydrogen bonds act in concert. They tend to be highly non-linear and involve unconventional donors and acceptors, such as C-H groups, the π-systems of aromatic rings or alkynes, or even transition metals and transition metal hydrides.

Table 4.5 Properties of hydrogen bonding interactions (A-H = hydrogen bond acid, B = hydrogen bond base)

	Strong	Moderate	Weak
A-H⋯B interaction	Mainly covalent	Mainly electrostatic	Electrostatic
Bond energy/(kJ·mol^{-1})	60 – 120	16 – 60	< 12
Bond lengths/Å			
H⋯B	1.2 – 1.5	1.5 – 2.2	2.2 – 3.2
A⋯B	2.2 – 2.5	2.5 – 3.2	3.2 – 4.0
Bond angles	175° – 180°	130° – 180°	90° – 150°
Relative IR vibration shift (stretching symmetrical mode)/cm^{-1}	25%	10% – 25%	< 10%
Examples	Gas phase dimers with strong acids/bases	Acids	Minor components of bifurcated bonds
	Proton sponge	Alcohols	C-H hydrogen bonds
	HF complexes	Biological molecules	O-H⋯π hydrogen bonds

The properties of hydrogen bonding are summarized as follows:

(1) Solubility: Lower alcohols are soluble in water because hydrogen bonding can occur between water and alcohol molecule.

(2) Volatility: Since the compounds involving hydrogen bonding between different molecules have a higher boiling point, they are less volatile.

(3) Viscosity and surface tension: The substances which contain hydrogen bonding exist in the form of associative molecule. Therefore, their flow becomes comparatively difficult. They have higher viscosity and higher surface tension.

(4) The lower density of ice than water: In the case of solid ice, the hydrogen bonding forms a cage-like structure of water molecules. In fact, each water molecule is connected to four water molecules in a tetrahedron. The molecules are not as tightly packed as in a liquid state. When ice melts, the structure in this case collapses and the molecules move closer to each other. Thus, for the same mass of water, the volume decreases and the density increases. Therefore, at 273 K, ice has a lower density than water. That is why ice floats.

Hydrogen bonds are ubiquitous in supramolecular chemistry. In particular, hydrogen bonds are responsible for the overall shape of many proteins, recognition of substrates by many enzymes, and (along with π-π stacking interactions) the DNA double helix structure shown in Figure 4.41 (a). A real-life example of hydrogen bond is the double helix of DNA. There are many hydrogen bond donors and acceptors that hold base pairs together, as shown between the nucleobases cytosine (C) and guanine (G) in Figure 4.41 (b) and Figure 4.41 (c). The CG base pair has three primary interactions (i.e. traditional hydrogen bonds) and also has secondary interactions that are attractive and repulsive.

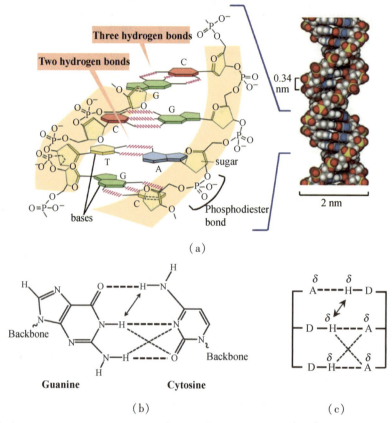

Figure 4.41 Hydrogen bonds in DNA: (a) double helix structure of DNA; (b) primary and secondary hydrogen bonding interactions between guanine and cytosine base pairs in DNA; (c) a schematic representation

iv. π-π stacking interaction

Aromatic π-π interactions (sometimes called π-π stacking interactions) occur between aromatic rings, usually in situations where one is relatively electron rich and the other is electron poor. There are two general types of π-π interactions: face-to-face and edge-to-face, although a wide variety of intermediate geometries are known [Figure 4.42 (a)]. Face-to-face interactions are responsible for the smooth feel of graphite and its useful lubricant properties. Similar π-π stacking interactions between the aromatic rings of nucleobase pairs also help to stabilize the DNA double helix. Edge-to-face interactions may be thought of as weak forms of hydrogen bonds between the slightly electron deficient hydrogen atoms of one aromatic ring and the electron rich π-cloud of another. Strictly speaking, they should not be called π-stacking, since there is no stacking of the π-electron surfaces. Edge-to-face interactions are responsible for the characteristic herringbone packing in a series of small aromatic hydrocarbon crystal structures, including benzene [Figure 4.42 (b)].

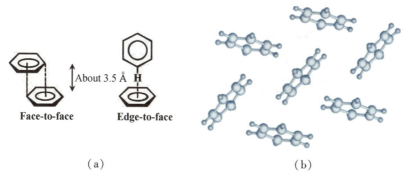

(a)　　　　　　　　　　　　　　　　(b)

Figure 4.42 (a) Limiting types of π-π interaction (Note the offset to the face-to-face mode, direct overlap is repulsive); (b) X-ray crystal structure of benzene showing herringbone motif arising from edge-to-face interactions

π-π interactions happen when aromatic π systems bind face to face with one another and involve a combination of dispersion and dipole-induced dipole interactions. Since the electron distribution in aromatic systems is relatively easily distorted, they can participate in atypically strong dispersion and dipole-induced dipole interactions known as π-π stacking interactions. In either case, the rings can be attracted to one another through a combination of dispersion and electrostatic interactions. The latter can be either attractive or repulsive, depending on the specific bonding parameters of the system. The overlapping arrangement of the benzene dimer is electrostatically unfavorable, whereas the attractive force between the positive hydrogen substituents and the π system of an adjacent ring is attractive in the overlapping arrangement.

When negatively charged substituents are used, the polarity of the ring system is reversed, resulting in favorable electrostatic interactions in overlapping benzene and hexaflourobenzene heterodimers [Figure 4.43 (a) and Figure 4.43 (b)]. In the heterodimer of benzene and hexaflourobenzene, we finally have an example of a π-stacking system that can more readily be described as involving a Lewis acid-base interaction. Specifically, the relatively electron-rich benzene π system acts as a Lewis base in donating "electron pairs" to the electron-deficient hexaflourobenzene π system.

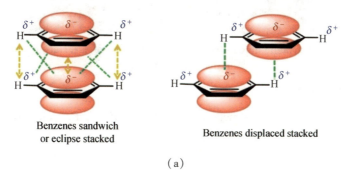

(a)

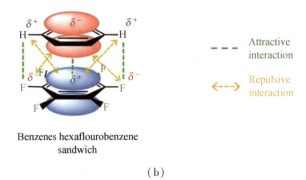

Benzenes hexaflourobenzene sandwich

(b)

Figure 4.43 π-π interactions in benzene system

As an attractive, non-destructive and non-covalent interaction, π-π stacking interactions have been widely explored in the fields of modern chemistry, molecular biology, and supramolecular armamentarium, among which their bio-applications have attracted tremendous attention due to the unique advantages such as strong binding force, non-destructive fabrication process, and simple operation. Impressively, great achievements have been made in the fields of nucleobase stacking, biosensing, drug-controlled release, protein folding, molecular recognition, self-assembly, template-directed synthesis. etc., shown in Figure 4.44.

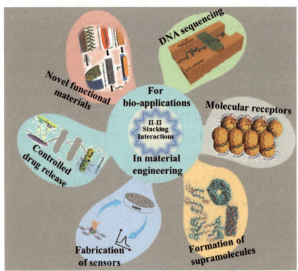

Figure 4.44 Schematic illustration for bio-applications of π-π stacking interactions in material engineering

v. Cation-π interaction

It is well known that transition metal cations such as Fe^{2+}, Pt^{2+} etc. form complexes with olefinic and aromatic hydrocarbons such as ferrocene $[Fe(C_5H_5)_2]$ and Zeise's salt $[PtCl_3(C_2H_4)]^-$. The bonding in such complexes is strong and in no way can be considered non-covalent, since it is intimately connected to the partially occupied d-orbitals of the metals. Even species such as $Ag^+ \cdots C_6H_6$ have important covalent component. However, the interaction of alkaline and alkaline earth metal cations with C = C double bonds is, a much more non-covalent "weak" interaction, and is therefore believed to play an important role in biological systems. For example, the interaction energy of K^+ and benzene in the gas phase is about 80 kJ · mol^{-1} (Figure 4.45). In contrast, the association of K^+ with a single water molecule is similar at 75 kJ · mol^{-1}. The reason why K^+ is more soluble in water than in benzene is related to the fact that many water molecules can interact with the potassium ion, while only a few bulkier benzene molecules can fit around it. The interaction of nonmetallic cations such as RNH_3^+ with double bonds may be considered as a form of X—H···π hydrogen bond.

Figure 4.45　Schematic drawing of the cation-π interaction showing the contact between the two (The quadrupole moment of benzene, along with its representation as two opposing dipoles, is also shown.)

The cationic interaction is a stable electrostatic interaction of a cation with the polarizable π electron cloud of an aromatic ring. Six-carbon aromatic rings occur in the side chains of 3 of the 20 standard amino acids: namely phenylalanine, tryptophan, and tyrosine. One p atomic orbital on each aromatic carbon overlaps in a π ("sideways") manner with its two neighboring atoms, forming a conjugated π orbital system. The π electrons of aromatic rings form delocalized annular clouds above and below the ring plane [Figure 4.46 (a)]. A positive charge close to one face of the ring attracts and polarizes the π electron cloud. In the appropriate conformation, this proximity forms an energetically significant cation-π interaction [Figure 4.46 (b)].

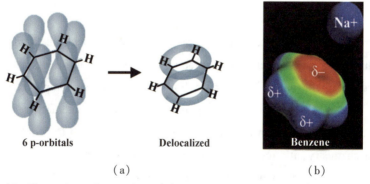

Figure 4.46　The cation-π interaction: (a) the π electrons of aromatic rings comprising delocalized annular clouds above and below the ring plane; (b) cation-π interaction (Surfaces colored by electrostatic potential)

The chemical community now recognizes that the cation-π interaction is a major force in molecular recognition, incorporating the hydrophobic effect, the hydrogen bond, and the ion pair in determining macromolecular structure and drug-receptor interactions. The cation-π interaction is now appreciated as an important factor in molecular recognition and catalysis in chemistry and biology.

vi. van der Waals force

van der Waals (VDW) force, also known as London dispersion energy, is probably the most basic type of interaction imaginable. Any two molecules experience van der Waals force. Even macroscopic surfaces experience VDW force, but more on this later. As part of molecular physics, this force is derived from the name of a Dutch scientist, Johannes Diderik van der Waals. He first discovered the van der Waals bond in 1873 while working on a theory of real gasses.

The following pointers show some of the characteristics of van der Waals force:

(1) It contains relatively weak electricity forces compared to ionic, metallic or covalent bonds.

(2) When a large number of molecules are present, the interactions can be addictive. These are still present when molecules get placed afar.

(3) van der Waals force is omnipresent and responsible for the attraction of atoms and molecules to each other.

(4) It remains unaffected due to temperature changes except in the case of dipole-dipole interactions.

(5) Although it is present in most of the materials, its influence is overpowered by the primary bonds.

(6) Moreover, it cannot get saturated.

van der Waals force is caused by the polarization of an electron cloud near an adjacent nucleus, resulting in a weak electrostatic attraction. They are non-directional and thus possess only limited scope in the design of specific hosts to selectively complex specific guests. In general, van der Waals force provides generally attractive interaction for most "soft" (polarizable) substances, with an interaction energy proportional to the contact surface area. In supramolecular chemistry, it's most important in formation of "inclusion" compounds, in which usually small organic molecules are loosely incorporated within crystalline lattices or molecular cavities. For example, toluene is contained within the molecular cavity of the p-tert-butylphenol-based macrocycle, p-tert-butylcalix[4]arene (Figure 4.47). Strictly speaking, van der Waals force can be divided into dispersion (London) and exchange exclusion terms. The dispersion interaction is the attractive component arising from the interactions between fluctuating multipoles (quadrupole, octupole etc.) in adjacent molecules. The attraction decreases very rapidly with distance (r^{-6} dependence) and adds to each bond in the molecule, contributing to the overall interaction energy. The exchange-repulsion defines molecular shape and balances dispersion over short range, decreasing with the 12th power of interatomic separation.

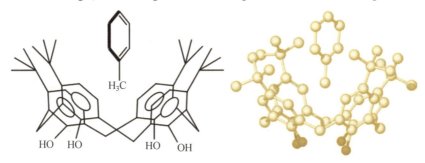

Figure 4.47   X-ray crystal structure of a typical van der Waals inclusion complex p-tert-butylcalix[4] arene toluene

van der Waals force has a variety of applications in molecular science: This intermolecular force allows Gecko lizards to move efficiently across surfaces. Likewise, some other species of spiders have these biological patterns. This force is responsible for the interactions of proteins with other atoms. It also affects various properties of gases, adhesion, and colloidal stability. This force plays a fundamental role in research in supramolecular chemistry, nanotechnology, surface, and polymer science, and more. For a bit of entertainment, and to show you that VDW force is real and very powerful, here is a picture of a Gecko and its foot (Figure 4.48). Geckos gain the ability to stick to almost any surface through VDW force. The trick is to get the area of two contacting surfaces close enough for the VDW force to start to become effective. The interaction energies fall off—drop by r^{-6}, which means close to within a fraction of an Angstrom. The gecko does this with superfine, flexible bristles on its feet that press very small

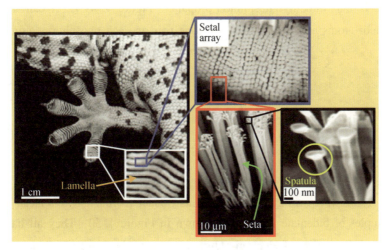

Figure 4.48   Geckos and van der Waals force

protrusions, called spatula, against surfaces, even if those surfaces are quite rough at the molecular level. It has been calculated that if all of those scrapers make full contact to the

Fundamentals of Photophysics

surface, VDW force would be strong enough to suspend a 200 pound weight from a single Gecko toe.

The factors that affect van der Waals force are as follows:

(1) The number of electrons present in an atom: The amount of electrons present is what causes the temporary dipoles. The strength of van der Waals force is dependent on the number of dipoles. Therefore, increasing the number of dipoles increases the bonds of van der Waals.

(2) Size of the atoms: The strength of attractive bonds of this force varies with changes in the size of atoms. An intermolecular force increases with the size of atoms, such as helium, radium, krypton, etc. In contrast, the boiling and melting points of materials change as a result of the force.

(3) Nature of the elements: The properties of an element or a non-metals are related to the strength of van der Waals force. Most non-metals that exist in a liquid or gaseous state consist of this force, while some metals include strong, cohesive forces.

(4) The shape of atoms: The form of an atom is directly related to the strength of the force. A thinner molecule has the potential to create more temporary dipoles than short, fat ones.

4.2.3 Entropy Increasing Factor

i. Chelation effect

Monodentate ligands are bound by only one donor atom. Monodentate means "one-toothed". The halides, phosphines, ammonia and amines seen earlier are monodentate ligands. Bidentate ligands bind via two donor sites. Bidentate means "two-toothed". An example of a bidentate ligand is bis(dimethylphosphino) propane (Figure 4.49). It can bind to a metal through two donor atoms at once: It uses one lone pair on each phosphorus atom.

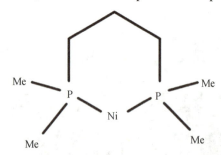

Figure 4.49 An example of a bidentate donor coordinating to a metal

More examples of bidentate ligands are shown in Figure 4.50. They all have at least two distinct atoms with lone pairs of electrons. In some cases, there are additional atoms with lone pairs, but only two of them can face the metal at the same time. Oxalate and glycinate would act as bidentate donors, donating up to two lone pairs simultaneously.

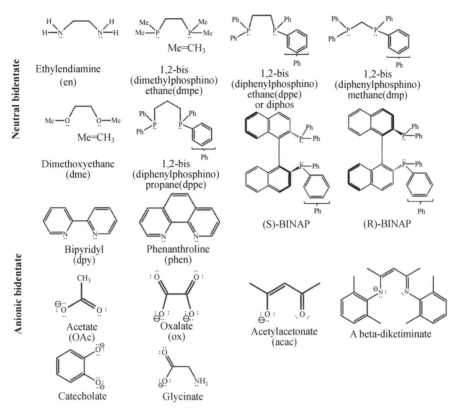

Figure 4.50 Some common neutral bidentate ligands and anionic bidentate

Bidentate binding allows the ligands to bind more tightly. Tridentate ligands bound by three donors, can bind even more tightly, and so on. This phenomenon is often referred to as "chelate effect". This term comes from the Greek chelos, which means "crab". A crab has no teeth at all, but it does have two claws to hold onto something. A very simple analogy is that, if you are holding something with two hands instead of one, you are less likely to drop it.

Chelating ligands have higher affinity for a metal ion than similar monodentate ligands. The chelating effect is the enhanced affinity of a chelating ligands for metal ions compared with their monodentate counterparts. For example, ethylenediamine ($H_2NCH_2CH_2NH_2$) is a bidentate ligand that binds metal ions more strongly than monodentate amine ligands such as ammonia (NH_3) and methylamine (CH_3NH_2).

The chemical cause of the chelation effect involves relative enthalpy and entropy changes upon binding of a multidentate ligand, which can be explained using thermodynamic principles. In terms of enthalpy, for complete removal of a bidentate ligand, both coordination bonds must be broken. That takes more energy than breaking one coordinate bond of a monodentate ligand. In terms of dealing with the entropy of the energy distribution within a system, it is generally considered that the cost of combining two molecules (a bidentate ligand and a metal complex) is lower than the cost of combining three molecules (two monodentate ligands and a metal complex). That's because individual molecules are free to move around, tumble and vibrate independently. Once they come together, they have to do all these things together. Since these different types of motion represent different ways of distributing energy, if the system becomes more constrained, energy cannot be distributed in as many states as possible.

ii. Macrocyclic effect

Many supramolecular host-guest complexes are even more stable than would be expected from cooperation/chelation effects alone. The hosts in these species are often macrocyclic (large ring) ligands which in turn sequester their guests, through a number of binding sites. Such compounds are also stabilized by a method traditionally known as the macrocyclic effect. This effect is not only related to the sequestration of the guest by multiple binding sites, but also to the spatial organization (i.e. preorganization) of those binding sites to guest binding so that binding energy is not consumed in the guest having to "wrap" the host about itself in order to benefit from the most chelation. Furthermore, the enthalpy losses associated with bringing lone pairs of donor atoms close proximity to one another (with consequent unfavorable repulsion and desolvation effects) have been "paid in advance" during the synthesis of the macrocycle. This makes macrocycles difficult to fabricate, but stronger complexing agents than similar non-macrocyclic hosts (podands).

Related to the chelation effect, it is necessary to improve system stability in terms of energy and entropy factors. The macrocyclic effect makes cyclic hosts such as corands (e.g. crown ethers) up to a factor of 10^4 times more stable than closely-related acyclic pods with the same type of binding sites. The macrocyclic effect was first elucidated by Cabbiness and Margerum in 1969, who studied the Cu(II) complexes **4.14** and **4.15** (shown in Figure 4.51). Both ions benefit from the stability associated with the four chelated donor atoms. However, the macrocyclic complex **4.14** is approximately 10^4 times more stable than the acyclic analogue **4.15** due to a consequence of the additional preorganization of the macrocycle.

Table 4.6 Thermodynamic parameters for Zn^{2+} complexes of 4.14 and 4.15 (298 K)

Complex	**4.14**	**4.15**
lg K	15.34	11.25
$\Delta H°/(kJ \cdot mol^{-1})$	−61.9	−44.4
$−T\Delta S°/(kJ \cdot mol^{-1})$	−25.6	−19.8

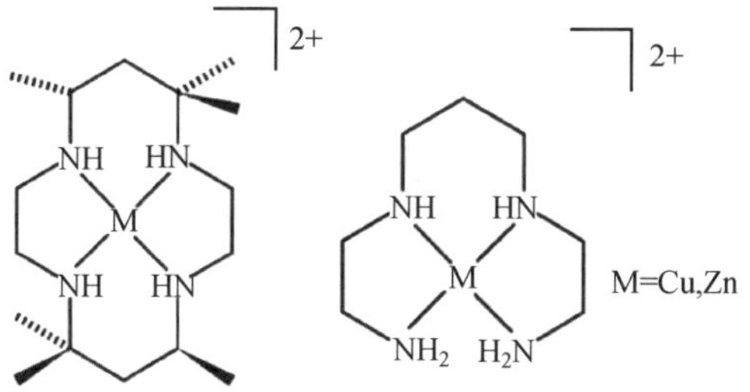

![Structures 4.14 and 4.15 with M=Cu,Zn, 2+ charge]

Figure 4.51 Structural schematic diagram of 4.14 and 4.15

Thermodynamic measurements on the analogous (unmethylated) Zn^{2+} complexes indicate that the stability of the macrocyclic pre-organization has both enthalpic and entropic contributions (Table 4.6). The enthalpy term arises from the fact that macrocyclic hosts are frequently less strongly solvated than their acyclic analogs. This is because they simply exhibit less solvent-accessible surface area. Therefore, there are fewer solvent-ligand bonds to break than in the extended, acyclic case. Entropy-wise, macrocycles are less conformationally flexible and so lose fewer degrees of freedom upon complexation. In general, the relative importance of the entropic and enthalpic terms varies depending on the system under study, although the enthalpy is frequently dominant due to other factors such as lone pair repulsions. For almost the same reason, bicyclic hosts such as crypts are found to be even more stable than monocyclic corands. Historically, this further additional stability is referred to as the macrobicyclic effect (Figure 4.52), which simply represents the more rigid, preorganized nature of the macrobicycle. The macrocyclic and macrobicyclic effects make an important contribution to the host of alkali metal binding (Figure 4.53).

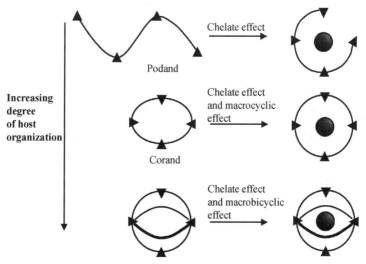

Figure 4.52 The chelate, macrocyclic and macrobicyclic effects

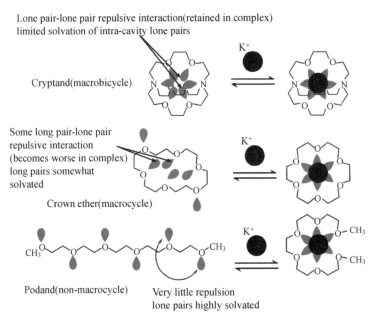

Figure 4.53 Comparison of preorganization effects in K binding by a macrobicycle, macrocycle and non-preorganized podand pentaethyleneglycol dimethyl ether

The macrocyclic effect is that when the multidentate ligands form a cyclic ring with central metal atom or ion, it enhances the stability of the complexes because it is a multidentate ligand. The macrocyclic effect is the high affinity of metal cations for macrocyclic ligands, compared with their acyclic analogs. A macrocyclic ligand is a macrocycle of at least nine ring sizes (including all hetero atoms) and three or more donor sites. Macrocyclic ligands are multidentate, and because they are covalently constrained to their cyclic form, they allow less conformational freedom. Macrocyclic ligands cannot be dissociated by a simple mechanism because there is no end group. Examples of the macrocyclic ligands are crown ethers and porphyrins. The magnitude of the macrocyclic effect depends on the ligand and metal center. The enhanced stability of a macrocyclic ligand complex may stem from its slow dissociation rate compared with its open chain analogue. Macrocyclic complexes are synthesized by combining macrocyclic ligands and metal ions. Macrocyclic ligands are present in many cofactors of proteins and enzymes. The natural occurrences of these complexes are Heme, the active site in the haemoglobin (the protein in blood that transports oxygen). Chlorophyll is a green photosynthetic pigment found in plants that contain a chlorine ring. For example, as shown in Figure 4.54, Heme b is a tetradentate cyclic ligand that strongly complexes transition metal ions, including (in biological systems) Fe^{2+}.

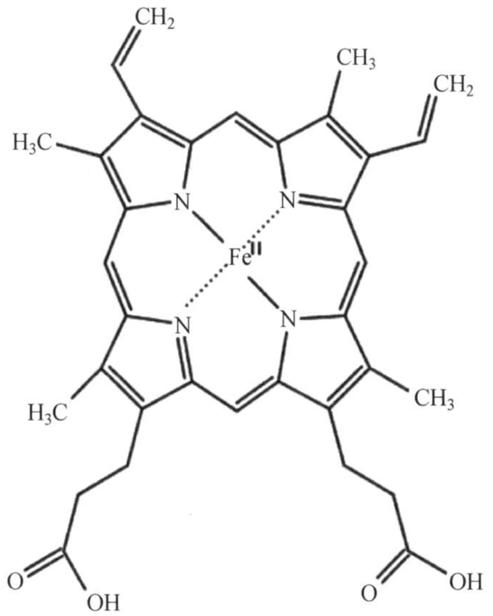

Figure 4.54 Molecular structure of Heme b

Macrocyclic effects can be seen in coordination compounds. It forms cyclic compounds to maintain the stability of the complex. In this multidentate ligand, the ligand forms complexes with metal ions. Different compounds exhibit different properties. Macrocyclic complexes have higher stability than acyclic complexes because they form cyclic compounds. The dissociation of these complexes requires a large amount of energy because it is a complex mechanism. These compounds are widely used for different purposes in various industries. The macrocyclic effect follows the same principle as the chelation effect, but the effect is further enhanced by the cyclic conformation of the ligand. Ligands are macrocyclic, and because they are covalently confined in their cyclic form, they also allow fewer degrees of conformational freedom. The ligand is said to be bound "preorganized" and the ligands wrap around the metal ions with little entropy loss. The order of stability is: macrocyclic > chelate > monodentate.

iii. Hydrophobic effect (Cavity effect)

The hydrophobic effect (cavity effect) refers to the hydrophobic effect or entropy enhancement effect of a hydrophobic cavity. In the hydrophobic cavity, water molecules are held together in a relatively orderly manner through hydrogen bonds. When there is a hydrophobic guest molecule (G), the guest molecule will spontaneously enter the cavity and squeeze out the water molecules. At this time, the water molecules are in a free state, and the disorder degree increases, that is, the entropy increases. Although occasionally mistaken for a force, hydrophobic effects are often associated with the exclusion from polar solvents, particularly water, of large particles or those that are weakly solvated. The effect is obvious in the immiscibility of mineral oil and water. Essentially, the water molecules are strongly attracted to one another, leading to a natural agglomeration of other substances (such as nonpolar organic molecules), because of the way they are squeezed out of the strong inter-solvent interactions.

This can produce effects resembling attraction between one organic molecule and another, although there are also van der Waals force and π-π stacking attraction between the organic molecules themselves.

American chemist Walter Kauzmann found that nonpolar substances such as fat molecules tend to clump up together rather than distributing itself in a water medium, as this allows the fat molecules to have minimal contact with water.

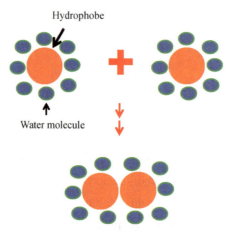

Figure 4.55 Schematic illustration for the hydrophobic effect

Figure 4.55 shows that when the hydrophobes come together, they will have less contact with water. They interact with a total of 16 water molecules before coming together and there are only 10 atoms after the interaction.

Hydrophobic interactions are relatively stronger than other weak intermolecular forces (i.e. van der Waals force or hydrogen bonds). The strength of hydrophobic interactions depends on several factors, including (in order of strength of influence):

(1) Temperature: As the temperature increases, the strength of hydrophobic interactions also increases. However, at an extreme temperature, hydrophobic interactions can denature.

(2) Number of carbons on the hydrophobes: Molecules with the highest number of carbons will have the strongest hydrophobic interactions.

(3) The shape of the hydrophobes: Aliphatic organic molecules have stronger interactions than aromatic compounds. Branching in a carbon chain will reduce the hydrophobic interaction of that molecule, while linear carbon chain can produce the largest hydrophobic interaction. This is because carbon branches create steric hindrance, so it is harder for two hydrophobes to have very close interactions with each other to minimize their contact to water.

The hydrophobic effect is very important in biological systems in the production and maintenance of protein and polynucleotide structure, and in the maintenance of phospholipid bilayer cell walls. Hydrophobic effects are crucial for the organic-guest binding of cyclodextrins and cyclophane hosts in water, and may be divided into two energetic components: enthalpy and entropy. The enthalpic hydrophobic effect involves the stabilization of water molecules driven from the host cavity upon guest binding. Because host cavities are generally hydrophobic, intracavity water does not interact strongly with the host walls and is therefore highly energetic. After being released into the bulk solvent, it is stabilized by interactions with other water molecules. The entropic hydrophobic effect stems from the fact that the presence of two (often organic) molecules in solution (host and guest) creates two "holes" in the structure of bulk water. Combining host and guest to form complex results in less disruption to the solvent structure and hence an entropic gain (resulting in a lowering of overall free energy). Figure 4.56 shows the process schematically.

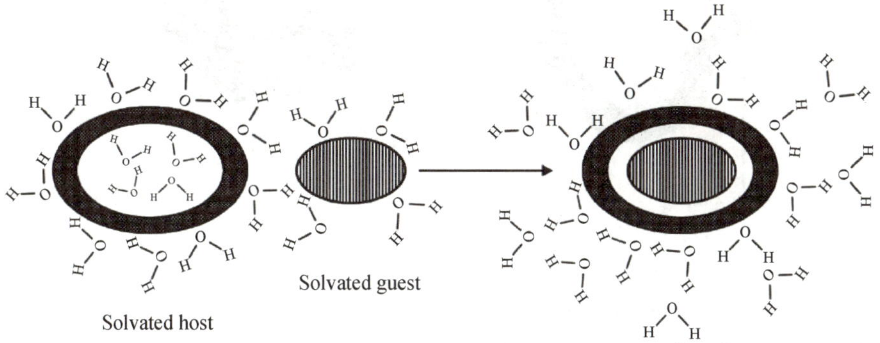

Figure 4.56 Hydrophobic binding of organic guests in aqueous solution

For example, consider the binding of the guest p-xylene to the cyclophane host **4.16** (Figure 4.57). The binding constant in water is 9.3×10^3 mol^{-1}. At 293 K, the complexation free energy, ΔG, is -22 kJ · mol^{-1}, divided into a favorable enthalpic stabilization, $\Delta H = -31$ kJ · mol^{-1}, and an unfavorable entropic component, $T\Delta S = -9$ kJ · mol^{-1}. In this case, it is mainly the enthalpic contribution to the hydrophobic binding. The enthalpic contribution is too large to cause attractive forces between hosts and guests (which experience only weak π-stacking and van der Waals force) and thus must arise from specific solvent-solvent forces. In methanol solvent, the enthalpic component is greatly reduced due to weaker solvent-solvent interactions.

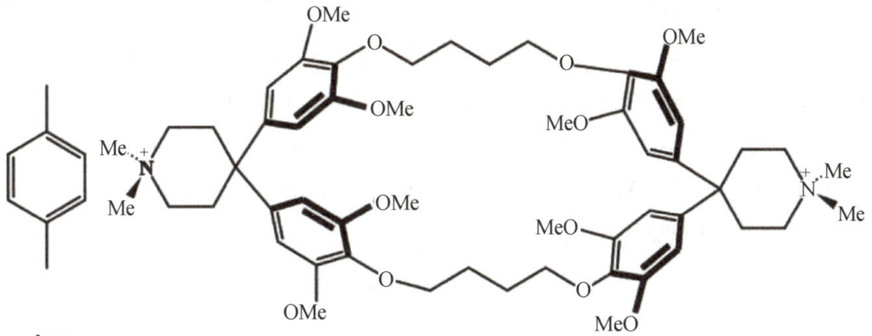

Figure 4.57 Structural schematic diagram of 4.16

Fundamentals of Photophysics

Hydrophobic interactions are important for protein folding. This is important for maintaining a protein stability and biological activity, as it allows the protein to decrease in surface and reduce the undesirable interactions with water (Figure 4.58). In addition to proteins, there are many other biological substances that depend on hydrophobic interactions for its survival and functions, such as the phospholipid bilayer membranes in every cell of your body.

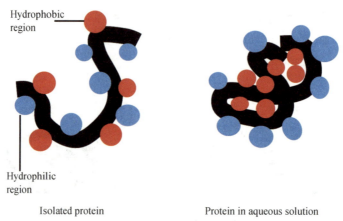

Isolated protein   Protein in aqueous solution

Figure 4.58 Illustration of how protein changes shape to allow polar regions (blue) to interact with water while prevent nonpolar hydrophobic regions (red) from interacting with water (CC BY-SA 3.0; Treshphrd)

4.3 Typical Molecules

4.3.1 Crown Ethers

Crown ethers, also known as "macrocyclic ethers", are a class of macrocyclic compounds found to contain multiple oxygen atoms. Crown ethers are among the simplest and most appealing macrocyclic (large ring) ligands, and are ubiquitous in supramolecular chemistry as hosts for both metallic and organic cations. Ethers are polar, relatively inert, and are not affected by alkalies, alkali metals and dilute acids as they do not contain any active functional group, and there is no double or triple bond. They simply consist of a cyclic array of ether oxygen atoms linked by organic spacers, typically -CH_2CH_2- groups. The typical crown ethers are shown in Figure 4.59. While the metal binding ability of unidentate ethers such as the common solvent diethyl ether is very poor, crown ethers are much more effective by virtue of the chelate effect and the partial preorganization arising from their macrocyclic structure.

>>> **Chapter 4** Supramolecules in Photophysics

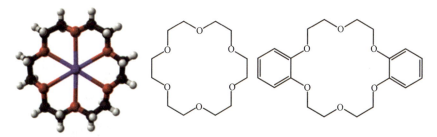

Figure 4.59 Schematic illustration of typical crown ethers

Crown ethers have their own unique naming scheme. According to the naming rule, the total number of atoms contained in the ring is marked before the word "crown", and the number of oxygen atoms contained in it is marked after the name, such as -crown (ether)-6, dicyclohexane and 18-crown (ether)-6.

The most important feature of crown ethers is their ability to complex with positive ions, especially with alkali metal ions, and complex with different metal ions depending on the size of the ring. Few complexes of the alkali metal cations with organic molecules had been discovered in the late 1960s. Specialized biological molecules such as valinomycin were known to complex selectively with the potassium cation K^+ for the transport across cell membranes, but rare reports were related to the synthetic ionophores (molecules that can form complexes with ions). All the alkali cations have a charge of +1 and, except for lithium, are chemically similar and rather inert. The only significant difference between one alkali cation and another is the size. In 1967, Charles J. Pedersen synthesized crown ethers, which provided size-selective cyclic molecules consisting of ether oxygens forming a ring or "crown" that could complex with a cation of the right size to fit into the hole in the center of the molecule. In some cases, two crown ether molecules can encapsulate a cation in a "sandwich" fashion. For example, K^+ just fits into the center of an 18-crown-6 ring (18 atoms in the ring, 12 of which are carbon atoms and 6 are ether oxygen atoms) to form a 1∶1 complex (that is, 1 cation∶1 crown ether), $K^+(18C6)$. Cs^+ is too large to fit into the ring but can be complexed on one side to form the $Cs^+(18C6)$ complex or can be sandwiched between two 18-crown-6 molecules to form the 1∶2 complex, $Cs^+(18C6)_2$. Therefore, the selectivity of a crown ether for a particular cation depends on the ring size. Common crown ethers are 15-crown-5, 18-crown-6 and 21-crown-7. These molecules are selective for Na^+, K^+ and Cs^+, respectively. Complexation between crown ether and alkali metal ions is displayed in Figure 4.60.

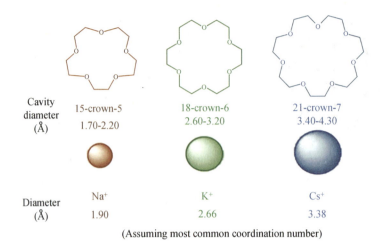

Figure 4.60　Complexation between crown ethers and alkali metal ions

The applications of crown ethers include:

(1) Crown ethers are used in the laboratory as a phase transfer catalyst.

(2) Crown ethers are used in esterification, saponification, displacement, elimination reaction, generation of carbenes, etc.

(3) Crown ethers are used in the supercritical fluid extraction of trace metals from solid and liquid materials.

(4) Crown ethers are used in photocyanation, resolution of ceramic mixtures, heterocyclization, etc.

(5) A typical kind of crown ether called dicyclohexyl-18-crown-6 is used in the determination of gold in geological samples.

Crown ethers allow many reactions that are difficult to react or do not occur under conventional conditions to proceed smoothly. Crown ethers complex with the positive ion in the reagent, so that the positive ion can be dissolved in organic solvent, and the corresponding negative ion with it also enters the organic solvent. Crown ethers do not complex with the negative ion, so that the free or bare negative ion reaction activities are very high, and they can react quickly. In this process, crown ethers bring the reagent into the organic solvent, known as a phase transfer agent or phase transfer catalyst, and the reaction that occurs in this way is called a phase transfer catalytic reaction. The reaction rate is fast. The condition is simple. The operation is convenient. The yield is high.

4.3.2　Cyclodextrin

Cyclodextrins (CDs) are cyclic molecules made up of glucose monomers coupled to form a rigid, hollow, tapering torus with a hydrophobic interior cavity. Due to the presence of the cavity, cyclodextrins are able to act as hosts, binding with small guest molecules held within the internal cavity. Cyclodextrin is a generic term for a series of cyclic oligosaccharides produced by amylose in a cyclodextrin glucosyltransferase produced by bacillus, usually containing 6-12 D-pyrans glucose unit. Among them, more studied and of practical significance are molecules containing 6, 7, or 8 glucose units, which are called alpha-, beta-, and gama-cyclodextrin, respectively (Figure 4.61). The schematic diagram of β-cyclodextrin is shown in Figure 4.62.

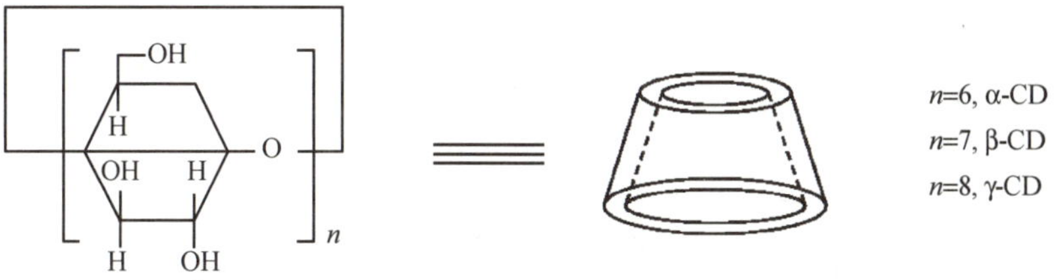

$n=6, \alpha\text{-CD}$
$n=7, \beta\text{-CD}$
$n=8, \gamma\text{-CD}$

Figure 4.61 Molecular configuration of cyclodextrin

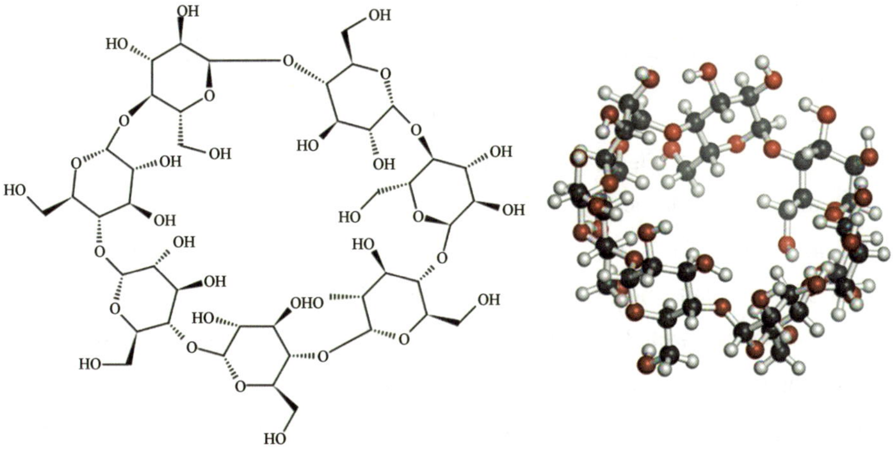

Figure 4.62 Schematic diagram of β-cyclodextrin

The shape of a cyclodextrin can often be represented as a tapering torus or truncated funnel and, like the upper and lower rims of calixarenes, there are two different faces to the cyclodextrins, referred to as the primary and secondary faces. The primary face is the narrow end of the torus, and is composed of the primary hydroxyl groups. The wider secondary face contains the -CH_2OH groups. The six-membered D-glucopyranoside rings are linked edge to edge, with their faces all pointing inwards towards a central hydrophobic cavity of varying dimensions. It is this cavity, coupled with the water solubility derived from the hydrophilic alcohol functionalities, that gives the cyclodextrins their unique complexation ability in aqueous solution. Figure 4.63 shows a functional scheme of cyclodextrin anatomy along with the cavity sizes.

Because the rim of the cyclodextrin is hydrophilic and the cavity is hydrophobic, it can provide a hydrophobic binding site like an enzyme, as a host to enclose a lot of appropriate guest, such as organic molecules, inorganic ions, and gas molecules. Its hydrophobic and externally hydrophilic properties make it possible to form inclusion complexes and molecular assembly systems with many organic and inorganic molecules by van der Waals force, hydrophobic interaction forces, and host-guest molecular matching, and become the interest of chemical and chemical research.

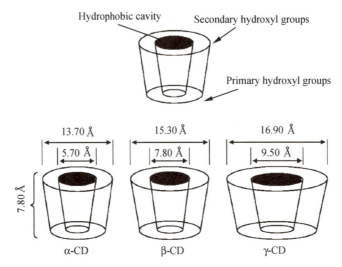

Figure 4.63 Anatomy of the cyclodextrins

Of particular interest in the application of cyclodextrins is the enhancement of luminescence from molecules when they are present in a cyclodextrin cavity. Polynuclear aromatic hydrocarbons show virtually no phosphorescence in solution. However, if these compounds in solution are encapsulated with 1,2-dibromoethane (enhances intersystem crossing by increasing spin-orbit coupling) in the cavities of β-cyclodextrin and nitrogen gas passed, intense phosphorescence emission can occur at room temperature. Cyclodextrins form complexes with guest molecules, which fit into the cavity so that the microenvironment around the guest molecule is different from that in the bulk medium. Once the phosphorescent guest molecule occupies the cavity, other molecules are excluded from simultaneously occupying it and collisional deactivation of the excited triplet state of the guest molecule is reduced or prevented.

As shown in Figure 4.64, cyclodextrins have a large range of industrial applications. The market for them is growing as a consequence of their unique inclusion properties and decomplexation kinetics in conjunction with their stability, non-toxicity and relative cheapness. Cyclodextrins are the main active ingredient in Procter and Gamble's deodorizing product Febreze, for instance, where their complexation ability binds molecules responsible for household odors.

Cyclodextrins are widely used in biomedicine, for example, biomimetic corneal implants. Cyclodextrins are favorable for the growth of collagen in vitro, making its structure similar to the one found in the cornea. Collagen fibrils in the cornea have very particular properties, being narrower than the fibrils in other connective tissues to ensure transparency. When cyclodextrins are added to collagen during fibrinogenesis, they interact with hydrophobic amino acid residues of collagen, thereby interrupting the crosslinking process (Figure 4.65). By using native

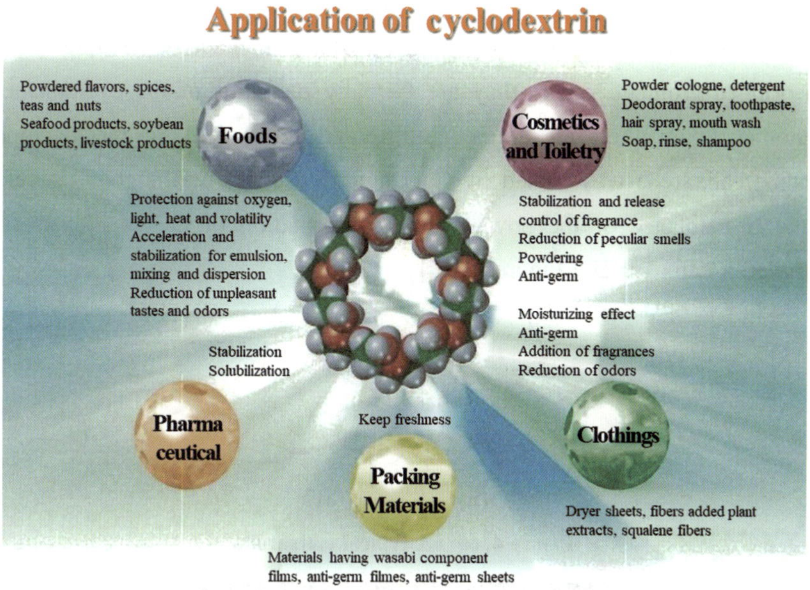

>>> **Chapter 4** Supramolecules in Photophysics

Figure 4.64 Applications of cyclodextrin

cyclodextrins to modulate the collagen I self-assembly process during vitrification, transparent and mechanically robust corneal substitutes can be engineered. The ultrastructure of these biomimetic corneal substitutes is similar to that of the native cornea. The saturation tests using the β-CD/collagen cornea also show that it is resistant enough and support re-epithelialisation and host tissue integration. These materials have huge potential as biomimetic cornea substitutes, and the technology is already protected by patents.

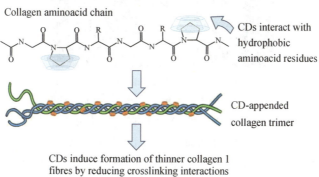

Figure 4.65 Mode of action proposed for in vitro modulation of collagen growth by cyclodextrins

4.3.3 Calixarene

Calixarene is a typical motif in supramolecular chemistry. Following crown ethers and cyclodextrins, calixarenes are third-generation host molecules. It is easy to synthesize calixarenes using condensation reactions of phenols with aldehydes in a desirable yield. The calixarenes are a popular and versatile class of macrocycle that is produced by the condensation of a p-substituted phenol (e.g. p-tert-butylphenol) with formaldehyde (Figure 4.66). Because they contain bridged aromatic rings, they are formally members of the cyclophane family. In cyclophane nomenclature they are termed substituted [1.1.1.1] metacyclophanes. The descriptive name calixarene was coined by C. David Gutsche (Washington University, USA) because of the resemblance of the bowl-shaped conformation of the smaller calixarenes to a Greek vase called a calix crater (Figure 4.67).

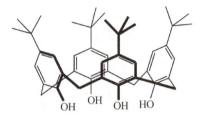

Figure 4.66 A typical calixarene

Figure 4.67 Greek Holy Grail (Calixarene is named "Calcylene" because of its similar structure to the Greek Holy Grail.)

In fact, calixarene was first discovered by Adolph von Bayer in 1872, and then Gutsche et al. improved the chemistry of phenol-formaldehyde products and coined the term "Calixarene" in the late 1970s. As a result of special geometry and the availability of active groups in calixarenes, it is possible to use these compounds for various purposes. The calixarene structure includes a polar rim, a non-polar rim, and a hydrophobic cavity. The rims can be selectively functionalized to provide analyte-selectivity or to facilitate polymerization. The internal cavity of calixarene can host different guest molecules. Therefore, calixarene-based polymers are suitable for manipulating selectivity in separation sciences. Because of these structural characteristics, calixarenes have been incorporated into polymers, but mostly as side-chain moieties.

Nowadays, the range of calixarene is gradually expanding, and similar structural substances formed by resorcinol and formaldehyde under acid-catalyzed conditions are also named as calixarene, and even other substances of the same structure are also named calixarene. According to the number of phenol units, it is named as a calix[n]arene. More common are calix[4]arene, calix[6]arene and calix[8]arene. In particular, the preparation of calixarene-containing polymers has two general synthetic methods. One is based on polymerization reactions of a functionalized calixarene monomers, and the other relies on the immobilization of the calixarene moiety on a polymeric matrix.

Once the concentration of iodine is too high, it will lead to an environmental pollution which has critical ecological and health effect. Iodine is also an interesting radionuclide that creates serious risks to the public. Therefore, it is important to sense and remove iodine from water and vapor phase and to this end, several calixarene-based polymers have been developed. These studies have introduced different types of adsorbents and molecules which have been examined to eliminate excess iodine from water and environment. For example, as shown in Figure 4.68, Shetty et al. introduced calix[4]arene-based 2D polymer 3 through Sonogashira-Hagihara cross-coupling reaction between tetra-bromo calix[4]arene tetrol 1 and 4,4'-diethynyl-1,1'-biphenyl 2 in anhydrous 1,4-dioxane. The results of the characterization indicated that 2D polymer 3 was porous, covalent, and isolated as few-layer thick (3.52 nm) nanosheet. The polymer was examined as an I2 vapor adsorbent, and the findings were supported with a maximum capacity of 114%. Moreover, it is easy to regenerate the material through washing with mild ethanol, and thus it can be recycled with a very small loss of the efficiency.

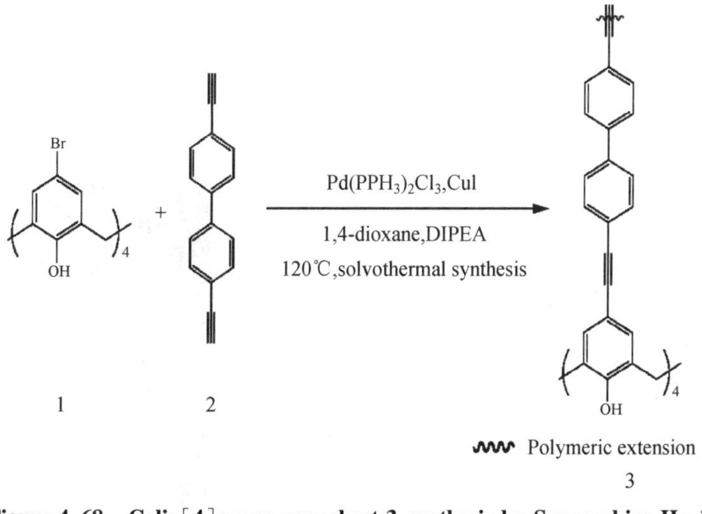

Figure 4.68 Calix[4]arene nanosheet 3 synthesis by Sonogashira-Hagihara cross-coupling reaction

Both calixarene and cyclodextrin have similar cavity structures, however, there is a big difference between them. Cyclodextrin is a ring of zero space liberation composed of glucose units, and is a sub-area in calixarene.

Calixarenes are characterized by a three-dimensional basket, cup or bucket shape. In calix

[4]arenes the internal volume is around 10 cubic nanometers. Calixarenes are characterized by a wide upper rim and a narrow lower rim and a central annulus. The phenol units attached to the group are free to rotate. Thus calixarenes have a variety of conformational isomers. The results of X-ray diffraction study show that the cups of aromatic hydrocarbons in the crystal adopt a conical conformation due to the intramolecular hydrogen bonding between the light bases. In calix[4]arene 4 up-down conformations exist: cone (point group C_{2v}, C_{4v}), partial cone Cs, 1, 2 alternate C_{2h} and 1,3 alternate D_{2d} (Figure 4.69). The 4 hydroxyl groups interact by hydrogen bonding and stabilize the cone conformation. This conformation is in dynamic equilibrium with other conformations. Conformations can be locked in place with proper substituents replacing the hydroxyl groups which increase the rotational barrier. Alternatively placing a bulky substituent on the upper rim also locks a conformation. The calixarene based on p-tert-butyl phenol is also a cone.

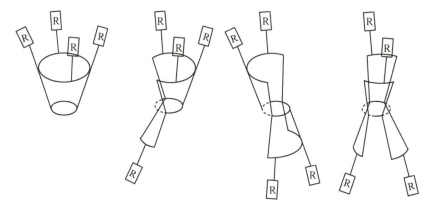

Figure 4.69 Four up-down conformations of calix [4] arene

4.3.4 Dendrimer

The word "dendrimer" originates from the Greek word "dendros", which means trees and branches. The branches on the tree grow to a certain length and then divide into two branches. This is repeated until it grows so dense that it grows into a spherical shape. (Figure 4.70) These hyperbranched molecules were first discovered by Fritz Vogtle in 1978, by Donald Tomalia and co-workers in the early 1980s, and at the same time, but independently by George R.

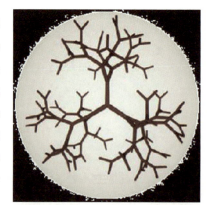

Figure 4.70 Schematic diagram of dendrimer

In physiology, a dendrite or dendron is a tree-like branched extension of a nerve cell. Dendrimers, or cascade molecules, are ideally monodisperse macromolecules with a highly branched, three-dimensional architecture. A dendrimer can be regarded as a many-component compound that has grown from a central core in a tree-like fashion. Therefore, a dendrimer begins with a multifunctional root or core structurally and often synthetically. From each core functional group springs another multifunctional repeat unit (often the same chemical entity as the core itself) to give an oligomer with a central core and a number of functionalized pendant arms. This is the first generation of dendrimer. If a second iteration of monomer units is added to the functional groups of the first-generation material, a much larger, more highly branched second-generation dendrimer results. In principle, this iterative process may be repeated indefinitely, ultimately resulting in a well-defined, three-dimensional polymer with each polymeric molecule being of identical molecular weight to all the others if each reaction step has been carried out perfectly. This concept is summarized in Figure 4.71 and examples of common, commercially available dendrimers showing the various generations are shown in Figure 4.71. In the case of poly (amido amine) (PAMAM) dendrimers the intermediate compounds with carboxylate surface groups are given half-integer generation names.

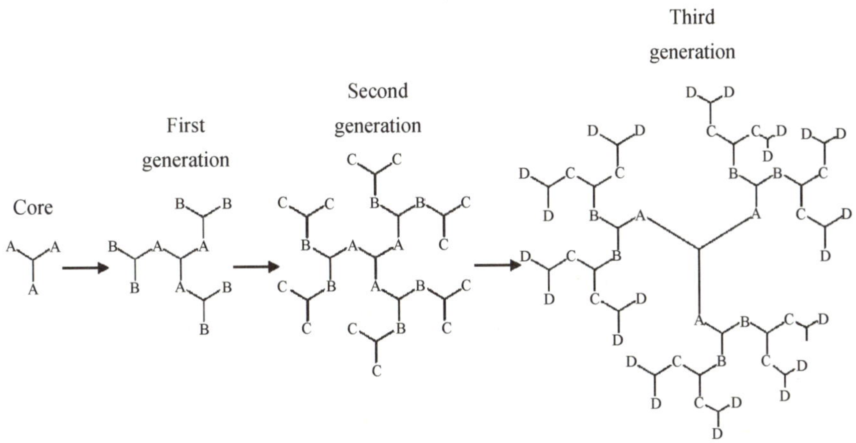

Figure 4.71 Schematic diagram showing the iterative preparation of dendrimers

Dendrimers are of interest in a variety of contexts, such as host-guest chemistry and catalysis in which the dendritic core region, which is often highly porous, can exhibit interesting host behavior, while the densely packed outer layer acts to shield the interior region from the surrounding medium. This can have great potential in mimicking the hydrophobic pocket regions in enzymes. In medicine, the feature of an outer surface comprising multiple identical binding sites is potentially pharamcologically important due to the amplified substrate binding. Dendrimers also show their great importance in the context of drug delivery or functional materials, as in the construction of structured molecular devices applied in fields such as light-harvesting technology. Additionally, the well-defined structural relationship of one layer to another, and the narrow molecular weight range (mono-dispersity) of well-designed dendrimers make their behavior easier to characterize than polymeric materials, even though they may reach very high molecular weights and display bulk properties.

Today, dendrimers have several medicinal and practical applications, which are summarized as follows:

i. Dendrimers in biomedical field

Dendritic polymers have advantages in biomedical applications. These dendritic polymers are analogous to protein, enzymes, and viruses, and can be easily functionalized. Dendrimers and other molecules can either be attached to the periphery or can be encapsulated in their interior voids. Modern medicine uses a variety of this material as potential blood substitutes, e. g. polyamidoamine dendrimers.

ii. Anticancer drugs

Unique pathophysiological traits of tumors will enable passive targeting, such as extensive angiogenesis resulting in hypervascularization, the increased permeability of tumor vasculature, and limited lymphatic drainage, and consequently, selective accumulation of macromolecules in tumor tissue. This phenomenon is known as enhanced permeation and retention (EPR). The drug-dendrimer conjugates exhibit high solubility, reduced systemic toxicity, and selective accumulation in solid tumors. Different strategies have been proposed to enclose within the dendrimer structure drug molecules, genetic materials, targeting agents, and dyes either by encapsulation, complexation, or by conjugation.

iii. Transdermal drug delivery

Transdermal delivery suffers poor rates of transcutaneous delivery because of the barrier function of the skin. Dendrimers have found applications in transdermal drug delivery systems. Generally, in bioactive drugs having hydrophobic moieties in their structure and low water solubility, dendrimers are a good choice in the field of efficient delivery system.

iv. Magnetic resonance imaging contrast agents

Dendrimer-based metal chelates act as magnetic resonance imaging contrast agents. Dendrimers are extremely appropriate and used as image contrast media because of their properties.

v. Dendritic sensors

Dendrimers, although are single molecules, can contain high numbers of functional groups on their surfaces. This makes them hard for applications where the covalent connection or close proximity of a high number of species is important. For example, Balzani and co-workers investigated the fluorescence of a fourth-generation poly(propylene amine) dendrimer decorated with 32 dansyl units at the periphery (Figure 4.72). Since the dendrimer contains 30 aliphatic amine units in the interior, suitable metal ions are able to coordinate. It was found that when a Co^{2+} ion is incorporated into the dendrimer, the strong fluorescence of all the dansyl units is quenched. Low concentrations of Co^{2+} ions (4.6×10^{-7} M) can be detected using a dendrimer concentration of 4.6×10^{-6} M. The many fluorescent groups on the surface serve to amplify the sensitivity of the dendrimer as a sensor.

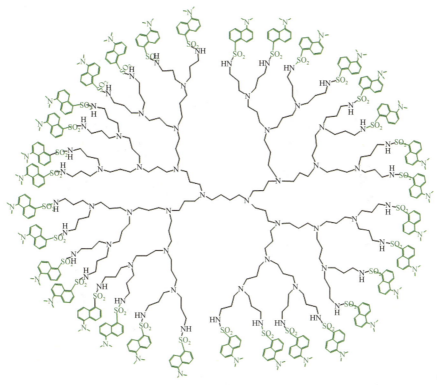

Figure 4.72　Poly (propylene amine) dendrimer, containing 32 dansyl units at its periphery

vi. *Photodynamic therapy*

Photodynamic therapy (PDT) relies on the activation of a photosensitizing agent with visible or near-infrared (NIR) light. Upon excitation, a highly energetic state is formed which, upon reaction with oxygen, affords a highly reactive singlet oxygen capable of inducing necrosis and apoptosis in tumor cells. Dendritic delivery of PDT agents has been investigated within the last few years in order to improve upon tumor selectivity, retention, and pharmacokinetics.

4.3.5　Metal Organic Framework

The term coordination polymer very broadly encompasses any extended structure based on metal ions linked into an infinite chain, sheet or three-dimensional architecture by bridging ligands, usually containing organic carbon. Recently, the term metal organic framework (MOF) has entered the literature. As depicted in Figure 4.73, a metal organic framework is a kind of coordination polymer that is a three-dimensional, crystalline solid that is both robust and porous. In general, the organic bridging ligands within MOFs are subject to some kind of synthetic choice and hence coordination polymers involving simple ligands such as cyanide are not generally considered MOFs. MOFs have aroused great research interest due to their special structure. As a porous material, it has special advantages compared with inorganic or organic porous materials. Metal organic framework porous materials have a wide range of applications in separation and purification, catalysis, microreactor, negative ion exchange and composite functional materials. Figure 4.74 summarizes the reported MOF structures.

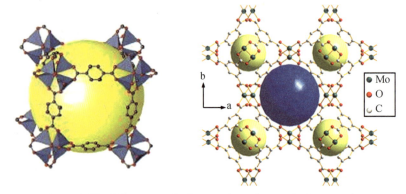

Figure 4.73 Diagrammatic figure of metal-organic frameworks

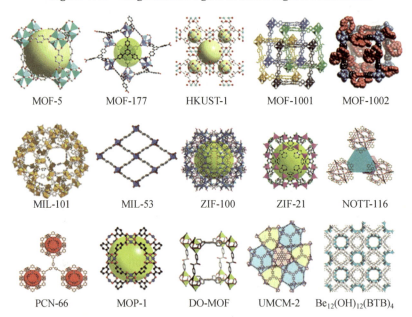

Figure 4.74 Metal organic framework (MOF) structures

Photothermal therapy (PTT) is an emerging treatment method with noninvasiveness, high selectivity, and tissue-penetrating depth. It is capable of converting near-infrared laser into hyperthermia by means of photothermal agents, resulting in photothermal ablation of tumor cells. Taking these into account, a multifunctional supramolecular nanohybrids constructed from core-shell MOFs and pillararene nanogates has been developed for targeted chemophotothermal treatment of cervical cancer (Figure 4.75). Such a system highly integrated a polypyrrole nanoparticle (PPy NP) core with ideal photothermal conversion capability. Specifically, Pymodified UiO-66-based drug reservoir with high drug-loading capacity is served as shell. pH/temperature-stimuli-responsive water-soluble pillar[6]arene (WP6) nanovalves work on the MOF surface. And folic acid (FA)-modified polyethyleneimine (PEI-FA) is served as active targeting group coated in the outermost layer by electrostatic interactions to meet multimodal synergetic treatment of tumors and controlled drug release.

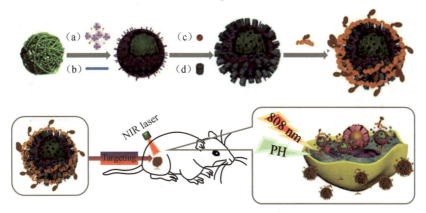

Figure 4.75 Schematic representation for the preparation of DDS based on WP6-capped core-shell MOFs and its application for synergistic chemophotothermal therapy

As shown in Figure 4.76, a series of conductive porous composites can be synthesized by polymerizing 3,4-ethylenedioxythiophene (EDOT, a conductive monomer) in the holes of MIL-101(Cr) MOF material. By controlling the content of EDOT in MOF, the conductivity and porosity of the composite can be readily regulated. Materials prepared by this method have appropriate electronic conductivity (1,100 S·cm^{-1}) while maintaining high porosity (specific surface area 803 m^2·g^{-1}). Therefore, it is expected that this strategy can be used to synthesize nanoconductive polymers capable of trapping small gas molecules for the preparation of chemical impedance sensors that can detect NO_2.

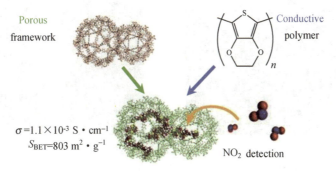

Highly porous and conductive composite

Figure 4.76 EDOT polymerizing into PCPs to form porous conductive composites used to prepare a chemical impedance sensor capable of detecting NO_2

4.3.6 Carbon Nanotube

A carbon molecule consisting of tubular coaxial nanotubes, now called the carbon nanotube (CNT), is also known as a bucky tube (Figure 4.77). The tubular body consists of a hexagonal carbon ring microstructure unit, and the end cap portion is a polygonal structure composed of a pentagon-shaped carbon ring, or a polygonal tapered multi-wall structure. Carbon nanotube is a one-dimensional quantum material with a special structure (the radial dimension is on the order of nanometers, the axial dimension is on the order of micrometers, and the ends of the tube are substantially sealed). Carbon nanotubes can either be open or capped at the end by carbon pentagons which give all fullerenes their closed curvature.

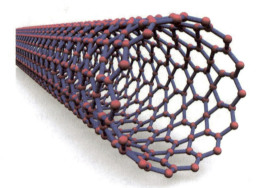

Figure 4.77 Schematic diagram of a ball-and-stick model of carbon nanotubes

As shown in Figure 4.78, according to the number of layers of graphene sheets, carbon nanotubes can be divided into single-walled nanotubes (SWNTs) and multi-walled nanotubes (MWNTs).

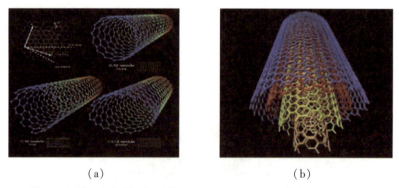

(a)   (b)

Figure 4.78 (a) Single walled carbon nanotube; (b) multi-walled carbon nanotube

Multi-walled nanotubes are formed at the beginning. At the time, the layers easily become the center of the trap to capture various defects, and thus the wall of the multi-walled nanotube is usually covered with small hole-like defects. Multi-walled nanotubes were first used as electrically conductive fillers in plastics. Nowadays, they are also used to enhance fiber composites. Examples include wind turbine blades and hulls for maritime security boats. 50% of lithium batteries incorporated carbon nanofibers by 2005, which are wires spun from CNTs. Carbon nanotubes are even being used to enhance sporting goods like tennis rackets, baseball bats, and bicycle frames. Compared with the multi-walled nanotube, the single-walled nanotube is composed of a single-layer cylindrical graphite layer, which has a small distribution range of diameters, few defects, and higher uniformity. Single-walled nanotubes have a diameter of close to 1 nm, and can be many millions of times longer. The structure of a SWNT can be conceptualized by wrapping a one-atom-thick layer of graphite called graphene into a seamless cylinder. Single-walled carbon nanotubes are made of one atomic sheet of carbon atoms in a honeycomb lattice, and they are hollow, long cylinders with extremely large aspect ratios. SWNTs possess extraordinary thermal, mechanical, and electrical properties and are considered as one of the most promising nanomaterials for applications and basic research.

Carbon nanotubes have a higher tensile strength than steel and Kevlar. Their strength stems from the sp^2 bonds between the individual carbon atoms. This bond is even stronger than the sp^3 bond found in diamond. Under high pressure, individual nanotubes can bond together, trading some sp^2 bonds for sp^3 bonds. This makes it possible to produce long nanotube wires. Carbon nanotubes are not only strong, they are also elastic. You can press on the tip of a nanotube and cause it to bend without damaging the nanotube, and the nanotube will return to its original shape when the force is removed. The elasticity of a carbon nanotube does have a limit, and under very strong forces, it is possible to permanently deform to the shape of a nanotube. A nanotube's strength can be weakened by defects in the structure of the nanotube. Defects occur from atomic vacancies or a rearrangement of the carbon bonds. Defects in the structure can lead to a small segment of the nanotube to become weaker, which in turn causes the tensile strength of the entire nanotube to weaken. The tensile strength of a nanotube depends on the strength of the weakest segment in the tube.

As mentioned previously, the structure of a carbon nanotube determines how conductive the nanotube is. When the structure of atoms in a carbon nanotube minimizes the collisions between conduction electrons and atoms, a carbon nanotube is highly conductive. The strong bonds between carbon atoms also endow carbon nanotubes with higher electric currents than copper. Electron transport occurs only along the axis of the tube. Single-walled nanotubes can route electrical signals at speeds up to 10 GHz when used as interconnects on semi-conducting devices. Nanotubes also have a constant resistively.

The strength of the atomic bonds in carbon nanotubes allows them to withstand high temperatures. Therefore, carbon nanotubes exhibit very good thermal conductivity. When compared with copper wires, which are commonly used as thermal conductors, the carbon nanotubes can transmit over 15 times the amount of watts per meter per Kelvin. The thermal conductivity of carbon nanotubes depends on the temperature of the tubes and the outside environment.

CNTs are extensively used as carbon nanomaterials in numerous applications due to their exceptional optical, electrical, and mechanical characteristics. They are utilized in photocatalysts, catalysts, adsorbents, membranes, sensors, conductive coatings, batteries, supercapacitors, hydrogen storage devices, solar cells, and fuel cells. Additionally, they are applied in biomedicine as extremely sensitive biosensors (for sensing glucose, choline, and hydrogen peroxide) and drug delivery systems (particularly in cancer treatments). CNTs exhibit other outstanding properties such as open, meso-/macro-porous structures, large specific surface areas, high resistance against acidic and alkaline corrosions, high mechanical and thermal (in inert gas) stabilities, controllable porosity, and possibility of the surface modification allowing changing their hydrophobic characteristics. Such features make them appropriate support materials for photocatalysts and catalysts employed in diverse reactions.

In terms of fuel cells, the use of CNTs as a catalyst support can potentially reduce Pt usage by 60% compared with carbon black, and doped CNTs may ensure that fuel cells do not require Pt. For organic solar cells, ongoing efforts are leveraging the properties of CNTs to reduce undesired carrier recombination and enhance resistance to photooxidation. In future, photovoltaic technologies may incorporate CNT-Si heterojunctions and leverage efficient multiple-exciton generation at p-n junctions formed within individual CNTs. In the nearer term, commercial photovoltaics may incorporate transparent SWNT electrodes [Figure 4.79 (a)]. An upcoming application domain of CNTs is water purification. Here, tangled CNT sheets can provide mechanically and electrochemically robust networks with controlled nanoscale porosity. These have been used to electrochemically oxidize organic contaminants, bacteria, and viruses. Portable filters containing CNT meshes have been commercialized for purification of contaminated drinking water [Figure 4.79 (b)].

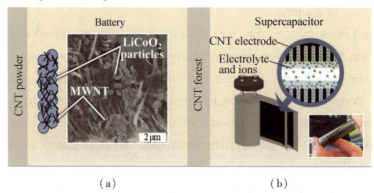

Figure 4.79 Energy-related applications of CNTs: (a) mixture of MWNTs and active powder for battery electrode; (b) concept for supercapacitors based on CNT forests

>>> **Chapter 4** Supramolecules in Photophysics

4.4   Basic Functions of Supramolecules

Supramolecular chemistry is one of the new areas of chemistry which deals with secondary interactions rather than covalent bonds in molecules and focuses on the chemical systems made up of a discrete number of assembled molecular subunits or components. The forces responsible for the spatial organization may vary from weak intermolecular forces to strong covalent bonding. The weak intermolecular forces are hydrogen bonding, metal coordination, hydrophobic forces, van der Waals force, π-π interactions and electrostatic effects. These weak interactions endow supramolecules with three basic functions, i. e. molecular recognition, molecular transportation, and chemical reaction. The relationship between the molecules (or components) involved in the three functions is the non-covalent connection. At the same time, in the implementation of these functions, the complementary and key-locking principles are widely used.

4.4.1   Molecular Recognition

Molecular recognition is the specific binding of a guest molecule to a complementary host molecule to form a host-guest complex. Normally, the definition of which species is the "host" and which species is the "guest" is arbitrary. The molecules are able to identify each other using non-covalent interactions.

In molecular recognition, the recognition subject (the acceptor) and the identified object are identified by non-covalent interactions. Due to the lock-key relationship between the two, it can form a strict specific identification. A complex such as R-A can be formed between the acceptor and the guest with a certain complexation constant. If R can also form a complex with B and its complexation constant is greater than the former, B can drive A out to form a new complex R-B. From the size of the complexation constant, it can be clearly seen that a suitable combination has a large complexation constant, and an inappropriate combination has a small complexation constant.

Molecular recognition plays a crucial role in biological systems and is observed in between receptor ligand, antigen-antibody, DNA-protein, sugar-lectin, RNA ribosome, etc. It should be noted that the best known molecular recognition function in biological systems is the naturally occurring DNA, which exists in the form of a double helical. The two single strands are held together by a number of hydrogen bonds, involving acidic hydrogen atoms (hydrogen bonding donor), oxygen (hydrogen bonding acceptor), and nitrogen atoms (hydrogen bonding acceptor) of the purine and pyrimidine bases in order to maintain the double helical structure

[Figure 4.80 (a)]. In this double helix, guanine (G) forms triple hydrogen bonds with cytosine (C) and adenine (A) forms double hydrogen bonds with thymine (T).

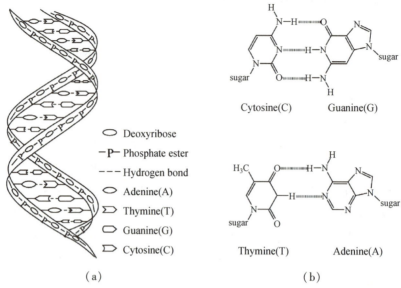

Figure 4.80 (a) Complementary base pairing in DNA helical structure; (b) base pairing in DNA (Guanine and cytosine form triple hydrogen bonds; adenine and thymine form double hydrogen bonds.)

Guanine selectively interacts with cytosine because G-C complex is much more stable than G-T complex that forms only one hydrogen bond [Figure 4.80 (b)]. Similarly, adenine exclusively forms complex with thymine because adenine will form no hydrogen bonds with cytosine. The X-ray diffraction studies have shown that the hydrogen bonds holding G-C and A-T complexes are about the same length (2.9 ± 0.1 Å).

One typical application of molecular recognition is the complex substitution function in cyanide poisoning. The mechanism of cyanide poisoning is that cyanide can form complexes with many metal ions. For instance, it can combine with Fe(Ⅲ) ions in the cytochrome oxidase-respiratory enzyme in the cell mitochondria, therefore, the oxygen function of the cells is inhibited and death occurs. The rescue method can be done by an antidote called Kelocyanor, which is a dicobalt salt of EDTA. The detailed formula is shown as follows:

$$2Co^{2+} + 4EDTA \rightarrow Co_2EDTA \ (K_{Formation} = 2.0 \times 10^{16}) \quad (4\text{-}9)$$

CN- can be used as a ligand for Co(Ⅱ) and has a large complexation constant, as shown by the following formula:

$$2Co^{2+}(aq) + 6CN^-(aq) \rightarrow Co(CN)_6^{4+}(aq) \ (K_{Formation} = 2.0 \times 10^{16}) \quad (4\text{-}10)$$

Thus, cyanide can be detached from Fe(Ⅲ) ions and combined with Co(Ⅱ), making people out of danger.

Key applications of this field are the construction of molecular sensors and catalysis. The specific interaction between host and guest molecules occurs through non-covalent bonding such as hydrogen bonding, metal coordination, hydrophobic forces, van der Waals force, π-π interactions, halogen bonding, electrostatic and/or electromagnetic effects. In addition to these direct interactions solvent can play a dominant indirect role in driving molecular recognition in solution as well. The host and guest involved in molecular recognition exhibit molecular complementarity.

4.4.2 Molecular Transportation

Molecular transportation can be used to study the role of supramolecules and their mechanisms when different chemical species cross certain obstacles, such as molecular membranes (or cell membranes). Because the cell membrane consists of a bilayer lipid membrane (BLM), the hydrophilic phase is on both sides, and the membrane itself is composed of a hydrocarbon chain. The hydrophilic chemical species dissolved in the water outside the membrane should pass over the lipophilic BLM layer.

A common mode of transportation is shown in Figure 4.81. Metal cations are not at all lipophilic. They cannot effectively diffuse through the cell wall unless something makes them lipophilic or a nonlipophilic pathway is created for them. The two main possible methods of such passive cation transport along a concentration gradient are: transport by some kind of lipophilic carrier, or controlled passage through a hydrophilic channel in the membrane. Transport of metal ions via the carrier mechanism is related to a carrier ligand that is able both to bind selectively to the metal cation and to shield it from the lipophilic region of the membrane. Such ion carriers are termed ionophores.

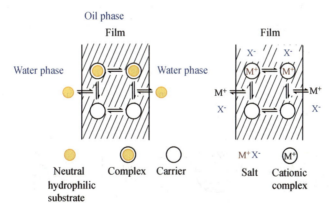

Figure 4.81 Schematic diagram of transport through biological membrane

Another typical example of molecular transportation function is the cell membranes in biological systems. Like synthetic lipid bilayers, cell membranes enable water and nonpolar molecules to permeate by simple diffusion. However, cell membranes also have to allow the passage of various polar molecules, such as ions, sugars, amino acids, nucleotides, and many cell metabolites that cross synthetic lipid bilayers only very slowly. Special membrane transport proteins are responsible for transferring such solutes across cell membranes. These proteins have been found in many forms and in all types of biological membranes. Each protein transports a particular class of molecule (such as ions, sugars, or amino acids) and often only certain molecular species of the class. The specificity of membrane transport proteins was first indicated in the mid-1950s by studies where single gene mutations were found to abolish the ability of bacteria to transport specific sugars across their plasma membrane. Similar mutations have now been discovered in humans suffering from a variety of inherited diseases that affect the transport of a specific solute in the kidney, intestine, or many other cell types. Individuals with the inherited disease cystinuria, for example, are unable to transport certain amino acids (including cystine, the disulfide-linked dimer of cysteine) from either the urine or the intestine into the blood. The resulting accumulation of cystine in the urine may cause the formation of cystine stones in the kidneys. In the process of molecular transportation above, supramolecules play an important role.

All membrane transport proteins that have been studied in detail have been found to be multipass transmembrane proteins, which means that their polypeptide chains traverse the lipid bilayer multiple times. With the formation of a continuous protein pathway across the membrane, these proteins enable specific hydrophilic solutes to cross the membrane without coming into direct contact with the hydrophobic interior of the lipid bilayer.

Carrier proteins and channel proteins are two major classes of membrane transport proteins. Carrier proteins (also called carriers, permeases, or transporters) bind the specific solute to be transported and undergo a series of conformational changes to transfer the bound solute across the membrane [Figure 4.82 (a)]. In contrast, channel proteins interact with the solute to be transported much more weakly. They form aqueous pores that extend across the lipid bilayer. When these pores are open, they allow specific solutes (usually inorganic ions of appropriate size and charge) to pass through them and thereby cross the membrane [see Figure 4.82 (b)]. It is not surprising that the transport through channel proteins occurs at a much faster rate than the transport mediated by carrier proteins.

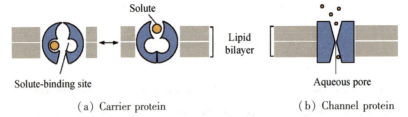

(a) Carrier protein (b) Channel protein

Figure 4.82 Carrier proteins and channel proteins: (a) A carrier protein alternates between two conformations, so that the solute-binding site is sequentially accessible on one side of the bilayer and then on the other; (b) A channel protein forms a water-filled pore across the bilayer through which specific solutes can diffuse

All and many carrier proteins allow solutes to cross the membrane only passively ("downhill"), this process is called passive transport or facilitated diffusion. In the case of transport of a single uncharged molecule, it is simply the difference in its concentration on the two sides of the membrane—its concentration gradient—that drives passive transport and determines its direction (Figure 4.83).

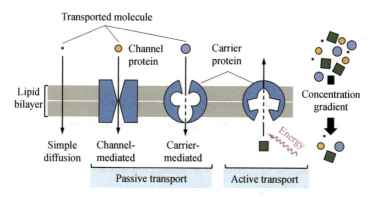

Figure 4.83　Passive and active transport compared

Passive transport down an electrochemical gradient occurs spontaneously, either by simple diffusion through the lipid bilayer or by facilitated diffusion through channels and passive carriers. In contrast, active transport requires an input of metabolic energy and is always mediated by carriers that harvest metabolic energy to pump the solute against its electrochemical gradient.

However, if the solute carries a net charge, both its concentration gradient and the electrical potential difference across the membrane, the membrane potential, influence its transport. A net driving force can be calculated with the combination of the concentration gradient and the electrical gradient. Cells also require transport proteins that will actively pump certain solutes across the membrane against their electrochemical gradient ("uphill"). This process is known as the active transport, and is mediated by carriers which are also called pumps. In active transport, the pumping activity of the carrier protein is directional because it is tightly coupled to a source of metabolic energy, such as ATP hydrolysis or an ion gradient. Therefore, transport by carriers can be either active or passive, whereas transport by channel proteins is always passive.

4.4.3　Chemical Reaction

Chemical reactions in supramolecules refer to the promotion of the reaction due to the presence of supramolecular structure. This is based on the understanding of organisms, especially the catalytic function of enzymes. It involves intermolecular reactions, including promotion of redox-electron transfer, generation of transition states-environmental effects and the existence of a certain reaction "field".

Supramolecular chemistry refers to the area of chemistry beyond the molecules and focuses on the chemical systems consisting of a discrete number of assembled molecular subunits or components. The forces responsible for the spatial organization can be either weak (intermolecular forces, electrostatic or hydrogen bonding) or strong (covalent bonding), provided that the degree of electronic coupling between the molecular components remains small with respect to relevant energy parameters of the component. The research on non-covalent interactions is crucial to understanding many biological processes from cell structure to vision that rely on these forces for structure and function. Biological systems are often the inspiration for supramolecular research.

As nature has served as a dominant source of inspiration in the area of supramolecular chemistry, it is not surprising that enzymes can serve as natural prototypes for the design of supramolecular catalysts [Figure 4.84 (a)]. As a result, much effort has been made in creating systems in which a receptor or a cavitand is connected to an active site, with the aim of mimicking enzyme catalysis, generally involving typical reactions that are carried out by enzymes. This approach has led to several elegant examples of cage-driven reactions that display enhanced selectivity and/or activity. For example, the special micro-environment in the cage compounds of Raymond, formed by an assembly of metal-ligand, increases dramatically the acid-catalyzed hydrolysis reaction of orthoesters, even in a basic reaction medium. Fujita applied his typical M_6L_4 cage compounds in Diels-Alder reactions to obtain products that can otherwise not be obtained in this reaction, demonstrating the cage-directed selectivity effect. In addition to this enzyme-inspired approach, recent breakthroughs demonstrate that implementation of supramolecular strategies into traditional homogeneous catalysis approaches can be extremely powerful [Figure 4.84 (b)]. In this approach the aim is no longer to mimic enzyme behavior, but supramolecular chemistry tools are applied to solve problems in the area of homogeneous catalysis.

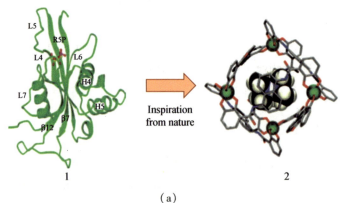

(a)

(b)

Figure 4.84 Supramolecular catalysis inspired by nature, and transition metal catalysis: (a) micro-environment of enzymes mimicked by the cage compounds 2 of Raymond (depicted enzyme: human ADP-ribose hydrolase, 1); (b) bidentate ligand Xantphos 3 as a model for the hydrogen bonded bidentate ligand assembly 4 of Br

Take enzymatic reactions for example. Early on in the history of supramolecular chemistry, people realized that an enzyme and its substrate were a type of host-guest system. Therefore, due to the post-World War II advances in synthetic and physical organic chemistry, attempts were made to simulate enzymes by constructing catalytic hosts. Success in this "biomimetic" branch of supramolecular chemistry has, unfortunately, been modest at best.

In enzyme catalysis, the lock-and-key image has been replaced by the "induced fit" theory of Daniel Koshland where both enzyme and substrate (host and guest) undergo significant conformational changes upon binding to one another (Figure 4.85). Due to these conformational changes, the enzymatic catalytic rate is accelerated since the substrate is commonly more like the reaction transition state in its bound form than in its unbound form. The occurrence of a conformational change upon guest binding is actually a very common phenomenon both in biological chemistry, where it lies at the heart of "trigger" processes such as muscle contraction and synaptic response, and in supramolecular chemistry.

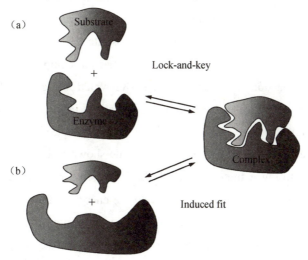

Figure 4.85 (a) Rigid lock and key; (b) induced fit models of enzyme-substrate binding

Fundamentals of Photophysics

The overall enzyme catalysis process may be represented by Equation 4-11, where E, S and P represent enzyme, substrate and product, respectively. Note that the enzyme is regenerated after the reaction, as required in a catalytic process.

$$E + S \rightleftharpoons ES \xrightarrow{k_{cat}} P + E \tag{4-11}$$

The specificity and selectivity of a particular enzyme for competing substrates depend on both rate constants k_1 and k_{-1} for each substrate (and hence on the equilibrium constant), and k_{cat}. Therefore, the most specific enzymes carry out their catalysis rapidly, without the need for particularly strong binding. A lot of enzymes adhere to Michaelis-Menten kinetic parameters (named after Leonor Michaelis and Maud Menten who first described a model of enzyme reactivity crucially dependent on the existence of the ES complex). The Michaelis-Menten model describes that the rate of catalysis v (defined as the number of moles of product formed per second) increases with increasing concentration of substrate, S. For a constant enzyme concentration, v is linearly proportional to the concentration of S (denoted [S]) when [S] is small. At high [S] (i.e. when S is in vast excess compared with the enzyme concentration), v is nearly independent of [S]. This gives a plot of v against [S] of the type shown in Figure 4.86 (a). Combining experimental observation with mathematical principles gives the famous Michaelis-Menten equation, as shown in Equation 4-12:

$$v = v_{max} \frac{[S]}{[S] + K_M} \tag{4-12}$$

The quantities K_M and v_{max} are important parameters that can be used to characterize and understand enzymatic reactivity. The K_M value is the Michaelis constant defined by the substrate concentration at which the reaction rate is half of its maximal value. Therefore, K_M (which varies considerably) is a measure of the relative affinity of an enzyme for its substrate, the higher the K_M value, the lower the affinity for the substrate.

Linus Pauling stated that enzymes are molecules that are complementary in structure to the transition states of the reactions they catalyze in 1948, which is regarded the key to understand how enzymes are able to achieve such large rate accelerations in the reactions they catalyze. This simple, but profound assertion relates to the overall transition state theory in chemical reactions, as exemplified by the type of reaction coordinate profile shown in Figure 4.86 (b). The role of the enzyme catalyst is to lower the energy of the activated enzyme substrate complex [ES]‡ and hence the free energy of activation $\Delta G_{cat}^{\ddagger}$ relative to the energy of the activated substrate, [S]‡, in the uncatalyzed reaction. Therefore, the enzyme must become more complementary to the activated substrate than it is to the ground-state species. In fact, the non-covalent forces involved in substrate binding should be sufficient to distort the substrate such that it proceeds some distance from left to right along the reaction coordinate. In other words, it becomes more like the transition state, hence lowering the activation energy required to form [ES]‡. This is a further manifestation of the induced fit model.

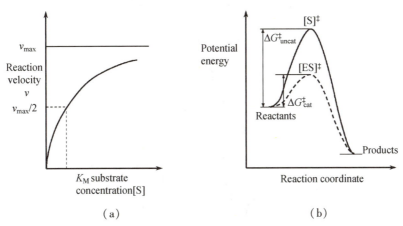

Figure 4.86 (a) **A plot of reaction velocity v against substrate concentration [S] for an enzyme that obeys Michaelis-Menten kinetics (v_{max} is the maximum reaction velocity and K_m is the Michaelis constant); (b) Chemical reactions profiles for catalyzed (---) and uncatalyzed reactions (—)**

The application of a self-assembled cage to supramolecular catalysis is shown below as an example where the characteristics of the reaction in the confined cavities of enzymes are emulated. Raymond and co-workers used a tetrahedral metal-ligand assembly as a catalyst that can function either for TS stabilization or for substrate preorganization. The self-assembled M_4L_6 cage 20 [M = Ga^{III}, L = N, N'-bis (2, 3-dihydroxybenzoyl)-1, 5-diaminonaphthalene] preferentially hosts cationic guests over neutral ones because of the negatively charged cage. It was expected that 20 could catalyze reactions containing a cationic TS through electrostatic stabilization of the TS. By using 20 as a catalyst, the hydrolysis of orthoformates in basic solution was achieved (Figure 4.87). The reaction rate of the hydrolysis of triisopropyl formate was accelerated by 890-fold compared with the uncatalyzed reaction. Mechanistic studies were performed to show that the neutral substrate first forms a host-guest complex with 20 and then undergoes encapsulation-driven protonation followed by acid-catalyzed hydrolysis inside 20 even in basic solution at pH 11 (Figure 4.88). The electrostatic environment in the cage is preorganized to accommodate the charged TS and effectively facilitates the transformation of the less polar substrate to the strongly polar TS.

Fundamentals of Photophysics

$$HC(OR)_3 + H_2O \xrightarrow[pH=11]{1\ mol\%\ 20} \underset{H}{\overset{O}{\underset{\|}{C}}}\text{—}OR + 2ROH \xrightarrow{OH^-} \underset{H}{\overset{O}{\underset{\|}{C}}}\text{—}O^- + 3ROH$$

(R=Me, Et, *n*-Pr, *i*-Pr, *n*-Bu, *i*-Bu)

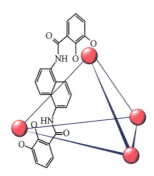

Figure 4.87 Hydrolysis of orthoformates in the presence of the self-assembled cage 20

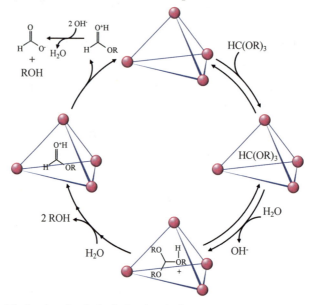

Figure 4.88 Mechanism for hydrolysis of orthoformates by the self-assembled cage 20

In another example of enzymatic catalysis, supramolecular interactions help to orient substrates around the enzyme's active site and to control subsequent reactivity. Therefore, the substrate reacts at the site that is held the closest to the reaction center, and not at nonadjacent sites, regardless of whether they are intrinsically more reactive. Recently, Crabtree and co-workers have reported an elegant catalysis example where reaction acceleration, together with selective molecular activation, were achieved. In this case, molecular recognition is combined through hydrogen bonding and C-H activation to obtain high turnover numbers in the catalytic regioselective functionalization of sp^3 C-H bonds remote from the -COOH recognition group in ibuprofen (Figure 4.89). The catalyst consists of a di-μ-oxido dimanganese catalytic core that is originally developed for its water oxidation activity, but it also proves to be active for C-H bond hydroxylation. The terpyridine ligand of the dimanganese catalyst is substituted with a Kemp's triacid unit through a phenylene linker. The Kemp's triacid unit provides a well-known U-turn motif having a -COOH group suitably oriented for the molecular recognition of the substrate COOH group.

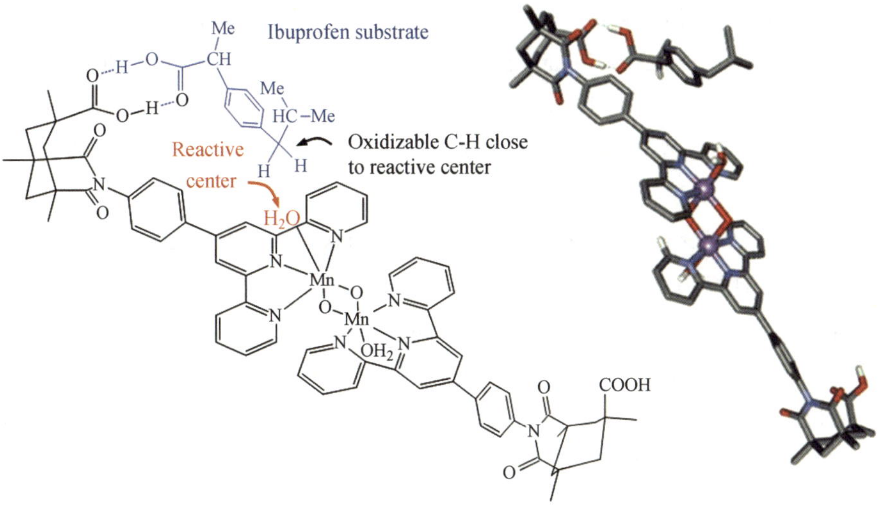

Figure 4.89 Model for the selective oxidation of ibuprofen (The 3D structure of the catalyst is generated with CAChe)

Molecular modeling studies have shown that, in the proposed geometry of the hydrogen-bonded -COOH⋯COOH- complex with ibuprofen [2-(4-isobutylphenyl) propanoic acid], the remote methylene group is positioned close to the metal-active site, and as a result, should be preferentially oxidized. When treating ibuprofen with the catalyst, the selectivity for oxidation of the benzylic position is >98%. However, with an analogous catalyst that lacks the -COOH groups, the selectivity is only 76%. The unselective product may be formed by decarboxylation followed by ketonization of the benzylic position. When the molecular recognition site is blocked by the addition of an excess of acetic acid, ibuprofen can still be oxidized, but all selectivity is lost. With a catalyst to substrate ratio of 0.1 mol%, the total turnover is 580, without the loss of regioselectivity.

References

[1] Ariga, K. & Kunitake, T. *Supramolecular Chemistry—Fundamentals and Applications* [M]. Heidelberg: Springer, 2006:45 - 74.

[2] Korlepara, D. B. & Balasubramanian, S. Molecular modelling of supramolecular one dimensional polymers [J]. *RSC. Adv.*, 2018(8): 22659 - 22669.

[3] Fung, S. Y., et al. *12-Self-assembly of Peptides and Its Potential Applications* [M].

Cambridge: Woodhead Publishing, 2005: 421 – 471.

[4] Ringsdorf, H., Schlarb, B. & Venzmer, J. Molecular architecture and function of polymer oriented system: models for the study of organization, surface recognition, and dynamics of biomembranes [J]. *Angew. Chem. Int. Ed. Engl.*, 1988(27): 113 – 158.

[5] Chen, T., Li, M. & Liu, J. π-π stacking interactions: a nondestructive and facile means in material engineering for bioapplications [J]. *Cryst. Growth Des.*, 2018(18): 2765 – 2783.

[6] Dougherty, D. A. The cation-π interaction [J]. *Acc Chem Res.*, 2013(46): 885 – 893.

[7] Autumn, K., et al. Evidence for van der Waals adhesion in gecko setae [J]. *PNAS*, 2002(99): 12252 – 12256.

[8] Braga, S. S. Cyclodextrins: emerging medicines of the new millennium [J]. *Biomolecules*, 2019(9): 801 – 819.

[9] Abbasi, E., et al. Dendrimers: synthesis, applications, and properties [J]. *Nanoscale Res. Lett.*, 2014(9): 247 – 256.

[10] Kampouraki, Z. C., et al. Metal organic frameworks as desulfurization adsorbents of DBT and 4,6-DMDBT from fuels [J]. *Molecules*, 2019(24): 4525 – 4547.

[11] Thomas, S., et al. *Handbook of Carbon-based Nanomaterials* [M]. San Diego: Elsevier, 2021: 321 – 364.

[12] Michael, F. L. De Volder, et al. Carbon nanotubes: present and future commercial applications [J]. *Science*, 2013(339): 535 – 539.

[13] Alberts, B., et al. *Molecular Biology of the Cell* [M]. 4th ed. New York: Garland Science, 2002:212 – 219.

[14] Meeuwissen, J. & Reek, J. N. H. Supramolecular catalysis beyond enzyme mimics [J]. *Nat. Chem.*, 2010(2): 615 – 621.

[15] Pablo, B., et al. *Advances in Catalysis* [M]. San Diego: Elsevier Academic Press Inc, 2011: 63 – 126.

>>> **Chapter 5** Fluorescence Quenching, Energy Transfer, and Electron Transfer

Chapter 5

Fluorescence Quenching, Energy Transfer, and Electron Transfer

5.1　Fluorescence Quenching

5.1.1　Definition and Theory

Fluorescence quenching refers to any process that decreases the fluorescence intensity of a sample. A variety of molecular interactions can result in quenching. These include excited-state reactions, molecular rearrangements, energy transfer, ground-state complex formation, and collisional quenching. For example, when two quinine solutions are irradiated by an ultraviolet laser, the left solution is quenched due to the presence of chloride ions, and the right solution emits the blue fluorescence of quinine as it normally would (Figure 5.1).

Figure 5.1　Quinine fluorescent under a purple laser (right) [In the presence of Chloride ions, its fluorescence is quenched (left).]

It is easy to find that the luminescent molecules can glow efficiently in dilute solution, but in concentrated solution or a concentrated state, their luminescence will be reduced or even completely disappear. For example, the phenomenon shown in Figure 5.2 is called self-quenching.

Fundamentals of Photophysics

Self-quenching is a special type of fluorescence quenching in which fluorophore and quencher molecules are the same. Fluorescence self-quenching is particularly evident in highly concentrated solutions of fluorophores at right-angle geometry.

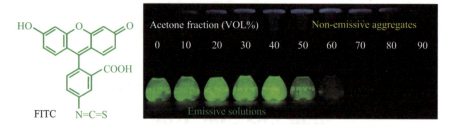

Figure 5.2　An example of fluorescence self-quenching

Fluorescence quenching has been widely studied as a fundamental phenomenon and as a source of information about biochemical systems. Due to the molecular interactions that result in quenching, many biochemical applications have been developed. Static and dynamic quenching require molecular contact between the fluorophore and quencher. In the case of collisional quenching, the quencher must diffuse to the fluorophore during the lifetime of the excited state. Upon contact, the fluorophore returns to the ground state without the emission of a photon. In general, quenching occurs without any permanent change in the molecules, that is, without a photochemical reaction.

Some phenomena of fluorescence quenching have been introduced above. Next, fluorescence quenching will be understood from the mechanism. Collisional quenching is illustrated on the modified Jablonski diagram in Figure 5.3. In this case, the fluorophore is returned to the ground state during a diffusive encounter with the quencher. The molecules are not chemically altered in the process. For collisional quenching, the decrease in intensity can be well-described by the Stern-Volmer equation:

$$\frac{F_0}{F} = 1 + K_{ST}[Q] \tag{5-1}$$

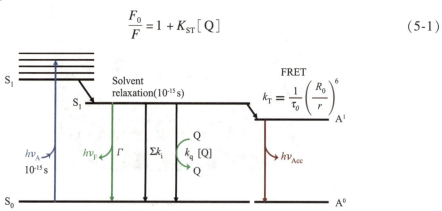

Figure 5.3　Jablonski diagram with collisional quenching and fluorescence resonance energy transfer (FRET) (The term Σk_i is used to represent non-radiative paths to the ground state aside from quenching and FRET.)

where F_0 is the fluorescence intensity in the absence of quencher. F is the fluorescence intensity in the presence of quencher. [Q] is the quencher concentration, and K_{ST} is the Stern-Volmer constant for quenching. If the fluorescence quantum yield is measured in the absence and presence of known concentrations of quencher, a Stern-Volmer plot of $\phi_f/^Q\phi_f$ against [Q] will give a straight line of slope K_Q and intercept 1 (Figure 5.4). Since fluorescence intensity and lifetime are both proportional to the fluorescence quantum yield, plots of $I_f/^Q I_f$ and $\tau_0/^Q\tau$ against [Q] are also linear with slope K_Q and intercept 1.

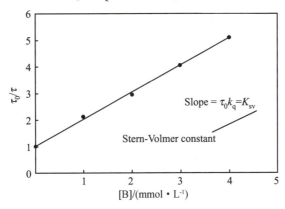

Figure 5.4 A representative of the Stern-Volmer plot fitting curve

5.1.2 Classification of Fluorescence Quenching

Fluorescence quenching can be divided into two types: dynamic quenching and static quenching.

Dynamic quenching occurs by the interaction of a quencher molecule (Q) with an excited molecule of the fluorescing substance (M*). Here, the interaction results in the dissipation of excitation energy by a non-radiative energy transfer from M* to Q without or with less fluorescence. The chemical equation therefore is:

$$M^* + Q \rightarrow M + Q^* \tag{5-2}$$

The exchange of energy between donor and acceptor fluorophore happens after excitation—acceptor may either quench the energy or release a lower energy photon. Dynamic quenching methods can be further explained as: FRET, Dexter energy transfer, and exciplex.

The Stern-Volmer plot of dynamic quenching is shown in Figure 5.5. Higher temperature results in a faster collision and hence larger amounts of dynamic quenching.

Fundamentals of Photophysics

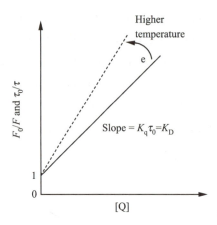

Figure 5.5 Stern-Volmer plot of dynamic quenching

Static quenching occurs due to the formation of a nonfluorescent ground-state complex between the fluorophore (M) and quencher (Q). The static quenching occurs when the molecules form a complex in the ground state, i.e. before excitation occurs. If the temperature increases (chemical stability will change), the rate of static quenching will decrease. The complex has its unique properties, such as changing the absorption spectrum and being nonfluorescent. The chemical equation can be expressed as M + Q→M · Q (Figure 5.6).

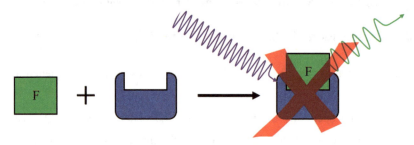

Figure 5.6 Schematic diagram of quenching

The chemical equilibrium (K_s) between the fluorophore, the quencher, and the complex FQ is formed by the law of mass action and equal to the Stern-Volmer constant (K_{sv}). [FQ] stands for the concentration of the complex FQ, [F] stands for the concentration of the loose fluorophore, and [Q] stands for the loose quencher:

$$K_{sv} = K_s = \frac{[FQ]}{[F][Q]} \tag{5-3}$$

The Stern-Volmer plot of static quenching is shown in Figure 5.7. The measurement of fluorescence lifetime is the most definitive method to distinguish static and dynamic quenching (Table 5.1).

>>> **Chapter 5** Fluorescence Quenching, Energy Transfer, and Electron Transfer

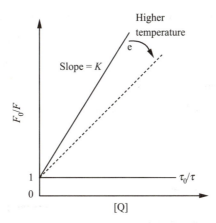

Figure 5.7 Stern-Volmer plot of static quenching

Table 5.1 Differences between static quenching and dynamic quenching

Topic	Static quenching	Dynamic quenching
Occurrence	It occurs in the ground state.	It occurs in the excited state of the molecule.
Effect of temperature	If the temperature increases (chemical stability will change), the rate of static quenching will decrease.	If the temperature increases, the dynamic quenching rate will increase.
Measurement	It can be measured by using a spectrophotometer.	It can only be measured by using Spectrofluorimeter.
Lifetime	There is no change in excited state lifetime.	There is a change in an excited state lifetime.
Fluorophore absorption spectrum	Fluorophore absorption spectrum distorted.	Fluorophore absorption spectrum unchanged.

5.1.3 Quencher

Any substance that increases the deactivation rate of an electronically excited state is known as a quencher and is said to quench the excited state. A wide variety of substances act as quenchers of fluorescence. Common quenching agents are molecular oxygen, halogen, aromatic, aliphatic amines, purines, pyrimidines, N-alkyl pyridinium, and picolinium salts.

i. Molecular oxygen

Molecular oxygen is a very efficient quencher, which quenches almost all known fluorophores. In any quantitative work, it is necessary to exclude oxygen either by bubbling oxygen-free nitrogen through the solution or by carrying out a number of freeze-pump-thaw cycles.

Molecular oxygen is unusual because it is a ground-state triplet. It is an efficient quencher of excited triplet states because it has two accessible excited singlet states, as shown in Figure 5.8. It is also capable of quenching excited singlet states; however, since the quenching occurs by a Dexter mechanism, the overlap integral between excited singlet states and ground state molecular oxygen tends to be very small, leading to tiny quantum yields of singlet oxygen formation.

$$O_2(^3\Sigma_g^-) + {}^*Dye(T_1) \rightarrow {}^*O_2(^1\Delta_g) + Dye(S_0) \quad (5\text{-}4)$$

The molecular oxygen can be excited into the Σ_g^+ state, which rapidly decays to the $^1\Delta_g$ state. The $^1\Delta_g$ is usually referred to as singlet oxygen and is relatively stable because of the spin forbidden deactivation, having a lifetime of a few μs in aqueous solvent and hours in the gas phase.

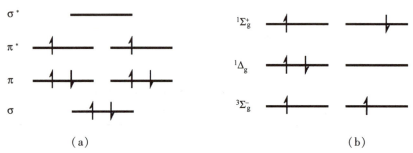

(a)   (b)

Figure 5.8 (a) The valence set of molecular orbitals in molecular oxygen, showing the full electronic arrangement for the $^3\Sigma_g^-$ triplet ground state; (b) The π^* HOMO orbital of molecular oxygen, with the $^3\Sigma_g^-$ triplet ground state and the singlet excited states $^1\Delta_g$ and $^1\Sigma_g^+$ (The relative energies of the two excited states are indicated.)

Chemically singlet oxygen formation is fascinating because it is highly reactive and has been shown to be an important factor in a number of mechanistic pathways leading to damage of DNA and proteins.

The mechanism by which oxygen quenches has been a subject of debate. It is most likely that the paramagnetic oxygen causes the fluorophore to undergo intersystem crossing to the triplet state. In fluid solutions, the long-lived triplets are completely quenched so that phosphorescence is not observed.

ii. Halogen

As early as 1869, George Stokes first described the fluorescence quenching of a fluorophore by halide. He observed that the fluorescence of quinine in dilute sulphuric acid was reduced after the addition of hydrochloric acid, i.e. chloride ions. The process that he observed is now commonly referred to as "dynamic fluorescence quenching", where both the lifetime and intensity of fluorescence are reduced in the presence of a quencher. This process is known to follow Stern-Volmer kinetics. Figure 5.9 shows a typical Stern-Volmer plot of a quinolinium-type dye quenched by aqueous halide ions.

>>> **Chapter 5** Fluorescence Quenching, Energy Transfer, and Electron Transfer

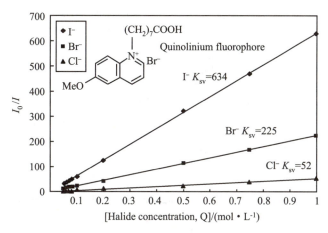

Figure 5.9 A typical Stern-Volmer plot, the respective K_{sv} values (units $L \cdot mol^{-1}$) and molecular structure for a quinolinium type dye quenched by aqueous halide ions at 21 ℃, pH = 10

Halogenated compounds such as trichloroethanol and bromobenzene can also function as collisional quenchers. Quenching by the larger halogens such as bromide and iodide may be caused by the intersystem crossing to an excited triplet state, promoted by spin-orbit coupling of the excited (singlet) fluorophore and the halogen. Since emission from the triplet state is slow, the triplet emission is highly quenched by other processes. The quenching mechanism is probably different for chlorine-containing substances. Indole, carbazole, and their derivatives have been found to be quite sensitive to quenching by chlorinated hydrocarbons and electron scavengers such as protons, histidine, cysteine, NO_3^-, fumarate, Cu^{2+}, Pb^{2+}, Cd^{2+}, and Mn^{2+}. Quenching by these substances probably involves a donation of an electron from the fluorophore to the quencher. In addition, indole, tryptophan, and its derivatives are quenched by acrylamide, succinimide, dichloroacetamide, dimethylformamide, pyridinium hydrochloride, imidazolium hydrochloride, methionine, Eu^{3+}, Ag^+, and Cs^+. Quenchers of protein fluorescence have been summarized in several insightful reviews. Hence, a variety of quenchers are available for studies of protein fluorescence. Especially, quenchers usually are used to determine the surface accessibility of tryptophan residues and the permeation of proteins by the quenchers.

A typical application is optical halide sensing using fluorescence quenching. Figure 5.10 shows a sensor implant incorporated just under human skin. In practice, it should be possible to adopt this principle to most other quenching systems, which have intrinsic or extrinsic fluorescent probes.

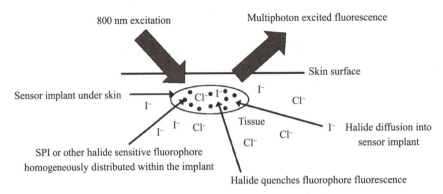

Figure 5.10 The principle of multiphoton transdermal halide sensing

iii. Aromatic

Aromatic and aliphatic amines are also efficient quenchers of most unsubstituted aromatic hydrocarbons. For example, anthracene fluorescence is effectively quenched by diethylaniline. Figure 5.11 shows the fluorescence images before and after anthracene quenching.

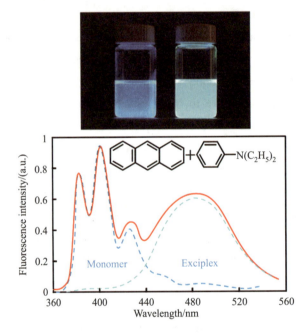

Figure 5.11 Emission spectrum of anthracene in toluene containing 0.2 M diethylaniline

In terms of anthracene and diethylaniline, the mechanism of quenching is the formation of an excited charge-transfer complex. The excited state fluorophore accepts an electron from the amine. In nonpolar solvents, fluorescence from the excited charge-transfer complex (exciplex) is frequently observed, and one may regard this process as an excited state reaction rather than quenching.

iv. Additional quenchers

Additional quenchers include purines, pyrimidines, N-methylnicotinamide, N-alkyl pyridinium, and picolinium salts. Some examples are shown in Figure 5.12.

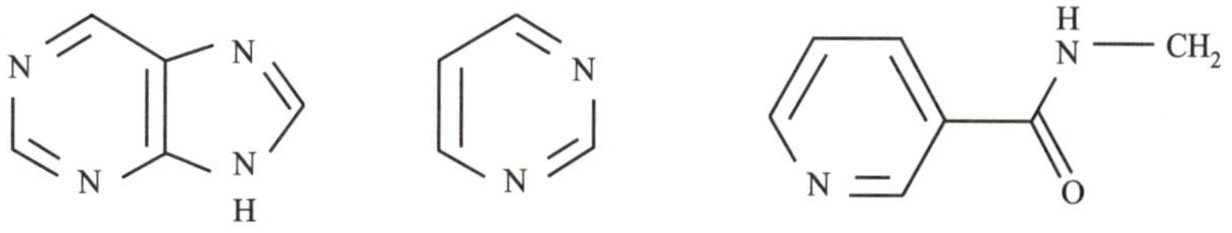

Figure 5.12 Purine, pyrimidine, and N-methylnicotinamide

Typically, the fluorescence of flavin adenine dinucleotide (FAD) and reduced nicotinamide adenine dinucleotide (NAD) are both quenched by the adenine moiety. Flavin fluorescence is quenched by both static and dynamic interactions with adenine, whereas the quenching of dihydronicotinamide appears to be primarily dynamic. These aromatic substances appear to be quenched by the formation of charge-transfer complexes. The ground-state complex can be reasonably stable depending upon the precise structure involved. As a result, both static and dynamic quenching are frequently observed.

5.1.4 Quenching Process

In materials science, quenching is the rapid cooling of a workpiece in water, oil, or air to obtain certain material properties. A type of heat treating, quenching prevents undesired low-temperature processes, such as phase transformations, from occurring. In photochemistry and photophysics, quenching refers to any process that decreases the fluorescence intensity of a given substance. A variety of processes can result in quenching (Figure 5.13), such as excited state reactions, energy transfer, complex-formation, and collisional quenching. Besides, there are common chemical quenchers like molecular oxygen, iodide ions, and acrylamide. According to different photophysical or photochemical processes, quenching processes can be divided into the following five categories. We will introduce them in detail in the following sections:

- Collision with a heavy atom or a paramagnetic species:

$$M^* + Q \rightarrow M + Q + heat \tag{5-5}$$

- Excimer formation:

$$^1M^* + {}^1M \rightarrow {}^1(MM) \tag{5-6}$$

- Exciplex formation:

$$^1M^* + Q \rightarrow {}^1(MQ)^* \tag{5-7}$$

$$^1M^* + Q \rightarrow {}^1(MQ^*) \tag{5-8}$$

- Energy transfer:

$$^1D^* + A \rightarrow {}^1D + A^* \tag{5-9}$$

- Electron transfer:

$$^1D^* + A \rightarrow D^{\cdot+} + A^{\cdot-} \tag{5-10}$$

Fundamentals of Photophysics

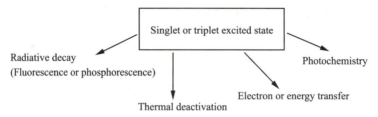

Figure 5.13　Different processes of fluorescence quenching

5.2　Collision Process

Collisional quenching occurs when the fluorophore and another molecule diffuse in the solution and collide with each other. In this case, the two molecules do not form a complex. Fluorescence parameters allow determining of quenching constants dependent on the dynamics of the studied system. Oxygen, acrylamide, iodide, and cesium ions are the most common quenchers. The collision process contain is related to the collision with a heavy atom (halogen) and collision with a paramagnetic species. The collision process could be expressed as follows:

$$M^* + Q \rightarrow M + Q + \text{heat} \tag{5-11}$$

5.2.1　Collision with a Paramagnetic Species

The processes can be related to O_2 or NO. Molecular oxygen is a well-known quencher of fluorescence, which is a small and uncharged molecule; thus it can diffuse easily. Therefore, the bimolecular diffusion constant k_q observed for oxygen in solution is the most important of all the excited quenchers. It deserves special attention because it is ubiquitous in solutions. Its ground state is a triplet state, and it has two low-lying singlet states. Quenching of singlet or triplet states via energy transfer to produce singlet oxygen is thus possible. The specific process could be described as follows:

$$^1A^* + {}^3O_2({}^3\Sigma) \rightarrow {}^1A + {}^1O_2({}^1\Delta) \tag{5-12}$$

$$^3A^* + {}^3O_2({}^3\Sigma) \rightarrow {}^3A + {}^1O_2({}^1\Delta) \tag{5-13}$$

From another point of view, molecular oxygen is a very efficient quencher, such that in any quantitative work, it is necessary to exclude oxygen either by bubbling oxygen-free nitrogen through the solution or by carrying out a number of freeze-pump-thaw cycles (cooling in liquid nitrogen, evacuating to remove any gas, sealing from the atmosphere, and thawing to release dissolved gases).

5.2.2　Collision with a Heavy Atom

As in the case of the intermolecular heavy atom effect, the intersystem crossing is favored by collision with heavy atoms. Cesium and iodide ions will quench Trp residues at or near protein's surface. Iodide ion is more efficient than cesium ion, i.e. each collision with the fluorophore results in a decrease in the fluorescence intensity and lifetime, which is not the case when cesium is used. Also, since cesium and iodide ions are charged, their quenching efficiency will depend on the charge of the protein surface. For a free Trp and Tyr in solution, the highest value of K_{sv} we found with iodide are 16 L·mol^{-1} and 19 L·mol^{-1}, respectively; thus, the corresponding k_q values are 5.8×10^9 L·mol^{-1}·s^{-1} and 5.3×10^9 L·mol^{-1}·s^{-1}, respectively. Another common example is acrylamide, an uncharged polar molecule; thus, it can diffuse within a protein and quenches fluorescence emission of Trp residues. The quencher should be able to collide with tryptophan on the surface or in the interior of a protein. Nevertheless, Trp residues, mainly those buried within the core of the protein, are not all reached by acrylamide. For a fully exposed tryptophan residue or a Trp free solution, the upper value of k_q found with acrylamide is 6.4×10^9 L·mol^{-1}·s^{-1}.

5.3 Excimer Formation

Excimer refers to an excited dimer formed by the association of excited and unexcited molecules, which can remain dissociated in the ground state. A photon is delocalized between two molecules that are weakly bound together in the excited state. Excimers are relatively long-lived species and were first noticed when increasing the concentration of certain fluorophores. It has been found that upon increasing the concentration of the chromophore, the fluorescence will not increase linearly, and upon reaching higher concentrations, the fluorescence decreases; however, this decrease in fluorescence can occur with an increase in a new emission band of lower energy.

When a dilute solution of the polynuclear aromatic hydrocarbon pyrene in toluene is illuminated by ultraviolet light, it emits a violet fluorescence. In a concentrated solution, the violet fluorescence is replaced by an intense sky-blue fluorescence. The fluorescence spectra of the two solutions are shown in Figure 5.14.

Fundamentals of Photophysics

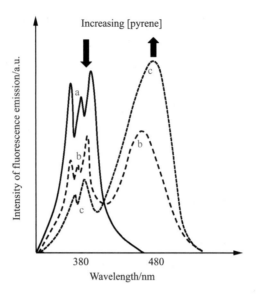

Figure 5.14 Fluorescence spectra of solutions of pyrene in toluene: (a) 10^{-6} mol · dm^{-3}; (b) 10^{-4} mol · dm^{-3}; (c) 10^{-3} mol · dm^{-3}

The following effects can be observed with increasing the pyrene concentration:
- The emission due to pyrene at a lower wavelength decreases in intensity.
- A new fluorescent emission appears at a longer wavelength and increases in intensity.

The isoemissive point indicates the involvement of only two species in the observed fluorescence. The explanation for this behavior is that at the higher concentration of pyrene, an excited-state dimer, or excimer, is formed through an interaction between the electronically-excited pyrene (M^*) and the ground-state pyrene (M).

The schematic diagram of the excimer and the formation of an excited dimer is displayed in Figure 5.15. It should be noted that there is no bonding between the chromophores in the ground state; excimers only occur at high concentrations as an excited state molecule has to interact with a second ground state molecule in the lifetime of the excited state. There is no change in the absorption of the sample (as there is only excitation of monomer species), only the formation of a new band in the emission spectrum.

$$^1M^* + M \rightarrow {}^1[MM]^*$$
$$\downarrow \qquad\qquad \downarrow$$
$$M + h\nu \qquad M + M + h\nu'$$

M: pyrene

Pyrene excimer

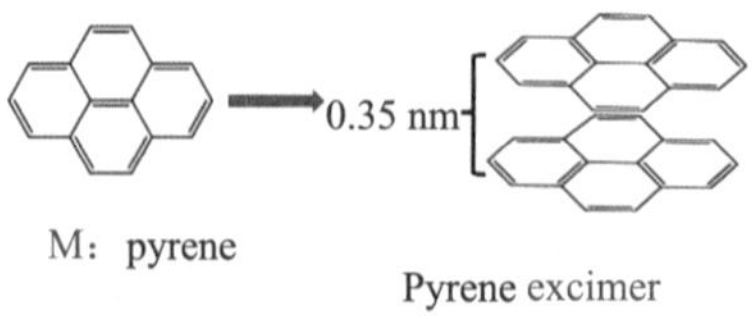

Figure 5.15 Schematic diagram of the formation of an excited dimer
(The structure of the pyrene excimer is sandwich-shaped, with the distance between the planes of the two rings being of 0.35 nm.)

>>> **Chapter 5** Fluorescence Quenching, Energy Transfer, and Electron Transfer

The formation of such excimers, which only exist in the excited state, is commonplace among polynuclear aromatic hydrocarbons, the simple potential energy diagram for which is shown in Figure 5.16.

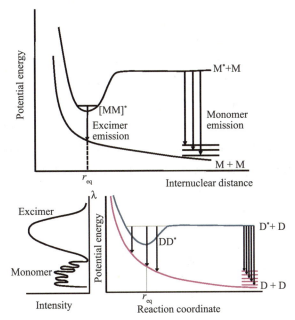

Figure 5.16 Potential energy diagram for a typical excimer, showing the ground-state and first excited singlet state

- The lower two potential energy curves in the figure, corresponding to the ground-state molecules, is characteristic of a dissociative system in which no bound dimer is formed.
- Bringing the M^* and M molecules together, however, results in a potential energy minimum at an internuclear distance (shown as r_{eq}). This minimum corresponds to the formation of the bound excimer.
- The energy change associated with the excimer emission is smaller than that for the monomer emission, so the excimer emission will occur at longer wavelengths than the monomer emission.

The vibrational fine structure is absent from the excimer emission because the Franck-Condon transition is to the unstable dissociative state where the molecule dissociates before it is able to undergo a vibrational transition. In the case of the monomer emission, all electronic transitions are from the $v = 0$ vibrational level of M^* to the quantized vibrational levels of M, resulting in a fine vibrational structure.

5.4 Exciplex Formation

Excimers consist of an excited state shared between two identical monomer units, whereas exciplexes are excited state complexes of two different chromophores. For instance, if N,N-diethylaniline is added to a solution of anthracene in a nonpolar solvent such as toluene, the anthracene fluorescence is quenched and replaced by a structureless emission at a longer wavelength (Figure 5.17). An excited complex, or exciplex, is formed by a reaction between the electronically excited anthracene molecule (M^*) and the ground-state N,N-diethylaniline molecule, which can effectively act as a quencher (Q). The schematic diagram of exciplex is displayed in Figure 5.18.

Figure 5.17 Fluorescence spectra of anthracene in toluene in the presence of N,N-diethylaniline of varying concentrations

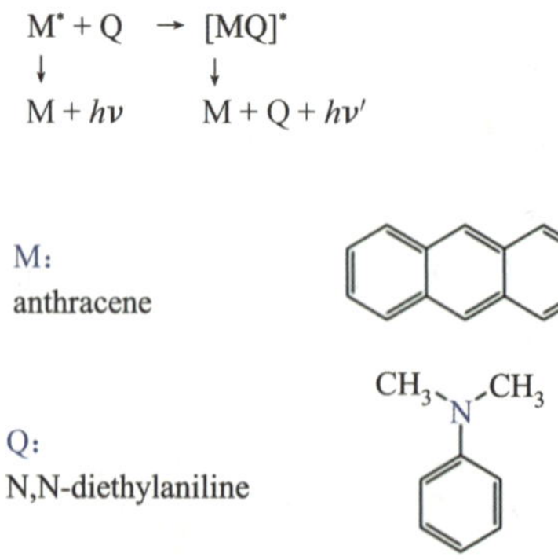

$$M^* + Q \rightarrow [MQ]^*$$
$$\downarrow \qquad \qquad \downarrow$$
$$M + h\nu \qquad M + Q + h\nu'$$

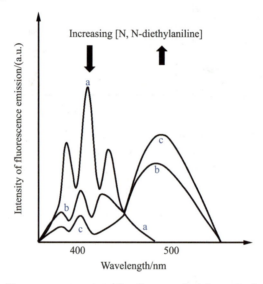

M: anthracene

Q: N,N-diethylaniline

Figure 5.18 Schematic diagram of the formation of an exciplex

>>> **Chapter 5** Fluorescence Quenching, Energy Transfer, and Electron Transfer

The chemical association of the exciplex results from an attraction between the excited state molecule and the ground-state molecule that is caused by a transfer of electronic charge between the molecules. Thus exciplexes are polar species, whereas excimers are nonpolar.

The exciplex emission can also be affected by solvent polarity, where an increase in the solvent polarity results in the decline of the energy level of the exciplex, simultaneously allowing stabilization of charged species formed by electron transfer (Figure 5.19). Therefore, in polar solvents, the exciplex emission is shifted to an even higher wavelength and accompanied by a decrease in the intensity of the emission due to competition between exciplex formation and electron transfer.

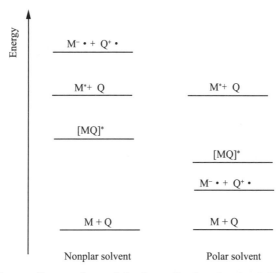

Figure 5.19 Energy diagram for exciplex formation in solvents of differing polarities

Exciplexes are usually observed in organic solvents as nonpolar solvents are not good at stabilizing ions in solution; as the polarity of the solvent increases, then the charge-separated state becomes more favored. Exciplexes and other charge separated species are of interest in chemistry for use in molecular wires and energy storage devices. The development of efficient combustion systems with reduced emissions can be brought about by having a clear understanding of the fuel: air mixing ratios in the combustion chamber. By doping the fuel with a chemical that has similar properties to the fuel, concentration distributions of the liquid and vapor phases of the fuel can be imaged. Such chemicals are designed to fluoresce at different wavelengths in each fuel phase when irradiated with ultraviolet light. Exciplex formation can only occur in the liquid phase, where a green emission occurs on irradiation, while in the vapor phase, a blue emission is observed because of the monomer. It is possible to perform accurate concentration measurements of the fuel/air mixtures using careful calibration techniques.

5.5 Energy Transfer

5.5.1 Introduction to Energy Transfer

From the photochemical point of view, intermolecular electronic energy transfer may be regarded as the transfer of energy from the excited state of the donor (D^*) to the acceptor (A). The acceptor is known as a quencher, and the donor is known as a sensitizer.

The excitation energy transfer from the donor to the acceptor will occur when an energy acceptor molecule is placed in the proximity of an excited energy donor molecule.

In the presence of a molecule of a lower energy excited state (acceptor), the excited donor (D^*) can be deactivated by a process that is known as energy transfer which can be represented by the following equation:

$$D^* + A \rightarrow D + A^* \tag{5-14}$$

The excited state of the donor is deactivated to a lower energy state by transferring energy to the acceptor, which is thereby raised to a higher electronic energy state. Since the excited state donor molecules are initially produced by photoexcitation and the energy is transferred to A, energy transfer is also known as the photosensitization of A or the quenching of D^*. Therefore, for energy transfer to occur, the energy level of the excited state of D^* has to be higher than that for A^*, and the time scale of the energy transfer process must be faster than the lifetime of D^*.

The nomenclature relating to electronic energy transfer is such that it is named as D^*-A^* energy transfer according to the multiplicity of D^* and A^*, common examples being:
- Singlet-singlet energy transfer:

$$^1D^* + {}^1A \rightarrow {}^1D + {}^1A^* \tag{5-15}$$

- Triplet-triplet energy transfer:

$$^3D^* + {}^1A \rightarrow {}^1D + {}^3A^* \tag{5-16}$$

Energy transfer is a widespread and important process in photochemistry and photophysics, and its importance is reflected in the following four aspects.

i. Produce the forbidden transition excited state

The excitation of some compounds is transition forbidden, and their excited states are difficult to obtain through direct excitation. Instead, the excitation of these compounds can be realized by the energy transfer process. A typical example is the excitation of O_2 molecules. Generally, the first excited state of oxygen ($^1O_2$) is difficult to achieve by direct excitation from the ground state ($^3O_2$). However, $^1O_2$ could be obtained through energy transfer, as shown in

>>> **Chapter 5** Fluorescence Quenching, Energy Transfer, and Electron Transfer

Figure 5.20. That means that energy transfer could produce the forbidden transition excited state.

$$S_0(\downarrow\uparrow)+h\nu \longrightarrow {}^1S^*(\downarrow\uparrow) \xrightarrow{ISC} {}^3S^*(\downarrow\uparrow)$$

$$^3S^*(\downarrow\downarrow)+O_2(\downarrow\downarrow) \longrightarrow S_0(\downarrow\uparrow)+O_2(\downarrow\uparrow)$$

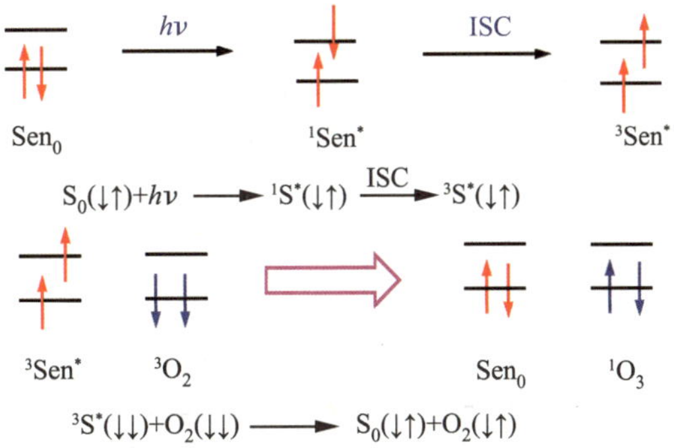

Figure 5.20 Producing the forbidden transition excited state $^1O_2$ through efficacious energy transfer

ii. Remove unnecessary excited state

The energy transfer process can be used to remove the influence or interference of an excited state. For example, an excitation process produces two excited states, $^1M^*$ and $^3M^*$. This process could be described as follows:

$$M + h\nu \rightarrow {}^1M^* + {}^3M^* \tag{5-17}$$

If we want to remove $^3M^*$ and keep $^1M^*$, a quencher of $^3M^*$ could be introduced to remove unnecessary excited state $^3M^*$:

$$^3M^* + Q \rightarrow M + {}^3Q^* \tag{5-18}$$

Similarly, if the $^1M^*$ is an unnecessary excited state, the removal process is as follows:

$$S_0 + h\nu \rightarrow {}^1S^* \rightarrow {}^3S^* \rightarrow {}^3M^* + S$$

iii. Influence photophysical process

For excited state deactivation of a single molecule, its lifetime τ is given by:

$$\tau = 1/(k_f + k_{ic} + k_{ET}) \tag{5-19}$$

The rate constants k_f, k_{ic}, and k_{st} respectively correspond to the photophysical processes of fluorescence, internal conversion, and intersystem crossing.

In the case of donor-acceptor energy transfer, τ can then be expressed by:

$$\tau = 1/(k_f + k_{ic} + k_{st} + k_{ET}) \tag{5-20}$$

where k_{ET} is the rate constant of energy transfer. The detailed process is shown in Figure 5.21.

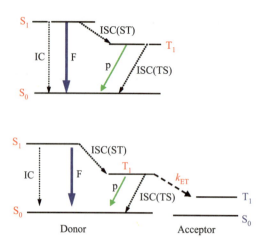

Figure 5.21 The photophysical processes with (bottom) and without (top) energy transfer

iv. Influence photochemical process

The energy transfer process can also alter reaction pathways, reaction mechanisms, and products, as shown in Figure 5.22.

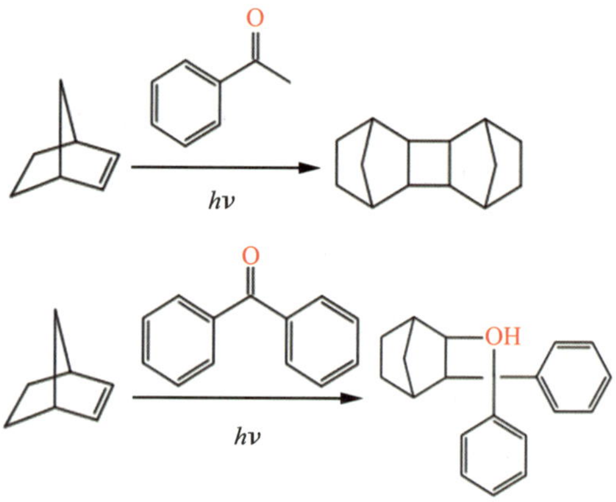

Figure 5.22 An example of energy transfer that influences the photochemical process

Energy transfer has also been extensively studied to understand the dynamic processes that occur in biological cells and determine the size and shape of biological macromolecules. Research at the forefront of cancer diagnosis using molecular beacons also relies on an understanding of energy-transfer processes.

Two possible types of energy transfers are known namely radiative and non-radiative (radiationless) energy transfer, which will be considered in more detail in Chapter 5.5.2 and Chapter 5.5.3, respectively. The principal mechanisms by which electronic energy transfer occurs are shown in Figure 5.23. All three mechanisms require overlap between the fluorescence spectrum of the donor and the absorbance spectrum of the acceptor.

>>> **Chapter 5** Fluorescence Quenching, Energy Transfer, and Electron Transfer

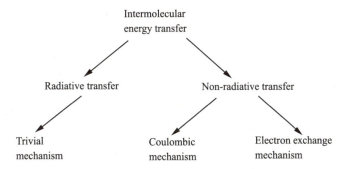

Figure 5.23 The principal mechanisms of electronic energy transfer

5.5.2 Radiative Transfer

Radiative transfer occurs when the extra energy of D^* is emitted in the form of luminescence, and this radiation is absorbed by the acceptor (A). This process involves the emission of a quantum of light by an excited donor, followed by the absorption of the emitted photon by the acceptor.

$$D^* \rightarrow h\nu + D \tag{5-21}$$

$$h\nu + A \rightarrow A^* \tag{5-22}$$

Thus, this mechanism requires that A must be capable of absorbing the photon emitted by D^*, which means that the acceptor absorption spectrum must overlap with the donor emission spectrum. Radiative energy transfer can operate over vast distances because a photon can travel a long way, and A simply intercepts the photon emitted by D^*. In other words, for this to be effective, the wavelengths of the light that D^* emits have to overlap with those where A absorbs. This type of interaction operates even when the distance between the donor and acceptor is large (100 Å). However, this radiative process is inefficient because luminescence is a three-dimensional process in which only a tiny fraction of the emitted light can be captured by the acceptor.

Experimental information on this term can be obtained from the overlap integral between the emission spectrum of the donor and the absorption spectrum of the acceptor: FC^{en} is negligible if A^* lies at lower energy than B^* [Figure 5.24 (a)], quite small if the two excited states are isoenergetic [Figure 5.24 (b)], and large if $^*A$ is higher than $^*B$ in energy [Figure 5.24 (c)]. Therefore, the radiative transfer has the following two features:

(1) This process does not affect the lifetime of the D^* excited state; however, it can lead to distortions of the emission spectrum from A^* (inner filter effect of fluorescence).

(2) Nevertheless, this radiative process is inefficient because luminescence is a three-dimensional process in which only a tiny fraction of the emitted light can be captured by the acceptor.

Fundamentals of Photophysics

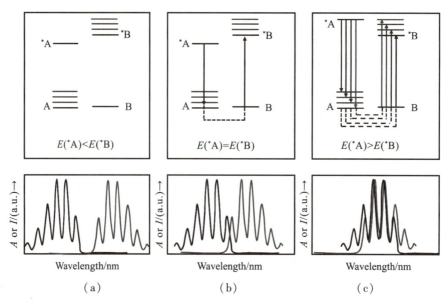

Figure 5.24 Schematic energy level diagram (top) and spectral overlap (bottom) between emission of the donor A (black line) and absorption of the acceptor B (gray line) for three different cases: (a) $E(^*A) < E(^*B)$, (b) $E(^*A) = E(^*B)$, and (c) $E(^*A) > E(^*B)$

Since the photon emitted by D^* is absorbed by A, the same rules will be applied to radiative energy transfer as to the intensity of absorption. Because singlet-triplet transitions are spin-forbidden, and singlet-triplet absorption coefficients are usually extremely small, it is impossible to build up a triplet state population by radiative energy transfer. For this reason, radiative energy transfer involving forming a triplet excited state of $^3A^*$ does not occur.

The mechanism of energy transfer from the donor molecule to the acceptor is a two-step process. Firstly, the Donor molecule from its excited state returns to a ground state by the release of a photon. In the second step, the photon released from the donor molecule excites the acceptor molecule. Now the acceptor molecule is transferred to the excited state. This mechanism is called trivial form energy transfer. The efficiency of this process depends on the following four parameters:

(1) The quantum yield of D^* when emitting a photon.

(2) The number density of ground state (A) molecules that can absorb the emitted photon.

(3) The absorption cross-section or the probability that A will absorb an emitted photon.

(4) The overlap of the fluorescence emission spectrum of D^* and the absorption (or excitation) spectrum of A.

5.5.3 Non-radiative Transfer

A mechanism of non-radiative transfer of this energy can also happen. In the transfer of energy from the donor molecule (D) to the acceptor molecule (A), the energy can also be transferred in a non-radiative way. This can occur in two ways:

(1) By collision (more precisely: by exchange interaction) (Dexter energy transfer).

(2) By Coulomb interaction (FRET).

i. Coulombic/Förster mechanism: long-range dipole-dipole energy transfer

The Coulombic mechanism is also known as the Förster mechanism or dipole-induced dipole interaction. It was first observed by Förster. The Coulombic mechanism is a relatively long-range process in as much as energy transfer can be significant even at distances of the order of 10 nm.

Coulombic energy transfer is a result of mutual electrostatic repulsion between the electrons of the donor and acceptor molecules. As D^* relaxes to D, the transition dipole thus created interacts by Coulombic (electrostatic) repulsion with the transition dipole created by the simultaneous electronic excitation of A to A^* (Figure 5.25).

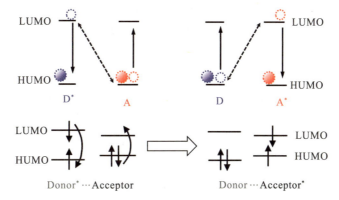

Figure 5.25 Electron movements occurring in long-range Coulombic energy transfer

This means that an electron of the excited donor placed in the LUMO relaxes to the HOMO, and the released energy is transferred to the acceptor via Coulombic interactions. As a consequence, an electron initially in the HOMO of the acceptor is promoted to the LUMO. This mechanism operates only in the singlet states of the donor and the acceptor. This can be explained based on the nature of the interactions (dipole-induced dipole) because only multiplicity-conserving transitions possess large dipole moments. This can be understood considering the nature of the excited state in both the singlet and the triplet states. The triplet state has a diradical structure, so it is less polar, making it hard to interact over long distances.

The rate of energy transfer (k_{ET}) according to this mechanism can be expressed by the following equation:

$$k_{ET} = k_D R_F^6 \left(\frac{1}{R}\right)^6 \qquad (5\text{-}23)$$

where k_D is the emission rate constant for the donor, R is the interchromophore separation, and R_F is the Förster radius, which means the distance between the donor and the acceptor at which 50% of the excited state decays by energy transfer, that is, the distance at which the energy transfer has the same rate constant as the excited state decay by the radiative and non-radiative channels ($k_{ET} = k_r + k_{nr}$). R_F can be calculated by the overlap of the emission spectrum of the donor excited state (D^*) and the absorption spectrum of the acceptor (A).

For the energy transfer to occur rapidly and efficiently by the Coulombic mechanism, there are several requirements:

(1) There should be good spectral overlap between the fluorescence spectrum of the donor and the absorption spectrum of the acceptor.

(2) The transition that corresponds to the donor emission should be allowed (large Φ_f).

(3) The relevant overlapping absorption band of the acceptor should be strong (large ε_A).

(4) A favorable mutual orientation of the transition dipoles should exist.

The Coulombic mechanism can only occur where spin multiplicity is conserved since it is only in such transitions that large transition dipoles are produced. Singlet-singlet energy transfer occurs by this mechanism as the donor (from excited singlet to singlet), and acceptor (from singlet to excited singlet) do not change in multiplicity, resulting in the creation of significant transition dipoles:

$$^1D^* + {}^1A \rightarrow {}^1D + {}^1A^* \tag{5-24}$$

However, triplet-triplet energy transfer cannot occur by this mechanism as this would require both donor and acceptor to change in multiplicity:

$$^3D^* + {}^1A \rightarrow {}^1D + {}^3A^* \tag{5-25}$$

Sometimes Coulombic energy transfer is called the resonance energy transfer because the energies of the coupled transitions are identical, or in other words, in resonance (Figure 5.26).

A detailed theory of energy transfer by the Coulombic mechanism was developed by Förster; therefore, fluorescence resonance energy transfer is usually named Förster resonance energy transfer. According to the Förster theory, the probability of Coulombic energy transfer falls off inversely with the sixth power of the distance between the donor and the acceptor. For donor-acceptor pairs, the efficiency of resonance energy transfer, E_T, increases with decreasing distance, R, according to:

$$E_T = R_0^6 / (R_0^6 + R^6) \tag{5-26}$$

where R_0 is the critical transfer distance, which is characteristic for a given donor-acceptor pair. R_0 is the donor-acceptor distance at which energy transfer from D^* to A and internal deactivation are equally probable; that is, R_0 is the donor-acceptor distance at which the efficiency of energy transfer is 50% (Figure 5.27).

>>> **Chapter 5**   Fluorescence Quenching, Energy Transfer, and Electron Transfer

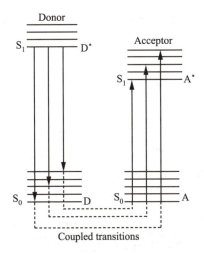

Figure 5.26   Energy-level diagram showing the coupling of donor and acceptor transitions of equal energy required for long-range non-radiative transfer

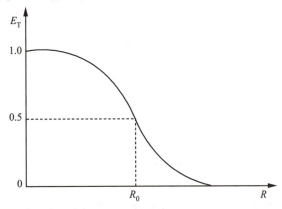

Figure 5.27   The dependence of the efficiency of energy transfer (E_T), on the donor-acceptor distance (R) according to the Förster theory

The efficiency of energy transfer can also be determined by the following three equations:

(1) The relative fluorescence intensity of the donor in the absence (F_D) and the presence (F_{DA}) of the acceptor:

$$E_T = 1 - (F_{DA}/F_D) \tag{5-27}$$

(2) The relative fluorescence quantum yield of the donor in the absence (Φ_D) and the presence (Φ_{DA}) of the acceptor:

$$E_T = 1 - (\Phi_{DA}/\Phi_D) \tag{5-28}$$

(3) The relative fluorescence lifetime of the donor in the absence (τ_D) and the presence (τ_{DA}) of the acceptor:

$$E_T = 1 - (\tau_{DA}/\tau_D) \tag{5-29}$$

When it is found experimentally that the rate constant for energy transfer is insensitive to solvent viscosity and significantly greater than the rate constant for diffusion, and then the Coulombic mechanism is confirmed. The process is equivalent to the energy being transferred across the space between donor and acceptor, like a transmitter-antenna system. As D^* and A are brought together in solution, they experience long-range Coulombic interactions due to their respective electron charge clouds.

The FRET mechanism can serve as an effective molecular ruler. Figure 5.28 shows a protein labeled with a blue fluorescent protein (BFP, the donor) and green fluorescent protein (GFP, the acceptor). The acceptor emission maximum (510 nm) will be observed when the complex is excited at the maximum absorbance wavelength (380 nm) of the donor, provided that the distance between the BFP and GFP will allow FRET to occur.

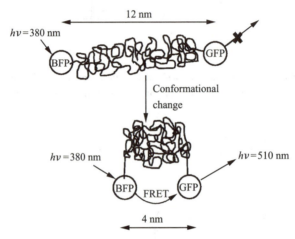

Figure 5.28 Labeling a protein molecule with a donor and acceptor group to show the dependence of FRET on distance

The development of pulsed lasers, microscopy and computer imaging, and the labeling techniques in which the donor and acceptor fluorophores become part of the biomolecules themselves have enabled the visualization of dynamic protein interactions within living cells.

ii. Dexter mechanism: short-range electron-exchange energy transfer

The Dexter mechanism is a non-radiative energy transfer process that involves a double electron exchange between the donor and the acceptor. Although the double electron exchange is involved in this mechanism, no charge separated-state is formed. From the viewpoint of molecular orbitals, electronic energy transfer by the exchange mechanism requires transfer of electrons between the donor and acceptor molecules, which is not like the Coulombic mechanism. Figure 5.29 shows the molecular orbital energy diagram of an electron exchange between D^* and A. The two-electron transfer processes co-occur so that the donor and the acceptor remain uncharged species throughout the exchange process.

The Dexter mechanism can be thought of as electron tunneling, by which one electron from the donor's LUMO moves to the acceptor's LUMO at the same time as an electron from the acceptor's HOMO moves to the donor's HOMO. In the Dexter mechanism, it is possible to realize both singlet-singlet and triplet-triplet energy transfers.

>>> **Chapter 5** Fluorescence Quenching, Energy Transfer, and Electron Transfer

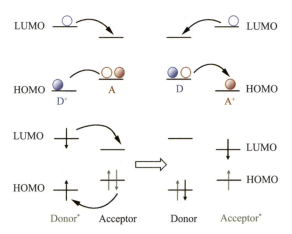

Figure 5.29 Mechanism of energy transfer action according to the Dexter mechanism

For this double electron exchange process to operate, there should be a molecular orbital overlap between the excited donor and the acceptor molecular orbital. For a bimolecular process, intermolecular collisions are required as well. This mechanism involves short-range interactions (6–20 Å and shorter). Because it relies on tunneling, it is attenuated exponentially with the intermolecular distance between the donor and the acceptor. The rate constant can be expressed by using:

$$k_{ET} = \frac{2\pi}{h} V_0^2 J_D \exp\left(-\frac{2R_{DA}}{L}\right) \tag{5-30}$$

where R_{DA} is the distance between the donor and the acceptor, J_D is the integral spectral overlap between the donor and the acceptor, L is the effective Bohr radius of the orbitals between which the electron is transferred, h is Plank's constant, and v_0 is the electronic coupling matrix element between the donor and acceptor at the contact distance.

In terms of triplet-triplet energy transfer:

$$^3D^* + {}^1A \rightarrow {}^1D + {}^3A^* \tag{5-31}$$

The Coulombic mechanism would require that both "$^3D^* \rightarrow {}^1D$" and "$^1A \rightarrow {}^3A^*$" were allowed transitions, but they are not. Thus, triplet-triplet energy transfer by the long-range Coulombic mechanism is forbidden.

When the donor and acceptor molecules approach each other so closely that their regions of electron density overlap, electrons can be exchanged between the two molecules. Therefore, this mechanism is called the exchange mechanism. The electron-exchange mechanism requires a close approach (1–1.5 nm), though not necessarily actual contact, between the donor and the acceptor to facilitate the physical transfer of electrons between the two.

Energy transfer by the exchange mechanism can happen if the spin states before and after overlap obey the Wigner spin conservation rule, providing the overall spin states before and after overlap have common components (Table 5.2).

Fundamentals of Photophysics

Table 5.2 Typical energy-transfer processes by the electron-exchange mechanism allowed according to the Wigner spin conservation rule

Energy Transfer	Spin States Before and After Energy Transfer
Singlet-singlet	$^1D^* + {}^1A \rightarrow {}^1D + {}^1A^*$ ↑↓ ↑↓ ↑↓ ↑↓ 0 0
Triplet-triplet	$^3D^* + {}^1A \rightarrow {}^1D + {}^3A^*$ ↑↑ ↑↓ ↑↓ ↑↑ 1 1
Singlet quenching by oxygen	$^1D^* + {}^3O_2 \rightarrow {}^3D^* + {}^1O_2$ ↑↓ ↑↑ ↑↑ ↑↓ 1 1
Triplet quenching by oxygen	$^3D^* + {}^3O_2 \rightarrow {}^1D^* + {}^1O_2$ ↑↑ ↓↓ ↑↓ ↑↓ 0 0

iii. Comparison between Coulombic/Förster mechanism and Dexter mechanism

As for the two energy transfer mechanisms, the Coulombic/Förster mechanism involves only dipole-dipole interactions, and the Dexter mechanism operates through electron tunneling. Another difference is their range of interactions. The Förster mechanism involves longer-range interactions (up to 5 – 10 nm). On the other hand, the Dexter mechanism focuses on shorter-range interactions (0.6 nm up to 2 nm) because the orbital overlap is necessary. Furthermore, the Förster mechanism describes interactions between singlet states, but the Dexter mechanism can be used for both singlet-singlet and triplet-triplet interactions. Hence for the singlet-singlet energy transfer, both mechanisms are possible.

Therefore, for the singlet-singlet energy transfer, both mechanisms are possible. Simulated graphs using reasonable values for the parameters for the two mechanisms have been proposed to distinguish between the zones where Förster and Dexter mechanisms are dominant. It has been found that the experimental values of the energy transfer rate in cofacial bisporphyrin systems agree with the theoretically constructed graphs (Figure 5.30).

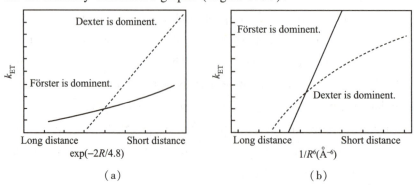

Figure 5.30 Qualitative theoretical plots for (a) and (b) k_{ET} versus $1/R^6$ (Förster) k_{ET} versus $\exp(-2R/4.8)$ (Dexter)

In these graphs, a Bohr radius value (L) of 4.8 Å (the value for porphyrin) is used in the Dexter equation. Additionally, the solid lines correspond to hypothetical situations where only the Förster mechanism operates; the dotted lines are hypothetical situations when the Dexter mechanism is the only process. The curved lines are simulated lines but transposed onto the other graph (i.e. Förster equation plotted against Dexter formulation and vice versa).

These plots clearly suggest the presence of a crossing point between the two mechanisms. There is a zone in which one mechanism is dominant and vice versa. All in all, the relaxation of an excited molecule via energy transfer processes will use all the pathways available. According to Figure 5.30, the dominant mechanism change is about 5 Å.

5.6 Electron Transfer

5.6.1 Definition and Classification of Electron Transfer

Electron transfer occurs when an electron relocates from an atom or molecule to another such chemical entity. ET is a mechanistic description of a redox reaction, wherein the oxidation state of reactant and product changes. As shown in Figure 5.31, during electron transfer, an electron jumps from one molecule to another:

$$D + A \rightarrow D^+ + A^- \quad (5\text{-}32)$$

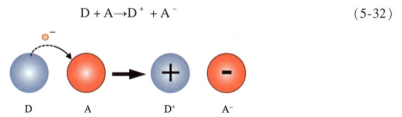

Figure 5.31 Schematic diagram of electron transfer

The energy contained in the excited state plays a crucial role in the ability of the excited state to donate or accept an electron. Electronically excited molecules are both better electron donors and electron acceptors than the ground-state species. As shown in Figure 5.32, light absorption promotes typically an electron from a lower to a higher orbital. The electron that has been promoted is thus more easily removed, which means that the excited state has a smaller ionization potential than the ground state. At the same time, the promotion of an electron leaves behind a low-lying vacancy that can accept an electron, which means that the excited state has a higher electron affinity than the ground state.

Fundamentals of Photophysics

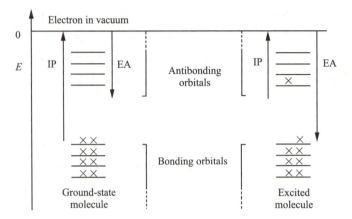

Figure 5.32 Orbital representation of changes in ionization potentials and electronic affinity of a molecule upon excitation

Electron transfer can be classified as intermolecular electron transfer and intramolecular electron transfer (Figure 5.33). During the intermolecular electron transfer, the donor and acceptor must diffuse together to form an encounter complex before forming D^* and A^-. Nonetheless, in intramolecular electron transfer, the inclusion of a rigid molecule bridge removes the need for diffusion. Electron transfer instead occurs at a fixed geometric distance.

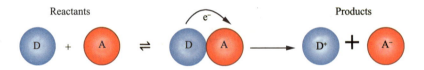

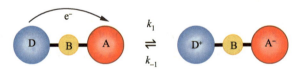

Figure 5.33 Schematic diagram of intermolecular electron transfer and intramolecular electron transfer

The factors that influence the electron transfer process include:

(1) The distance between the donor and acceptor. Electron transfer will become more efficient as the distance between donor and acceptor decreases.

(2) The reaction Gibbs energy. As the reaction becomes more exergonic, electron transfer will become more efficient.

(3) The reorganization energy. The energy cost incurred by molecular rearrangements of donor, acceptor, and medium during electron transfer. The electron transfer rate is predicted to increase because this reorgnization energy is matched closely by the reaction Gibbs energy.

5.6.2 Mechanism and Characteristics of Photoinduced Electron Transfer

Photoinduced electron transfer (PET) involves an electron transfer within an electron donor-acceptor pair. As shown in Figure 5.34, PET represents one of the most basic photochemical reactions, and at the same time, it is the most attractive way to convert light energy or store it for further applications. In Figure 5.34, it can be observed that a process occurs between a donor and acceptor after excitation, resulting in the formation of a charge-separated state, which relaxes to the ground state via electron-hole recombination (back-electron transfer).

Figure 5.34 PET process

Marcus developed a theory to study and interpret the PET in solution. In this theory, the electron transfer reaction can be treated by transition state theory, where the reactant state is the excited donor and acceptor, and the product state is the charge-separated state of the donor and acceptor ($D^+ A^-$), as illustrated in Figure 5.35.

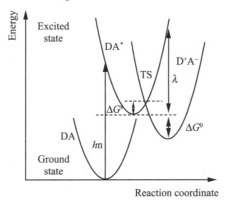

Figure 5.35 Potential energy surfaces for the ground state (DA), the excited state (DA*, reactant state), and the charge-separated state ($D^+ A^-$, product state), proposed by Marcus's theory (λ, total reorganization energy; TS, transition state)

According to the Franck-Condon principle, the photoexcitation triggers a vertical transition to the excited state followed by a rapid nuclear equilibration. The electron transfer process will be highly endothermic without donor excitation. However, after exciting the donor, electron transfer can happen at the crossing of the equilibrated excited state surface and the product state.

The change in Gibbs free energy associated with the electron transfer event can be expressed by:

$$\Delta G^\# = \frac{(\lambda + \Delta G^0)^2}{4\lambda} \tag{5-33}$$

In this equation, the total reorganization energy (λ) that is required to distort the reactant structure to the product structure without electron transfer is composed of solvent (λ_s) and internal (λ_i) components ($\lambda = \lambda_i + \lambda_s$). The reaction free energy (ΔG^0) is the difference in free energy between the equilibrium configuration of the reactant (DA^*) and of the product states (D^+A^-). The internal reorganization energy represents the energy change that occurs in bond length and bond angle distortions during the electron transfer step and is usually represented by a sum of potential harmonic energies. In the classical Marcus theory, the electron transfer rate is given by:

$$k_{ET} = \kappa_{et} \nu_n \exp\left(\frac{-\Delta G^{\#}}{k_B T}\right) \tag{5-34}$$

where ν_n is the effective frequency of motion along the reaction coordinate, and κ_{et} is the electronic transmission factor.

5.6.3 Classification of Photoinduced Electron Transfer

The absorption of a photon by a molecule promotes an electron to a higher energy level. Therefore, the molecule becomes a better electron donor (reducing agent) in its excited state than in its ground state. Because electronic excitation can also create an electron vacancy in the highest occupied molecular orbital, the molecule is also a better electron acceptor (oxidizing agent) in its excited state.

PET corresponds to the primary photochemical process of the excited state species ($R^* \rightarrow$ I), where R^* can be either an electron donor or an electron acceptor when reacting with another molecule, M.

When the electron donor is excited by light first, PET occurs between the excited state electron donor and electron acceptor. R^* acts as an electron donor and is therefore oxidized (oxidative electron transfer; Figure 5.36), which can be expressed by:

$$R^* + M \rightarrow R^{\cdot +} + M^{\cdot -} \tag{5-35}$$

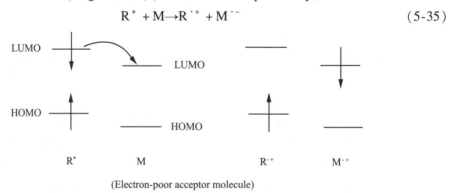

(Electron-poor acceptor molecule)

Figure 5.36　Molecular orbital representation of oxidative electron transfer

When the electron acceptor is excited by light, PET occurs between the excited state electron acceptor and electron donor. R^* acts as an electron acceptor and is therefore reduced (Figure 5.37):

>>> **Chapter 5** Fluorescence Quenching, Energy Transfer, and Electron Transfer

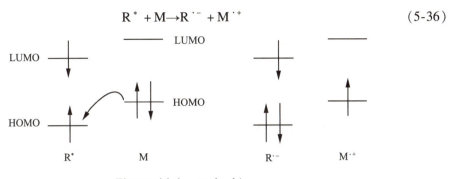

$$R^* + M \rightarrow R^{·-} + M^{·+} \tag{5-36}$$

(Electron-rich donor molecule)

Figure 5.37 Molecular orbital representation of reductive electron transfer

On the other hand, during the electron transfer process, when light is used as a reactant (photochemical reaction):

$$A + \nu \rightarrow {}^*A \tag{5-37}$$

$$^*A + B \rightarrow A^+ + B^- \tag{5-38}$$

There are two possible energetic situations for electron transfer reactions, as illustrated in Figure 5.38.

The scheme of Figure 5.38 (a) is related to an exergonic dark reaction, which is slow for kinetic reasons (high activation energy). Upon light illumination, the reductant A is excited to the much stronger reductant $^*A$, so the reaction between $^*A$ and B is much more exergonic than the reaction between A and B. Because the activation energy decreases with increasing exergonicity, the reaction involving the excited state will be much faster than that involving the ground state. In a system of this kind, light is simply used to overcome a kinetic barrier and, formally, plays the role of a catalyst.

The scheme shown in Figure 5.38 (b) corresponds to the case of a dark reaction "A + B → A$^+$ + B$^-$" that cannot take place because of thermodynamic reasons. Light excitation causes the formation of $^*A$, a reductant much stronger than A, so the reaction " $^*A + B \rightarrow A^+ + B^-$ " is thermodynamically allowed. The excitation by light can thus drive "A + B" to "A$^+$ + B$^-$" via " $^*A + B$", and in such a process, a fraction of the light energy is converted into the chemical energy of the products. The converted energy is released when A$^+$ and B$^-$ undergo the back electron-transfer reaction leading to "A + B".

Investigations on the conversion of light into chemical energy based on redox processes require the availability of compounds stable toward photodissociation and capable of undergoing reversible redox processes. For example, the metal complexes of the polypyridine family can meet both these requirements and cover a wide range of redox potentials and excited state energies.

In supramolecular systems, where the two partners of the photoredox reaction are linked together directly or through some chemical bridge, the conversion of light into chemical energy by PET reactions is widely used to power molecular artificial devices and machines.

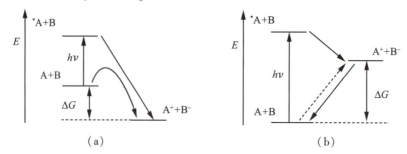

Figure 5.38 Schematic representation of the two possible energetic situations for electron transfer reactions involving an excited state reactant

5.6.4 Comparison Between Electron Transfer and Electron-exchange Energy Transfer

Electron transfer can occur between two ground-state molecules or between a ground-state molecule and an excited state molecule (Figure 5.39).

For a ground-state electron-transfer reaction in solution, the reaction can be expressed as:

$$D + A \rightarrow D^+ + A^- \tag{5-39}$$

the free energy change is obtained from the oxidation potential of the donor, $E^0(D^+/D)$, and the reduction potential of the acceptor, $E^0(A/A^-)$. Therefore, the free energy change is given by:

$$\Delta G^0 = \Delta E^0(D^+/D) - \Delta E^0(A/A^-) \tag{5-40}$$

For a reaction involving an excited state, the reaction then becomes:

$$D^* + A \rightarrow D^+ + A^- \text{ or } D + A^* \rightarrow D^+ + A^- \tag{5-41}$$

The extra amount of free energy carried by the excited state should then be considered. The free energy difference between $^*A$ and A consists of an enthalpic and an entropic term:

$$\Delta G(^*A/A) = \Delta H(^*A/A) - T\Delta S(^*A/A) \tag{5-42}$$

Specifically, PET involves the formation of ionic radical pairs. PET is also one of the deactivation ways of the excited states, which leads to the energy change of the donor and acceptor. The result of PET is transforming of light/photon energy into chemical energy.

>>> **Chapter 5** Fluorescence Quenching, Energy Transfer, and Electron Transfer

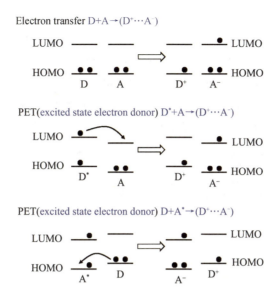

Figure 5.39 Schematic illustrations of the electron transfer process

Electron-exchange energy transfer occurs between an excited molecule and a ground-state molecule, and it is from excited state/donor to ground state/acceptor within a distance of about 0.6 nm to 2 nm:

$$D^* + A \rightarrow D + A^* \tag{5-43}$$

The exchange interaction can best be regarded as a double electron transfer process: one electron moves from the LUMO of the excited donor to the LUMO of the acceptor, and the other is from the acceptor HOMO to the donor HOMO. The important steps are illustrated in Figure 5.40.

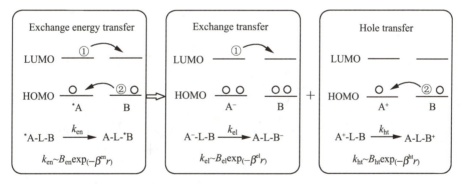

Figure 5.40 Exchange energy transfer represented as an electron and a hole transfer
(The relationships between the rate constants and the attenuation factors
of the three processes are also shown.)

5.6.5 Applications of Photoinduced Electron Transfer

One proper application of PET is the realization of fluorescence switching. The molecules used as fluorescent probes for cations, including H^+, are non-luminescent but become fluorescent when bonded to the cation.

The probe molecules used to consist of a fluorescent chromophore (fluorophore) bonded to a cation receptor group containing an electron-donating group. PET occurs from the receptor to the fluorophore, resulting in the quenching of the latter (Figure 5.41). The binding of the receptor group to the cation restores the fluorescence properties of the fluorophore as electron transfer is inhibited. Therefore, the PET sensor can act as a molecular switch.

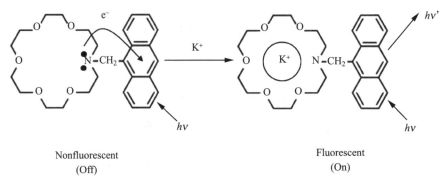

Figure 5.41 Action of a fluorescent PET potassium cation sensor as a molecular switch using a macrocyclic electron donor and anthracene fluorophore

As illustrated in Figure 5.42, excitation of the fluorophore promotes an electron from the HOMO to the LUMO. This enables PET to occur from the HOMO of the receptor to that of the fluorophore, resulting in fluorescence quenching of the latter. On binding with the K^+ cation, the redox potential of the electron-donating receptor is raised so that its HOMO becomes lower in energy than that of the fluorophore. Therefore, if PET is no longer possible, fluorescence quenching will not occur.

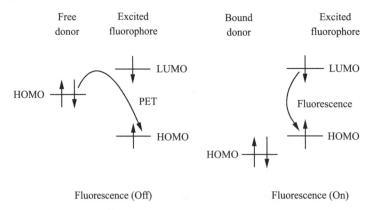

Figure 5.42 Principle of the PET cation sensor

PET quenching by DNA bases and nucleotides is also important in biological analysis. Fluorescently labeled oligomers are widely used in DNA analysis and biotechnology. Hence there is interest in understanding what factors influence the intensity of covalently bound fluorophores. The extent of quenching depends on the base and the structure of the fluorophore. Significantly, guanine appears to be the most efficient quencher among the bases, probably due to its highest tendency to donate an electron. Figure 5.43 shows the lifetime distributions for oligomers labeled with MR121. If the oligomer does not contain guanine, the decay of MR121 is a single exponential with $\tau = 2.72$ ns. The presence of a single guanine residue as the second base results in a short-lived component with $\tau = 0.43$ ns which is due to PET quenching. Therefore, one can expect the quantum yields and lifetimes of labeled oligomers to depend on the base sequence near the fluorophore.

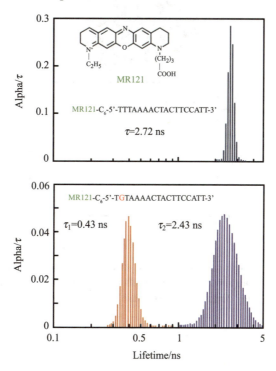

Figure 5.43 Lifetime distribution of MR121-labeled oligomers with T or G as the second base

PET has been used to study electron transfer along double-helical DNA. Figure 5.44 shows a DNA hairpin that contains a covalently bound stilbene residue at the bending site. A single guanine residue can be positioned next to the stilbene or spaced by several base pairs. The lifetime decreases dramatically as the guanine approaches the stilbene residue with $\beta = 0.64$ Å^{-1}. Therefore, one can imagine PET quenching being used to develop molecular beacons or DNA hybridization assays based on quenching by guanine residues.

Fundamentals of Photophysics

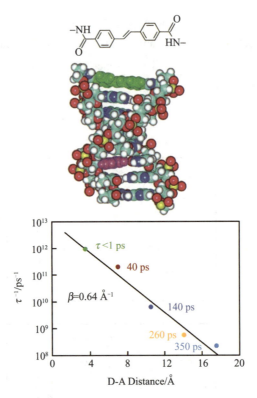

Figure 5.44 PET quenching of oligomers labeled with stilbene (green) and containing a single guanine residue (red base) (From left to right in the lower panel, this guanine is the 1st, 2nd, 3rd, 4th, or 5th base pair away from the stilbene.)

References

[1] Benson, C. R., et al. Plug-and-play optical materials from fluorescent dyes and macrocycles [J]. *Chem.*, 2020, 6(8): 1978 – 1997.

[2] Ghosh, M., et al. Fluorescence self-quenching of tetraphenylporphyrin in liquid medium [J]. *J. Lumin.*, 2013, 141: 87 – 92.

[3] Geddes, C. D. Optical halide sensing using fluorescence quenching: theory, simulations and applications—a review [J]. *Meas. Sci. Technol.*, 2001, 12(9): R53 – R88.

[4] Kalyanasundaram, K. *Photochemistry of Polypyridine and Porphyrin Complexes* [M]. London: Academic Press, 1992.

[5] Turro, N. J., Ramamurthy, V. & Scaiano, J. *Modern Molecular Photochemistry of Organic Molecules* [M]. Sausalito: University Science Books, 2010.

[6] Ceroni, P., et al. Luminescence as a tool to investigate dendrimer properties [J]. *Prog. Polym. Sci.*, 2005, 30(3 – 4): 453 – 473.

[7] Balzani, V., Credi, A. & Venturi, M. *Molecular Devices and Machines: Concepts and Perspectives for the Nanoworld* [M]. 2nd ed. Weinheim: Wiley-VCH Verlag GmbH, 2008.

[8] Cheng, Y. Y., et al. Improving the light-harvesting of amorphous silicon solar cells with photochemical upconversion [J]. *Energy Environ. Sci.*, 2012(5): 6953-6959.

Chapter 6

Fluorescence Chemical Sensors

6.1 Chemical Sensors and Fluorescence Chemical Sensors

Sensors provide a critical interface for humans to connect to the world around them. According to the Oxford English Dictionary, a sensor is "a device that detects or measures a physical property and records, indicates, or otherwise responds to it". A sensor responds to an external stimulus in the form of a signal which can be recorded. For example, the human body has at least five sets of sensors, i.e. nose, tongue, ears, eyes, and skin; human beings sense the world by their smell, taste, hearing, sight, and touch. Commercially available sensors include temperature sensors, pressure sensors, flow sensors, stress/strain sensors, conductivity sensors, etc.

A sensor contains typically three elements: a receptor, a signal transducer, and a read-out (Figure 6.1). The receptor is a critical element that interacts with the analyte in the media of interest. A selective receptor can recognize and distinguish a specific analyte from other interference. The signal converter is responsible for "translating" the physical or chemical change from the interaction of analyte and receptor to a different form, suitable for read-out. The read-out reveals the presence of the change happening in the analyte, which can be an electronic signal or optical change. The three elements of a sensor device are not always independent or physically separated; sometimes, two or more elements can be integrated into the same molecule.

>>> **Chapter 6** Fluorescence Chemical Sensors

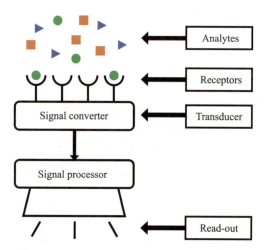

Figure 6.1 Schematic illustration of a sensor

There are many ways to categorize sensors. Based on the stimulus types, sensors can be separated into three main categories: physical property sensors, chemical sensors, and biosensors. Physical sensors can sense changes such as temperature, pressure, level, humidity, speed, and distance. Both chemical sensors and biosensors target specific chemical analytes, but the biosensors have a biological sensing recognition element, which is the main difference between them. Based on the energy transfer type that sensors detect, the classification can be summarized in Table 6.1.

Table 6.1 Classification of sensors

Sensor technology	Description
Optical sensors	Based on the principle of absorbance, reflectance, luminescence, fluorescence, refractive index, optothermal effect, and light scattering
Electrochemical sensors	Comprise voltammetric and potentiometric devices, chemically sensitized field-effect transistor, and potentiometric solid electrolyte gas sensors
Electrical sensors	Sensors with metal oxide and organic semiconductors as well as electrolytic conductivity sensors
Mass sensitive sensors	Include piezoelectric devices and those based on surface acoustic waves
Magnetic sensors	Based on paramagnetic gas properties (mainly for oxygen)
Thermometric sensors	Based on the measurement of the heat effect of a specific chemical reaction or adsorption that involves the analyte

6.1.1 Chemical Sensors

Chemical sensors can be realized by measuring a chemical or physical property of a particular analyte or a chemical transducer interacting with a particular analyte. A biosensor is a device that recognizes an analyte in an appropriate sample and interprets its concentration as an electrical signal via a substrate combination of a biological recognition system and a suitable transducer. A typical chemical/biological sensor configuration is shown in Figure 6.2.

Fundamentals of Photophysics

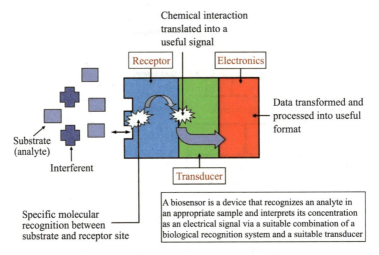

Figure 6.2 Chemical/biological sensor configuration

The ideal chemical sensor is an inexpensive, portable, foolproof device that responds with perfect and instantaneous selectivity to a particular target chemical substance (analyte) present in any desired medium to produce a measurable signal output at any required analyte concentration. Such ideal chemical sensors, however, are far from the requirement of practical application despite enormous advances over the past decades. Chemical sensors, in actuality, are complex devices universally optimized for a particular application. The complexity of a chemical sensor application is related to the technical difficulties associated with these determinations and the specific nature (i.e. elemental or molecular) of the chemical substance to be analyzed. Given the huge number ($>10^6$) of known molecular substances, molecular sensing primarily depends on the recognition of molecular structure or relevant reactivity. This capability of recognition is called selectivity.

The increasing desire to detect analytes (components of mixtures of compounds) in situ and in real-time, and to monitor the chemical changes uninterruptedly in industrial and biological processes, has promoted great developments in the field of chemical sensors, also referred to as chemosensors. Chemosensing can be accomplished by measuring a chemical or physical property of a particular analyte or a chemical transducer interacting with a particular analyte. In reality, the latter type of chemical sensor is the most important. Prominent characters in this context are highly fluorescent conjugated polymers that possess a large number of receptor sites for analytes, in fact, one receptor site per repeating unit. Noncovalent binding of analyte results in a shift of the maximum of the emission spectrum or causes quenching or enhancement of the fluorescence intensity.

6.1.2 3R and 3S

A somewhat different type of chemosensor comprises molecules, in some cases supramolecules, that recognize and signal the presence of analytes based on a 3R scheme—"recognize, relay, and report", which is schematically depicted in Figure 6.3. The sensor system contains a receptor site and a reporter site, commonly covalently linked. A noncovalent recognition event at the receptor site is communicated to the reporter site, which creates a measurable signal. Energy transfer, electron transfer, a conformational change in the molecular structure, or a combination of these processes constitutes the relay mechanism. Generally, chemosensor systems operating according to the 3R scheme consist of sensor molecules or groups that are physically blended or covalently linked to a polymer matrix.

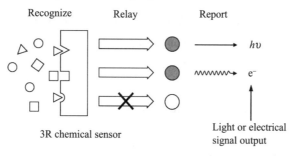

Figure 6.3 Schematic depiction of chemical sensor action (An optical or electrical signal reports the noncovalent binding of analyte to the receptor site.)

Chemical sensors have three important indicators—selectivity, sensitivity, and stability (3S). Sensitivity and detection limit relate to the quantity or concentration of the element or molecule to be analyzed (the analyte). The amount of analyte present in a sample can have a dynamic range of greater than 10^{23}, and chemical sensors are commonly required to detect 10^{-9} molar concentrations or less. Therefore, the challenge of attaining the needed sensitivity in chemical sensing is comparable to achieving the required selectivity. The sensitivity and selectivity of chemical sensing are influenced by the phase, dimensional, and temporal aspects of the desired determination. The analyte can appear in the form of a gas, liquid, or solid phase on various dimensional scales ranging from bulk volumes of liters to picoliters or surface layers from nanoscopic to monomolecular scales. It may also be persistent or discrete. A further set of requirements comes from repetitive measurements of the analyte over long times (e.g. days, months) or at multiple and perhaps remote locations, such as in environmental analysis and personal monitoring. The design of chemical sensors also requires an appreciation of the necessary degree of quantitative reliability (precision or accuracy). Stability (or repeatability) refers to the degree of inconsistency of the retention curve obtained when the sensor inputs are measured in the same direction at the full range many times. The more identical the output characteristic curves repeatedly tested under the same input conditions, the better their repeatability will be, and the error will be more minor. The non-repeatability of the output holdup of the sensor is mainly caused by the wear, clearance, looseness of the mechanical part of the sensor, internal friction of the parts, dust accumulation, and the naming and drift of the auxiliary circuit.

6.1.3 Fluorescence Chemical Sensors

Accurate detection and quantification of biological and chemical substances play a more significant role in many applications, such as environmental monitoring, clinical diagnostics, DNA sequencing, and even biological warfare agent detection. An efficient sensor should be highly sensitive and selective and capable of concurrently detecting and distinguishing multiple target analytes in a simple and rapid way. Optical sensing technology is one of the most promising approaches in this aspect because of the numerous benefits over other sensing methods. The main advantages of optical sensing include immunity to electromagnetic interference, durability under intense pressures and temperatures, and most importantly, high sensitivity and selectivity because the measurement is performed utilizing particular excitation and emission wavelengths specific to the target analytes.

With the benefits of optical sensing technologies, different optical detection techniques, such as absorbance, diffraction, reflection, scattering, chemiluminescence, and refractive index have been used and reported in developing portable biochemical sensors. Among various optical detection techniques, fluorescent sensing is considered highly useful in practical applications for its high sensitivity, specificity, and accuracy compared to other optical sensing techniques.

Molecules that conduct molecular interactions through fluorescence signals are called fluorescence chemical sensors. Figure 6.4 shows the schematic diagram of the fluorescent chemical sensor. The working principle of the fluorescent chemical sensor is succinctly summarized as follows: The surrounding environment, such as temperature, acidity, viscosity, solvent, especially foreign chemical or biological species, has an effect on the molecules with the characteristics of fluorescence emission. The fluorescence changes the constituent factor of emission, including spectrum and intensity. As a result, the characteristics of the surrounding environment or the information of certain species in the environment can be acquired.

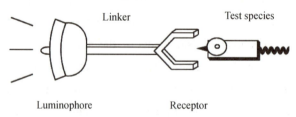

Figure 6.4 Schematic diagram of a fluorescent chemical sensor

The magnitude of the signal generated by the sensor usually is proportional to the concentration of the analyte. Regarding practical applications, optical chemosensors that monitor changes in fluorescence intensity, or to a lesser degree in optical absorption, are much more universal compared with chemosensors that observe changes in electrical conductivity or electrical current.

In many cases, optical chemosensor devices consisting of a probe are called an optode, in which modulation of the optical signal takes place, and an optical link connects the probe to the instrumentation. The main parts of the latter are a light source, a photodetector, and an electronic signal-processing unit. A schematic depiction of a typical optode is presented in Figure 6.5. This optode operates with the aid of two fluorophores that undergo a change in fluorescent light emission in the presence of O_2 or CO_2. Fluorophore I is admixed, and fluorophore II is chemically linked to the polymer.

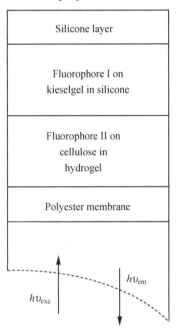

Figure 6.5 Structure of an optode for detecting molecular oxygen and carbon dioxide.
[Fluorophore I (O_2): tris (2,2'-bipyridyl) ruthenium (II) dichloride;
fluorophore II (CO_2): 1-hydroxypyrene-3,6,8-trisulfonate]

During the last few decades, significant advances in fluorescence-based instruments have greatly influenced the chemical and biological sensing. The sensitivity of a modern fluorometer is as low as a single photon level, and a fluorescence microscope can identify two different particles spaced less than 10 nm distance by utilizing state-of-the-art super-resolution technology.

Fluorescence-based instruments can be divided into several types based on measured fluorescence parameters. The most fundamental function of a fluorometer is to simply measure fluorescence intensity at fixed wavelength values of excitation and emission, often called a steady-state measurement (wavelength-based). Fluorescence lifetime is decided by emission intensity decay, a unique property of different fluorophores. Anisotropy is the utilization of polarized excitation light for characterizing the rotational motions of fluorophores through detecting fluorescence emission that has the same polarity as the excitation wavelength.

A spectrometer contains five necessary components as illustrated in Figure 6.6 (a): a light source for broad excitation spectrum, an excitation monochromator for wavelength selection, a sample container (usually cuvette and holder jig), an emission monochromator for fluorescence wavelength selection, and a photodetector. Light absorbance can be measured as an option (spectrophotometry function). Among all five main components, the quality of the measured fluorescence (resolution and intensity) usually depends on the monochromator components. A monochromator consists of a collimator, focusing mirror, a grating, slits, optomechanical components, and peripheral systems for controlling optics, as illustrated in Figure 6.6 (b). Monochromators and detectors are mainly responsible for making it hard to miniaturize a spectrometer system unless its performance is significantly compromised in terms of resolution and sensitivity.

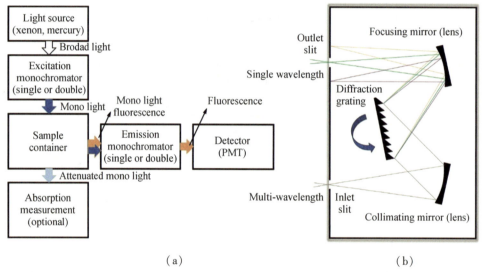

Figure 6.6 Basic spectrometer components: (a) block diagram of spectrometer components; (b) illustration of a basic monochromator for excitation and emission wavelength selection

A filter-based fluorometer is the oldest structure of all fluorescence-based instruments. Components of this type of instrument are similar to that of a spectrometer. Still some components, including excitation and emission wavelength selectors, are substituted by optical filters, as illustrated in Figure 6.7. These excitation and emission optical filters are either single or multiple layers of color filters or dichroic mirrors; thus, it is only possible to select fixed excitation and emission wavelengths. Nevertheless, multiple wavelengths can be chosen with a filter-wheel configuration which contains a limited number of optical filters to obtain various wavelengths of interest.

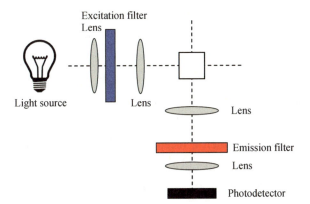

Figure 6.7 Illustration of the filter-based fluorometer (Compared to the conventional spectrometer, filter-based fluorometers are simpler, cheaper, compact, and more application-specific.)

A proper selection of excitation light is essential to improve the sensitivity, selectivity, and many other parameters of the fluorescent sensing system. Various excitation light sources (such as lamp, light-emitting diode, and laser) are applied for different applications. Each of them has its advantages and disadvantages. The spectral linewidth of excitation light should be as narrow as possible to reduce the interference with emission fluorescence.

6.1.4 Design of Fluorescence Chemical Sensors

The complexation of small species, such as inorganic/organic cations and anions and neutral molecules (with a receptor), influences the photochemical or photophysical characteristics and some photoactive units. Figure 6.8 shows the working process of the fluorescent chemical sensor.

There are three methods of designing the fluorescence chemical sensor: enhancement of fluorescence intensity, decrease of fluorescence intensity, and change in emission wavelength.

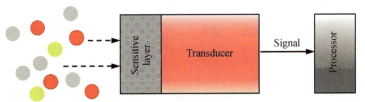

Figure 6.8 The working process of fluorescent chemical sensor

The classical design of a fluorescent indicator contains two parts, a receptor responsible for the molecular recognition of the analyte and a fluorophore utilized for signaling the recognition event. There are three main strategies to approach the design of fluorescent molecular indicators for chemical sensing in solution. The first consequence is intrinsic fluorescent probes, i.e. fluorescent molecules, where the mechanism for signal transduction involves the interaction of the analyte with a ligand that is part of the P system of the fluorophore. The second is extrinsic fluorescent probes, in which the receptor moiety and the fluorophore are covalently linked but are electronically independent. The extrinsic probes have also been denoted conjugate; nevertheless, for homogeneity reasons, we prefer to call them "extrinsic". In this case, different receptor molecules might be synthesized and then attached to a fluorophore to make the sensitive probe. Because of the covalent linking through a spacer, both moieties are in close proximity; the interaction of the analyte with the receptor induces a change in the fluorophore surroundings and changes its fluorescence. The third strategy is called chemosensing ensemble, based on a competitive assay in which a receptor-fluorophore ensemble is selectively separated by the addition of an appropriate competitive analyte able to interact efficiently with the receptor leading to a detectable response of the fluorophore.

After the production of a fluorescent indicator, the next step toward the preparation of a sensor is usually the production of the sensing material by incorporating the indicator in solid support. Until now, the most popular method of the immobilization step is the physical entrapment of the sensor probe in a polymer matrix. After inducing, the polymer is deposited on a device such as an optical fiber or the surface of a waveguide to create the working sensor. However, physical entrapment of the dyes in the polymer matrix produces inhomogeneity in the material. It brings stability problems because of the filtering of the fluorescent probe, decreasing the lifetime and reproducibility of the sensor. Thus, despite the easy preparation of these materials, they are rarely incorporated into commercial instruments. To enhance the stability of these materials, we take the alternative as the covalent attachment of the probes to the polymeric matrices. Parallel to the production of polymeric materials, a new tendency in material science for chemical sensing is emerging. Other materials have been developed, where the components of a sensing system (receptor and fluorophore) are directionally held in a physical space, i.e. they are covalently immobilized at a surface or form surfactant aggregates. Many materials such as silica particles, glass and gold surfaces, quantum dots, Langmuir-Blodgett films, vesicles, liposomes, and others combine with many chemical receptors to synthesize sensitive fluorescent materials.

6.2 Recognize

Generally, the types and utilities of light-related functions depend on the properties of the system that uses photons. From the photophysical or photochemical viewpoint, some molecules have outstanding properties by themselves [e.g. $[Ru(bpy)_3]^{2+}$ (bpy = 2,2'-bipyridine)]. However, more profitable use of photons is achieved when two or more molecular components are connected in an organized way in most cases. Molecular self-assembly and self-organization are dominant processes in the chemistry of biological systems. It is indeed amazing how nature is capable of mastering weak intermolecular forces to construct complex molecular devices and machines.

Prominent in fluorescence chemical sensors are highly fluorescent conjugated polymers that possess a large number of receptor sites for analytes, actually one receptor site per repeating unit. Noncovalent binding of an analyte results in a shift of the maximum of the emission spectrum or causes quenching or enhancement of the fluorescence intensity. A somewhat different type of chemosensor comprises molecules, in some cases supramolecules, that recognize and signal the presence of analytes on the basis of a 3R scheme "recognize, relay, and report", which is schematically depicted in Figure 6.9. The sensor system includes a receptor site and a reporter site, which are covalently linked commonly. A noncovalent recognition event at the receptor site is communicated to the reporter site, which could produce a measurable signal. Energy transfer, electron transfer, a conformational change in the molecular structure, or a combination of these processes constitutes the relay mechanism.

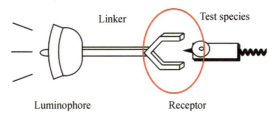

Figure 6.9 The recognize part in fluorescence chemical sensors

To improve the selectivity and specificity of the acceptor part for species identification, the following must be addressed in the process of designing:

(1) The identified object.

(2) The electrification and charge distribution of the object.

(3) The shape and size of the object.

(4) The existing functional groups of the object.

(5) Weak interaction in the recognition process.

In the process of recognizing, weak interaction between acceptor and substrate plays an important role in molecular recognition. It mainly includes six kinds: electrostatic interaction, hydrogen bonding, π-π stacking interaction, cation-π interaction, van der Waals Force, and hydrophobic effect. These six of weak interactions will be described briefly in the following sections because these interactions have been introduced in the enthalpy reduction factor of Chapter 4.

6.2.1 Electrostatic

The intermolecular forces of attraction in which complete or partial ionic species are attracted to each other are termed electrostatic interactions. These attraction forces do not include any sharing of electrons between atoms. Therefore, they are also named noncovalent bonds.

Fundamentals of Photophysics

The term electrostatic interactions comprise both attractive and repulsive forces ionic species, which means ions with opposite charges are attracted to each other, while similar charges repel each other. There are three types of electrostatic interactions: dipole-dipole interactions, London dispersion forces, and hydrogen bonds. Generally, there are three types of electrostatic interactions, ion-ion, ion-dipole, and dipole-dipole electrostatic. Corresponding examples are shown in Figure 6.10, respectively.

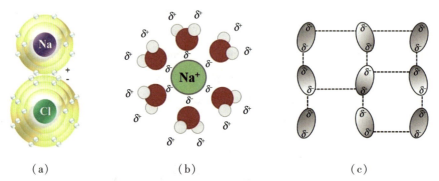

(a)　　　　　　　(b)　　　　　　　(c)

Figure 6.10　Three types of electrostatic interactions: (a) ion-ion interactions in NaCl; (b) Ion-dipole interaction between Na^+ and H_2O; (c) dipole-dipole interactions in H_2O or HCl

i. Ion-ion interactions are attractive forces between ions with opposite charges

They are also referred to as ionic bonds and are the forces that hold together ionic compounds. Similar charges repel each other, and opposite charges attract. These Coulombic forces operate over relatively long distances in the gas phase. The force depends on the product of the charges (Z_1, Z_2) divided by the square of the distance of separation (d^2). The strength of ionic bonding is comparable to covalent bonding (bond energy = 100 – 350 kJ · mol^{-1}). A typical ionic solid is sodium chloride, presented in Figure 6.10 (a), which has a cubic lattice in which each Na^+ cation is surrounded by six Cl^- anions. It would require a large stretch of the imagination to regard NaCl as a supramolecular compound; however, the ionic lattice of NaCl does illustrate how a Na^+ cation can organize six complementary donor atoms about itself in order to maximize noncovalent ion-ion interactions.

ii. Ion-dipole interaction results from an electrostatic interaction between a charged ion and a molecule with a dipole

The bonding of an ion with a polar molecule, such as Na^+ cation and water, respectively, is an example of an ion-dipole interaction. Figure 6.10 (b) is a schematic diagram of this. The range in strength of ion-dipole interaction is from ca. 50 – 200 kJ · mol^{-1}. This bonding could be seen both in the solid state and solution. A supramolecular analogue is readily apparent in the structures of the complexes of alkali metal cations with macrocyclic (large ring) ethers termed crown ethers, in which the ether oxygen atoms play the same role as the polar water molecules, although the complex is stabilized by the chelate effect and the effects of macrocyclic preorganization. The oxygen lone pairs are attracted to the positive cation charge.

iii. Dipole-dipole interactions are a type of intermolecular force between two molecules with net dipole moments (asymmetrical charge distributions, where polar molecules develop partial positive and partial negative charges)

Molecules tend to align themselves so that the positive end of one dipole is near the negative end of another, and vice versa. When a positive dipole and a negative dipole get close to each other, it creates an attractive intermolecular interaction, while two dipoles with the same type of charge, two positive dipoles or two negative dipoles, will create a repulsive intermolecular interaction presented in Figure 6.10 (c).

Electrostatic interactions play an important role for gases at high pressures as these are accountable for their observed deviations from the ideal gas law at high pressures. Such interactions are important in maintaining the 3D structure of larger molecules. Nucleic acids and proteins are very typical examples. Such interactions are involved in several biological processes where larger molecules bind specifically but transiently to one. Additionally, these interactions also deeply influence the crystallinity and design of materials, in general, for many organic molecules synthesis, particularly for self-assembly.

6.2.2 Hydrogen Bonding

A hydrogen bond is an attractive intermolecular force in which a hydrogen atom that is covalently bonded to a small, highly electronegative atom is attracted to a lone pair of electrons on an atom in a neighboring molecule. Compared to other dipole interactions, the hydrogen bond is a very strong interaction force. The strength of a typical hydrogen bond is about 5% of that of a covalent bond.

Hydrogen bonding appears only in molecules where hydrogen is covalently bonded to one of three elements: fluorine, oxygen, or nitrogen. The electronegativity of these three elements is so strong that they withdraw the majority of the electron density in the covalent bond with hydrogen, leaving the H atom very electron-deficient. The H atom nearly acts as a bare proton, leaving it very attracted to lone pair electrons on a nearby atom. A classic case is water; the attractive force between water molecules is a dipole interaction. The hydrogen atoms tend to bind with the highly electronegative oxygen atom, which also possesses two lone pair sets of electrons, making for a very polar bond. The partially positive hydrogen atom of one molecule is then attracted to the oxygen atom of a nearby water molecule [Figure 6.11 (a)]. The hydrogen bonding that occurs in water would result in some unusual but significant properties. For example, water molecules are able to stay condensed in the liquid state due to the strong hydrogen bonds, while most molecular compounds that have a mass similar to water are gases at room temperature. Figure 6.11 (b) shows how the bent shape and two hydrogen atoms per molecule allow each water molecule to be able to hydrogen bond to two other molecules.

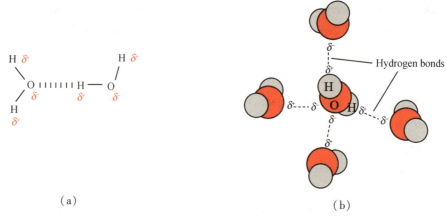

(a)   (b)

Figure 6.11 (a) A hydrogen bond in water occuring between the hydrogen atom of one water molecule and the lone pair of electrons on an oxygen atom of a neighboring water molecule; (b) Multiple hydrogen bonds occuring simultaneously in water because of their bent shape and the presence of two hydrogen atoms per molecule

Hydrogen bonding is vital in many chemical processes, especially in the field of life sciences. Hydrogen bonding is responsible for water's unique solvent capabilities. Hydrogen bonds hold complementary strands of DNA together and are responsible for determining the 3D structure of folded proteins including enzymes and antibodies.

6.2.3 π-π Stacking Interaction

Aromatic π-π interactions (sometimes called π-π stacking interactions) occur between aromatic rings, often in situations where one is relatively electron-rich and another one is electron-poor. This denomination denotes strong, attractive interactions between molecules carrying P systems, i.e. cyclic molecules with conjugated double carbon-carbon bonds. There are two general types of π-π interactions: face-to-face and edge-to-face, although a wide variety of intermediate geometries are known (Figure 6.12). Face-to-face π-stacking interactions shown in Figure 6.12(a), called sandwich stacking as well, are responsible for the slippery feel of graphite and its useful lubricant properties. Similar π-stacking interactions between the aryl rings of nucleobase pairs also help to stabilize the DNA double helix. Edge-to-face interactions (sometimes called T-shaped stacking interactions) may be regarded as weak forms of hydrogen bonds between the slightly electron deficient hydrogen atoms of one aromatic ring and the electron-rich π-cloud of another.

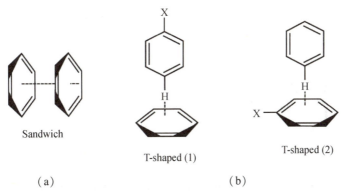

Figure 6.12 Two general types of π-π interactions: face-to-face (sandwich stacking) and edge-to-face (T-shaped stacking)

Many theoretical and experimental studies underline the importance of π-π interactions, which are fundamental to much supramolecular organization and recognition processes. Perhaps the simplest prototype of aromatic π-π interactions is the benzene dimer. Within dimeric aryl systems such as this, possible π-π interactions are the sandwich and T-shaped interactions, as shown in Figure 6.12. It has been shown that all substituted sandwich dimers bind more strongly than a benzene dimer, whereas the T-shaped configurations bind more or less favorably depending on the substituent. Electrostatic, dispersion, induction, and exchange-repulsion contributions are all significant to the overall binding energies.

6.2.4 Cation-π Interaction

Cation-π interaction refers to the stabilizing interaction between a cation and the face of a simple aromatic, a general, strong, noncovalent binding force used throughout nature. Figure 6.13 summarizes several studies of the cation-π interactions in the gas phase. These results have two important features. First, these are considerable binding energies for a clearly noncovalent binding interaction. Importantly, pioneering work by Kebarle in 1981 measured not only the interaction between benzene and K^+ but also the interaction between water and K^+. Everyone would agree that water is a potent ligand for ions, and it is, binding K^+ with a $-\Delta H$ of 18 kcal · mol^{-1}. Remarkably, benzene binds the K^+ ion more tightly than water, and this is the first indication that the cation-π interaction is a potentially important binding force. The second feature of Figure 6.13 is that the results follow a classical electrostatic trend, much like in aqueous solvation energies or crystal lattice energies. That means smaller ions with more focused charges have a larger affinity. These results, and many more, have led us to advocate a primarily electrostatic model for the cation-p interaction. While it is certainly true that van der Waals and polarization effects contribute to the cation-p interaction, the defining feature is electrostatic.

Fundamentals of Photophysics

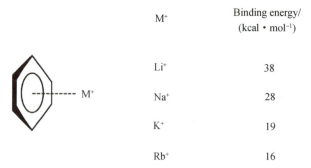

Figure 6.13 Gas-phase binding energies (2DH) for alkali metals to benzene

6.2.5 van der Waals Force

The van der Waals interaction is a long-range attraction between chemical species arising from a correlation between instantaneous electronic charge fluctuations on each. It is present even when there is no chemical bond and each species has no permanent multipole moment. In other words, van der Waals Forces are those bonds that play the role of attracting both molecules and atoms. The van der Waals interaction is much weaker in strength than a normal chemical bond, but it has important effects on the properties of materials. Moreover, The van der Waals interaction is the weakest intermolecular force (<5 kJ · mol^{-1}). The van der Waals interaction is pervasive in various kinds of materials. Some typical examples are shown in Figure 6.14. It is responsible for many phenomena such as sublimation of iodine, naphthalene, dry ice, and other organic molecules, high interlayer mobility of graphite, folding of long biomolecular chains such as DNA, RNA, and proteins, polymers.

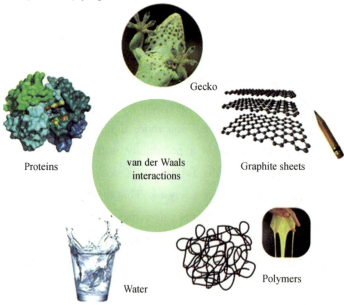

Figure 6.14 van der Waals interactions in nature

6.2.6 Hydrophobic Effect

Hydrophobic literally means "the fear of water". Hydrophobic molecules and surfaces repel water. Hydrophobic liquids, such as oil, will separate from water. Generally, hydrophobic molecules are nonpolar, meaning the atoms that make the molecule do not produce a static electric field. In polar molecules, these opposite regions of electrical energy attract water molecules. Without opposite electrical charges on the molecules, water cannot form hydrogen bonds with the molecules. The water molecules then form more hydrogen bonds with themselves, and the nonpolar molecules clump together.

The hydrophobic effect is caused by nonpolar molecules clumping together. Hydrophobic sections in large macromolecules will fold the molecule so they can be close to each other and away from water. Many amino acids in proteins are hydrophobic, which helps the proteins in life obtain their complicated shapes. The hydrophobic effect extends to organisms, as many hydrophobic molecules on the surface of an organism help them regulate the amount of water and nutrients in their systems.

Cell membranes are a typical example, which are made of macromolecules known as phospholipids. Phospholipids have phosphorous atoms in the heads of the molecules, which attract water. In contrast, the tail of the molecule is made of lipids, which are hydrophobic molecules. The hydrophilic heads point toward water, and the hydrophobic tails attract each other. In small groups, phospholipids form micelles, commonly small hydrophobic ball shown in Figure 6.15.

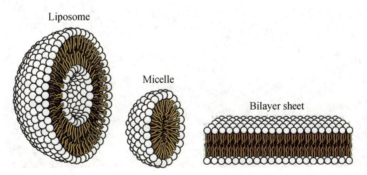

Figure 6.15 A micelle—a small hydrophobic ball

The hydrophobic tails expel water from the center of the ball. Cell membranes are composed of two phospholipid layers, known as the phospholipid bilayer. The middle of the sheet is made of hydrophobic tails, which expel water and can separate the contents of the cell from the outside environment. Cells have a variety of special proteins embedded into the membrane helping transport hydrophilic molecules like water and ions across the hydrophobic middle portion of the membrane. In eukaryotic cells, organelles are formed inside cells from smaller sacs created from phospholipid bilayers. The hydrophobic properties of phospholipids are utilized by scientists to create another structure to deliver medicine and nutrients to cells. As seen in the graphic above, liposomes are small sacs that can be filled with medicine. With the right proteins embedded into the membrane, the liposome will merge with the membrane of a target cell and deliver the medicine to the inside of the cell.

6.3 Relay

Fluorescence sensing is always based on the change of one or several fluorescence characteristics of a fluorophore (intensity, spectral shift, lifetime, and polarization) upon interaction with an analyte. These changes may only be due to the change of the ground state of the fluorophore during binding or may also be due to the disturbance of the analyte to the excited state process: electron (or proton) transfer, charge transfer, the formation or disappearance of excimer (or excimer complex), and energy transfer. These various mechanisms are illustrated in Figure 6.16 and will be successively presented in the subsequent sections.

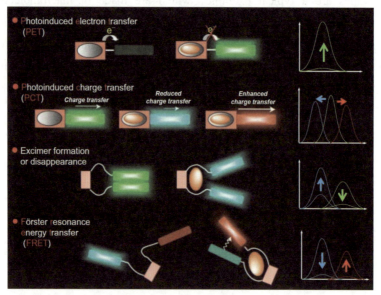

Figure 6.16 Various excited-state processes occurring in fluorescent molecular sensors for ions and molecules

6.3.1 Photoinduced Electron Transfer Mechanism

PET is a critical working mechanism for information, which has been widely used in molecular recognition. Generally, if electron donor and electron acceptor compounds can satisfy the result of calculated $\Delta G < 0$ (from the Weller formula) under illumination, the PET process between molecules can occur.

Weller formula can be abbreviated as follows:

$$\Delta G = E_{ox} - E_{red} - \Delta E_{0,0} - \text{Constant} \tag{6-1}$$

where ΔG is the free energy change of the reaction. E_{ox} and E_{red} represent the oxidation and reduction potential of the donor and acceptor. $\Delta E_{0,0}$ is the transition energy of the stimulating compounds.

Identified species are often designed to interfere with the PET process. If an electron donor group can be given through the coordination of protonated metal cations and the formation of hydrogen bonds, the "electron pair" of the original electron donor group of the system is stable. Thus, the driving force of the electron transfer process can be reduced so that the quenching process is not easy to occur, and fluorescence intensity can be improved.

In PET, a complex is formed between the electron donor D_P and the electron acceptor A_P, yielding $D_P^+ \cdot A_P^-$ (Figure 6.17). The subscript P is used to identify the quenching due to a PET mechanism. This charge transfer complex can return to the ground state without the emission of a photon, but exciton emission can be observed in some cases. Finally, the extra electron on the acceptor is returned to the electron donor.

The terminology for PET can be confusing because the excited fluorophore can be either the electron donor or acceptor. The electron transfer direction of the excited state is determined by the redox potential of the ground state and the excited state. When discussing PET, the term "donor" refers to the species that donates an electron to an acceptor. In PET, the terms "donor" and "acceptor" do not identify which species is initially excited.

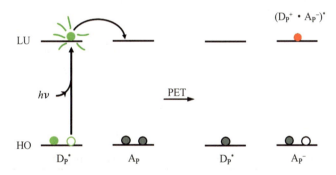

Figure 6.17 Molecular orbital schematic for PET

The nature of PET quenching is clarified by examining several examples. The more common situation is when the excited state of a fluorophore acts as an electron acceptor (Table 6.2). A typical example is an electron-rich species, such as dimethylaniline (DMA), which can donate electrons to a wide range of polycyclic aromatic hydrocarbons as electron acceptors. Electron transfer is even more favorable to electron-deficient species like cyanonaphthalenes. There are some unusual PET pairs, such as indole donating to pyrene and dienes donating to cyanoanthracene. This table is not exhaustive but provides only examples of PET pairs.

Fundamentals of Photophysics

Table 6.2 PET quenching where the fluorophore is the electron acceptor

$F^*-Q \rightarrow (F^- \cdot Q^+)^* \rightarrow$ heat or exciplex	
Fluorophore (PET acceptor)	Quencher (PET donor)
Polynuclear aromatic hydrocarbons	Amines, dimethylaniline
2-Cyanonaphthalenes	Dimethylaniline
7-Methoxycoumarin	Guanine monophosphate
Pyrene	Indole
9-Cyanoanthracene	Methyl indoles
9,10-Dicyanoanthracene	Dienes and alkenes, alkyl benzenes
Anthraquinones	Amines
Oxazine	Amines

PET quenching can also occur by electron transfer from the excited fluorophore to the quencher (Table 6.3). Examples include electron transfer from excited indoles to electron-deficient imidazolium or acrylamide quenchers. Quenching of carbon halides also occurs due to the transfer of electrons from fluorophores to electronegative carbon halides. Electron-rich dimethoxynaphthalene can donate electrons to pyridinium. And finally, there is the well-known example of electron transfer from excited $[Ru(bpy)_3]^{2+}$ to methyl viologen.

Table 6.3 PET quenching where the fluorophore is the electron donor

$F^*-Q \rightarrow (F^+ \cdot Q^-)^* \rightarrow$ heat or exciplex	
Fluorophore (PET donor)	Quencher (PET acceptor)
Indole or NATA	Imidazole, protonated
Indole	RCO_2H, but not RCO_2^-
Tryptophan	Acrylamide, pyridinium
Carbazole	Halocarbons; trichloroacetic acid
Indole	Halocarbons
$[Ru(bpy)_3]^{2+}$	Methyl viologen
Dimethoxynaphthalene	N-methylpyridinium

PET is the most widely employed mechanism for the design of fluorescence probes. In general, these molecules consist of a fluorophore, a spacer, and a receptor. The receptor bears free electron pairs, for example, on the nitrogen or oxygen atoms. One of these electrons can be transferred to the partially unoccupied HOMO of the photoexcited fluorophore. A back-electron transfer can now take place from the excited state of the fluorophore to the HOMO of the receptor. This leads to the non-radiation inactivation of the excited state and the quenching of fluorescence. PET is blocked, and the fluorescence of the molecule is turned on if a guest binds to the receptor (Figure 6.18). A quantitative approach to predict PET efficiency was developed by Weller. PET is fast and fully reversible.

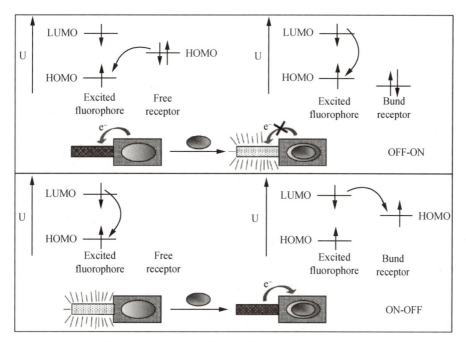

Figure 6.18 Principles of PET fluorescent sensors

PET sensors can either be a "turn-on" or a "turn-off" sensor (Figure 6.19). In the cases of "turn-off" sensors, the receptors take part in the photophysical process either directly or indirectly. When binding to the analyte, the energy level of the HOMO of the receptor/analyte pair moves somewhere between the fluorophore homo and the energy level of the lowest unoccupied molecular orbital, forming a non-radiative path, such as PET quenching, dissipating the excitation energy and a quenching of the fluorescence of the sensor molecule.

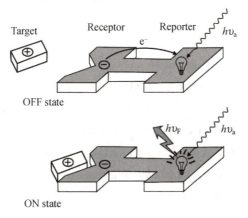

Figure 6.19 Signal transduction in the simplest PET-based sensor

"Turn-on" sensors work a little differently. The PET process is dominant before the binding event. The receptors usually contain a nonbonding electron pair with relatively high energy. In the absence of analytes, this electron pair quenches the emission by rapid intramolecular electron transfer from the receptor to the stimulated fluorophore. When this electron pair coordinates with electron-deficient analytes, the relative energy of the HOMO of the receptor is lowered. This prohibits the PET process, creates a direct pathway for energy to release through emitting a photon, and thus restores the fluorescence of the fluorophore.

Such simple realization of PET has mostly been applied in the sensing of ions. In a typical ion sensor, there are two electronic systems, one for ion recognition and the other for reporting, and they are connected by a short spacer. Cation binding produces the strong attraction of electrons with the suppression of PET.

Sensing neutral molecules in this way has also been made, though with limited success. The more complicated schemes allow excluding the target analyte binding from direct influence on PET. For instance, this can be done by influencing the ionization of an attached group and thus extending this transduction mechanism to a much broader range of analytes.

PET can occur not only through bonds but also through space, and for this process, the most important is the close distance between donor and acceptor. Therefore, donor and acceptor groups can be placed together or moved apart during sensing events due to conformational changes in the sensor.

One more possibility offered by the PET mechanism is the association or dissociation of nanoparticles: on association, they can play the role of PET donor or acceptor. Noble metal nanoparticles have proved to be the most efficient electron acceptors from different types of fluorescent donors, including semiconductor quantum dots. The latter can be reduced or oxidized at relatively moderate potentials, which often changes slightly with the physical size of these nanocrystals. Therefore, they can be used as efficient electron donors, and this possibility of signal transduction makes them efficient fluorescence reporters. Transition metal cations (such as Cu^{+2}) and free radicals quench fluorescence according to the PET mechanism. This property can be used for their detection.

The PET quenching in conjugated polymers deserves special attention. Deficient amounts of cationic electron acceptors quench the fluorescence of polyanionic conjugated polymer on a time scale shorter than picosecond so that compared with the so-called "molecular excited state", its sensitivity to quenching is more than one million times. Novel materials such as quantum dots suggest new strategies for optimal exploration of the PET phenomenon in sensing. The three important cases of PET frequently used in sensing are as follows:

(1) Electron transfer between molecular fragments of the same dye molecule.

(2) Intermolecular electron transfer.

(3) Quenching by spin labels.

Although, the mechanism of PET has been well discussed in the past three decades. A large number of metal ion PET probes have been reported, and PET probes may have some disadvantages. For example, although the NIR fluorescent probes are now more attractive than those of shorter excitation/emission wavelength due to their larger tissue penetration depth and lower phototoxicity, their higher fluorophore HOMO energy level decreases the energy gap between the HOMOs of the fluorophore and ionophore. It reduces the efficiency of the PET process to quench free probe emission. NIR probes with a PET mechanism may suffer from high background fluorescence and lower analyte-induced emission enhancement factor. Therefore, the design of turn-on NIR probes with low background is still challenging. Heavy metal ions like Hg^{2+} can quench fluorescence by several mechanisms, and metal ions with unpaired electrons in their d-orbitals, such as Cu^{2+} display a distinct ability to quench the emission of organic fluorophores via their unpaired electrons. Therefore, fine adjustment of metal coordination and reduction of metal paramagnetism and spin-orbit coupling by changing the structure of ion carrier and spacer are the key to the design of the PET probe. In addition, two-photon excitable probes with a PET mechanism for practical bioimaging applications are also currently attracting the interest of scientists.

6.3.2 Photoinduced Charge Transfer Mechanism

According to the nature of cation controlled photoinduction process, there are three main fluorescent molecular sensors for cation recognition:

(1) Sensors based on cation control of PET (PET sensors).

(2) Sensors based on cation control of photoinduced charge transfer (PCT sensors).

(3) Sensors based on cation control excimer formation or disappearance. In each class, the distinction is to be made according to the structure of the complexing moiety: chelators, podands, coronands (crown ethers), cryptands, and calixarenes.

A distinct advantage of PET sensors is the large change in fluorescence intensity usually observed upon cation binding, so the expression "off-on" and "on-off" fluorescent sensors is often employed. Another feature is no shift in the fluorescence or excitation spectrum, which eliminates the possibility of intensity ratio measurement at two wavelengths. In addition, the pet usually comes from tertiary amines, and its pH sensitivity may affect the reaction to cations.

In PCT sensors, the changes in fluorescence quantum yield on cation complexation are generally not very large compared to those observed with PET sensors. However, the absorption and fluorescence spectra are shifted upon cation binding, so a considerable change in fluorescence intensity can be observed by properly selecting the excitation wavelength and observation wavelength. Moreover, ratiometric measurements are possible: the ratio of the fluorescence intensities at two appropriate emission or excitation wavelengths provides a measure of the cation concentration, which is independent of the probe concentration (provided that the ion is in excess) and is insensitive to the intensity of incident light, scattering, inner-filter effects, and photobleaching. Ratiometric measurements are also possible with excimer-based sensors.

When a fluorophore contains an electron-donating group (often an amino group) conjugated to an electron-withdrawing group, it undergoes internal charge transfer (ICT) from the donor to the acceptor upon excitation by light. Then the change of dipole moment leads to Stokes displacement, which depends on the microenvironment of the fluorophore. The polarity probe can be designed on this basis. It can thus be anticipated that an analyte in close interaction with the donor or the acceptor moiety will change the efficiency of ICT and thus will affect the photophysical properties of the fluorophore.

Let us consider the important case of cation sensing. When a cation interacts with a group (like an amino group) playing the role of an electron donor within the fluorophore (Figure 6.20), it reduces the electron-donating properties of this group. Due to the decrease of conjugation, the blue shift of absorption spectrum and the decrease of the molar absorption coefficient is expected. Conversely, a cation interacting with the acceptor group enhances the electron-withdrawing character of this group; the absorption spectrum is thus redshifted, and the molar absorption coefficient is increased. The fluorescence spectra are, in principle, shifted in the same direction as the absorption spectra. In addition to these shifts, changes in quantum yields and lifetimes are often observed. All these photophysical effects are dependent on the charge and the size of the cation, and selectivity of these effects is expected.

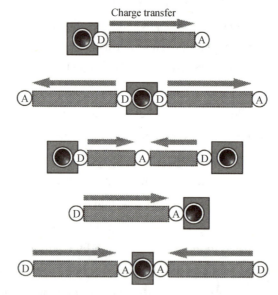

Figure 6.20 Topology of PCT fluorescent molecular sensors

The photophysical changes in cation binding can also be described in charge dipole interaction. Let us consider only the case where the dipole moment in the excited state is larger than that in the ground state. Then, when the cation interacts with the donor group, the destruction of the excited state by the ion is stronger than that of the ground state, and a blue shift of the absorption and emission spectra is expected. Conversely, when the cation interacts with the acceptor group, the excited state is more stabilized by the cation than the ground state, leading to a red shift of the absorption and emission spectra (Figure 6.21). Such spectral shifts offer the advantage of ratiometric measurements.

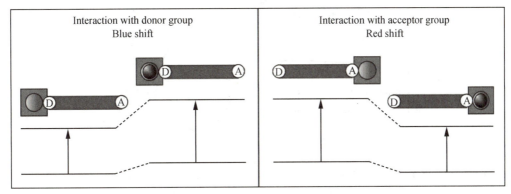

Figure 6.21 Spectral displacements of PCT sensors resulting from the interaction of a bound cation with an electron-donating or electron-withdrawing group

ICT fluorophores with a conjugated couple of electron-donating/electron-withdrawing groups (donor/acceptor, D/A) normally display large Stokes shift, visible light excitability and metal coordination-induced emission shift. Moreover, the ICT effect can decrease the basicity of the donor amine, which provides an opportunity to form pH-independent probes in near-neutral pH conditions for application in biological systems. Modification of D or A as a metal ionophore results in metal coordination-induced blue or red shift of excitation/emission via altering the photoinduced ICT excited state, which provides an effective strategy to devise ratiometric metal ion probes. As shown in Figure 6.22, metal coordination to the donor of an ICT fluorophore will decrease the HOMO energy and induce the hypsochromic shift of excitation or emission maxima. If its receptor is coordinated with metal, the opposite change will be observed. These sensing behaviors are desirable for ratiometric probes, whose self-calibration effect of two excitation/emission bands can eliminate the interference of photobleaching and deviated microenvironments, local probe concentration, and experimental parameters. These probes allow quantitative determination of target metal ions in living cells, tissues, and animals (Figure 6.22). Therefore, the PCT mechanism is an efficient strategy for the construction of ratiometric metal ion probes.

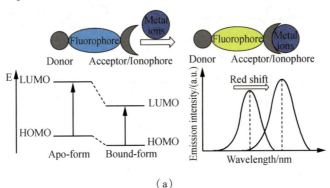

(a)

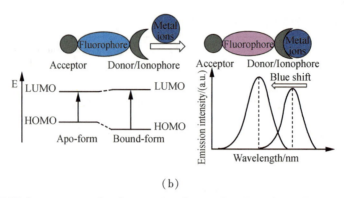

(b)

Figure 6.22 PCT fluorescent probes for metal cations and their ratiometric sensing mechanisms

However, many probes designed following this approach did not work or function as PCT probes, since the donor/acceptor modification may integrate electron donating atoms such as N or O into the ionophore, and the MCHEF effect will compete with the PCT response in the sensing system. If the PET process is dominant, the minor excitation/emission shift induced by metal coordination would be covered by the emission enhancement and the probe functions as a PET probe. If the PCT process induced by metal coordination is obvious and the intensity change is small, the probes are ratiometric (Figure 6.23). In addition, the PCT effect from donor to acceptor may decrease the metal coordination ability of the donor-derived ionophore, resulting in the photodisruption of metal coordination in the excited state and the absence of emission/excitation shift. Therefore, the design of ratiometric probes based on the PCT effect is complicated and challenging.

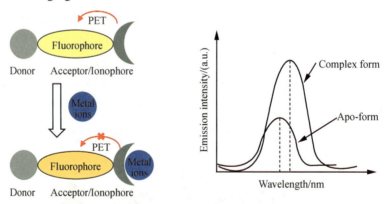

Figure 6.23 Competition of PET and PCT effects in fluorescent probes for metal cations

Many fluoroionophores contain an azacrown whose nitrogen atom is conjugated to an electron-withdrawing group (Figure 6.24). Compounds **6.1 – 6.7** exhibit a common feature: The blue shift of the absorption spectrum is much larger than that of the emission spectrum on cation binding. Similar photophysical effects were observed with compound **6.8** consisting of benzothiazyl group linked to polythiaazaalkane as a complexing unit in order to promote complexation of Ag^+. This tiny shift in the fluorescence spectrum—surprising at first glance—

can be explained as follows. The PCT reduces the electron density on the nitrogen atom of the crown, and this nitrogen atom becomes a noncoordinating atom because it is positively polarized. Therefore, excitation induces a photo-disruption of the interaction between the cation and the nitrogen atom of the crown. Therefore, the fluorescence spectrum is little affected because most of the fluorescence is emitted by species whose interaction between cations and fluorophores no longer exists or is much weaker.

The absence of fluorescence of **6.9** may be due to the formation of a nonfluorescent NCT state, and an acridinium type fluorescence is recovered upon binding of H^+ and Ag^+. A similar explanation applies to the low fluorescence of the **6.10** molecule.

ICT in conjugated donor-acceptor molecules may be accompanied with internal rotation leading to twisted ICT (TICT) states. A dual fluorescence may be observed as in **6.11** (which resembles the well-known DMABN containing a dimethylamino group instead of the monoaza-IS-crown-5): the short-wavelength band corresponds to the fluorescence from the locally excited-state and the long-wavelength band arises from an NCT state. The fluorescence intensity of the latter decreases upon cation binding because the interaction between bound cations and crown nitrogen is not conducive to the formation of the NCT state, which leads to a concomitant increase of the short-wavelength band.

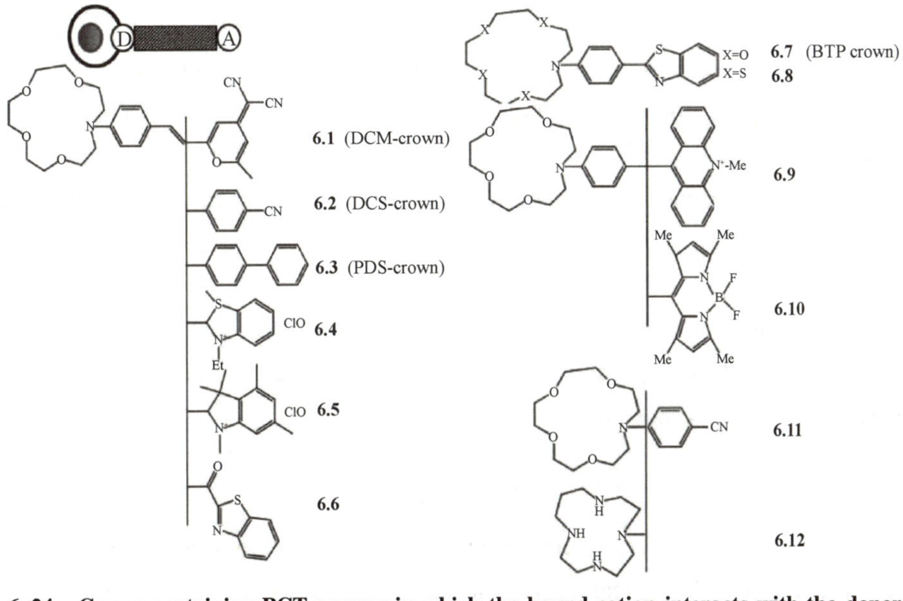

Figure 6.24 Crown-containing PCT sensors in which the bound cation interacts with the donor group

Compound **6.12** is an analog of **6.11** in which the monoazacrown has been replaced by a tetraazacrown (cyclam) in order to promote the complexation of transition metal ions. In this case, a triple fluorescence is observed: in addition to the emission from the locally excited and the NCT state, fluorescence is also emitted from an intramolecular exciplex (sandwich complex formed in the excited state thanks to the flexibility of cyclam). Such a triple fluorescence is

Fundamentals of Photophysics

solvent and pH-dependent and is perturbed by cation binding. And the relative changes of the bands depend on the nature of the cation.

6.3.3 Fluorescence Resonance Energy Transfer Mechanism

FRET is a non-radiative process in which the resonance energy transfers from an excited state of a donor fluorophore (D-F) to the ground state of an acceptor fluorophore (A-F) via a non-radiative "dipole-dipole coupling". The emission spectrum of D-F should have a certain overlap with the absorption spectrum of A-F, and a higher overlap will lead to a more effective FRET. Moreover, FRET processes are distance-dependent, and an effective FRET requires a distance between D-F and A-F from 10 Å to 100 Å (Figure 6.25).

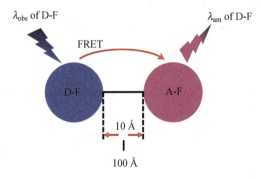

Figure 6.25 Schematic diagram of the FRET process

The extent of energy transfer is determined by the distance between the donor and acceptor and the extent of spectral overlap. For convenience, the spectral overlap (Figure 6.26) is described in terms of the Förster distance (R_0). The rate of energy transfer $k_T(r)$ is given by:

$$k_T(r) = \frac{1}{\tau_D}\left(\frac{R_0}{r}\right)^6 \quad (6\text{-}2)$$

where r is the distance between the donor (D) and acceptor (A), and τ_D is the lifetime of the donor in the absence of energy transfer. The efficiency of energy transfer for a single donor-acceptor pair at a fixed distance is:

$$E = \frac{R_0^6}{R_0^6 + r^t} \quad (6\text{-}3)$$

Hence the extent of transfer depends on the distance (r). Fortunately, the Förster distances are comparable in size to biological macromolecules: 30 Å to 60 Å. This is why energy transfer is used as a "spectral ruler" to measure the distance between protein sites. It should be noted that the value of R_0 for energy transfer should not be confused with the fundamental anisotropies (r_0).

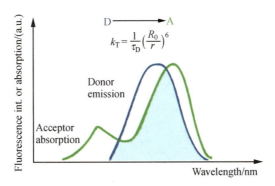

Figure 6.26 Spectral overlap for FRET

FRET involves long-range coupling of dipoles, resulting in an exchange of excitation energy through space without direct orbital overlap (Figure 6.27). The key factor for this mechanism is the overlap of the emission spectrum of the donor and the absorption spectrum of the acceptor. When the distance of donor and acceptor in space changes in the event of analyte binding, it changes the efficiency of FRET.

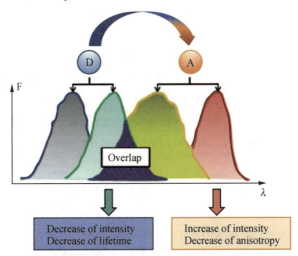

Figure 6.27 Absorption and fluorescence emission spectra of two FRET partners

The light-absorbing and emitting at shorter wavelengths fluorophore (donor, D) can transfer its excitation energy to another fluorophore (acceptor, A) absorbing and emitting at longer wavelengths.

The main difference between the energy transfer mechanisms of Förster and Dexter is that they depend on the distance between the donor and recipient sites. The rate of Dexter energy transfer is proportional to $\exp(-R_{DA})$, and the rate of FRET is proportional to $1/R_{DA}^6$ (Figure 6.28). Here R_{DA} denotes the distance between the donor and acceptor. The Förster mechanism is more likely to occur at extremely short and extremely long distances.

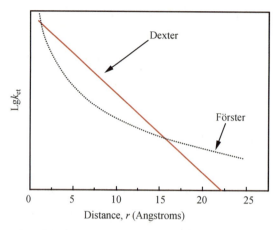

Figure 6.28 Plot of log(k_{ET}) versus distance, r, for both Dexter (solid red line) and Förster (dotted line) energy transfer mechanism, absent any criteria other than distance (k_{ET} denotes the rate of energy transfer.)

The FRET mechanism is commonly exploited for the design of ratiometric fluorescent probes with large pseudo-Stokes shifts. Three classes of FRET-based metal ion probes have been frequently reported. The first is developed by linking a non-emissive moiety such as rhodamine spirolactam derivatives to a donating fluorophore. FRET can be unlocked by binding metal ions to convert non-emitting receptors into fluorophores [Figure 6.29 (a)]. For instance, compound **6.13** (Figure 6.30) is a ratiometric fluorescent Cr^{3+} probe based on FRET, in which 1,8-naphthalimide and rhodamine derivatives were selected as the D-F and A-F, respectively. There is no energy transfer between the two because of the non-emissive nature of rhodamine spirolactam derivatives.

An efficient ring-opening reaction induced by Cr^{3+} generates fluorescent rhodamine, which induces an effective FRET process from naphthalimide to the switch-on rhodamine. In this case, the probe normally functions as an emission ratiometric probe. If the metal-induced switch on is devised for D-F, then the probe is normally an excitation ratiometric one. The sensing selectivity is clearly derived from the selective response of the fluorophore precursor. Tuning the overlap between the D-F emission spectrum and A-F absorption spectrum via metal coordination could also lead to a ratiometric response to metal cations. The ionophore of these types of probes is often conjugated directly with D-F or A-F [Figure 6.29 (b)]. In this case, the coordination of the metal with the ion carrier changes the absorption/emission spectra of either D-F or A-F. An exact example following this design rationale is probe **6.14** (Figure 6.30), which is composed of integrating 5-(4-methoxystyryl)-50-methyl-2, 20-bipyridine (bpy) with diamino-substituted naphthalimide (NDI) as D-F and A-F, respectively. It displays a specific Zn^{2+}-induced fluorescence enhancement via FRET. It was proposed that there should be a very small overlap between the emission spectrum of D-F and the absorption spectrum of A-F in the apo-form, and Zn^{2+} coordination makes its emission band undergo a red shift which enables a significant spectral overlap between the emission of the bpy/

Zn^{2+} complex and the absorption band of NDI. This facilitates the occurrence of FRET. Although only the emission enhancement is reported, a ratiometric response can be expected. The colorimetric or fluorescent response of the donor or acceptor fluorophore to certain metal cations determines the selective ratiometric response of these FRET probes.

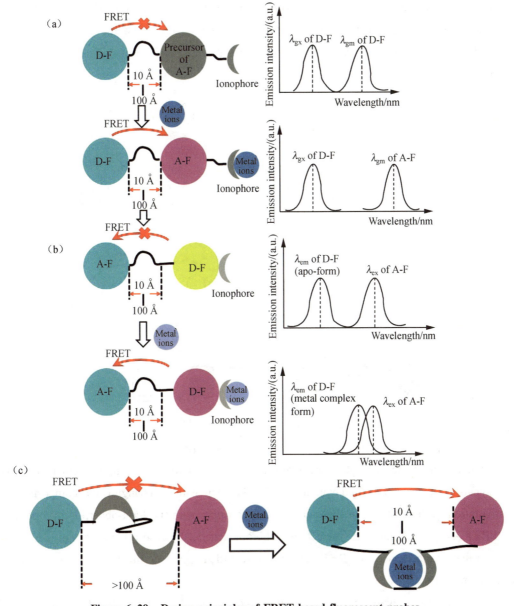

Figure 6.29 Design principles of FRET-based fluorescent probes

The third class of FRET probes for metal ions functions mainly by tuning the dipole-dipole distance between D-F and A-F via metal coordination. For these probes, a spacer acting as an ionophore was demanded. In the apo-form, the distance between D-F and A-F is longer than 100 Å, and FRET is switched off. With the addition of metals, metal coordination of the spacer alters the conformation of the spacer, resulting in a smaller distance between D-F and A-F in favor of FRET processes [Figure 6.29 (c)]. To date, most of the probes of this type are protein-based. For example, He and co-workers reported an Amt1-based Cu^+ fluorescent probe, called Amt1-FRET (**6.15**, Figure 6.30), which consists of a Cu^+-binding domain of Amt1 (residues 36–110) between a cyan fluorescent protein (CFP) and a yellow fluorescent protein (YFP). The coordination of Cu^+ with Amt1 changed the conformation of Amt1 and resulted in effective FRET from CFP to YFP. In this case, the selectivity of the spacer to the metal cation determines the selectivity of the probe, and elaborative design of this ionophore spacer is essential.

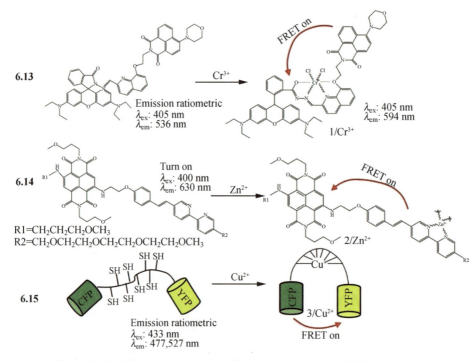

Figure 6.30 Fluorescent probes 6.13–6.15 based on a FRET mechanism

In some sensing technologies, especially in cellular imaging, it can be important to compare two signals or images, with and without FRET, keeping without changing the composition and configuration in the system. In these cases, one can play with the light-absorption properties of the FRET acceptor. If the dye is selectively photobleached, the absorption spectrum is lost. This leads to the disappearance of FRET that allows recorded signal or image as the reference with only donor emission.

The other very elegant approach was suggested and got the name of "photochromic FRET". The so-called photochromic compounds are those that change their absorbance reversibly in response to illumination at appropriate wavelengths. They serve as FRET receptors, with the ability to perform a reversible conversion between two different structural forms with different absorption (and, in some cases, fluorescence) spectra. Thus, they offer a possibility of reversible switching the FRET effect between "ON" and "OFF" states without any chemical intervention, just by light (Figure 6.31).

The best compounds serving this purpose are probably spiropyrans. These molecules exist in closed spiro forms absorbing at wavelengths shorter than 400 nm. They undergo a light-driven molecular rearrangement to an open merocyanine form with absorbance at 500 – 700 nm. A family of such photo-switchable acceptors has been extended and includes small organic dyes, conjugated polymers, fluorescent proteins, nanoparticles and their composites.

The sensing technology benefits from these possibilities. The optical switch quenching phenomenon provides the light activation control of the excited state to control any subsequent energy or electron-transfer process. In imaging the living cells, the FRET switching on and off allows obtaining the necessary controls on the distribution of donor molecules and their fluorescence parameters. Such reversible cycles of photoconversion allow reversible transitions between two (donor and acceptor) emission colors.

The ability to undergo numerous cycles of FRET switching is extensively used in super-resolution microscopy. This allows modulating the number of spatially resolvable emitters located in different positions in image for achieving nanoscale resolution.

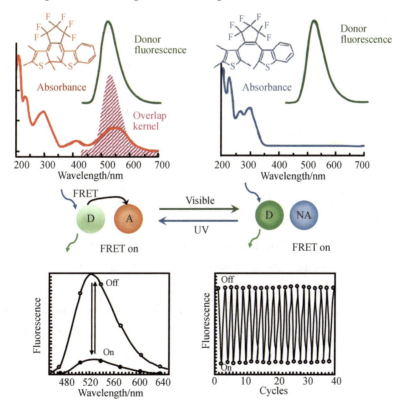

Figure 6.31 Photochromic FRET

The chemical structures depicted correspond to the photochromic dithienylethene in the colored closed form (left) and colorless open form (right). As we can see, the absorption spectrum of the latter overlaps well with the emission spectrum of the donor; the kernel of the overlap integral (striped) corresponds to the lucifer yellow dye as the donor. Ultraviolet light induces the photochromic transition to the closed-form (On), and visible (green) light reverses the process to the open form (Off).

6.3.4 Excimer Formation Mechanism

When a molecule absorbs light, its electronic properties change dramatically. Thus, it may participate in reactions that are not observable in the ground state. Particularly it can make a complex with the ground-state molecule like itself. These excited dimeric complexes are the excimers. The emission spectrum of excimer is very different from that of monemer; it is usually broad, shifted to longer wavelengths and does not contain vibrational structure. Particularly, this is the case of excimers of pyrene.

Excimer is a complex that weakly binds in the electronic ground state but strongly interacts in electronic excitation. Because they can be used as trap sites to limit exciton diffusion, they play an important role in the electronic relaxation of molecular aggregates and have a far-reaching impact on the functional performance of organic devices.

An excimer is formed by a fluorophore in the excited state with another fluorophore molecule in the ground state via weak interactions (e.g. π-π^* stacking). In chemical sensing, a number of molecular constructs have been suggested that employ excimer formation. The idea was to change the target binding the ground-state configuration of sensor molecules so that on excitation, they can form excimers but only in the presence of the target.

A typical metal ion probe functioning via excimer formation consists of two identical fluorophore moieties spaced by a flexible spacer which is also an ionophore. After coordinating with metal ions, the changed spacer makes the two fluorophore groups close (in van der Waals contact), resulting in effective weak interaction between the two fluorophores. In this way, the electronic excitation of one fluorophore causes an enhanced interaction with its neighbor, leading to the formation of an excimer. The excimer typically provides a redshifted and broad emission band compared to the monomer. In most cases, emission bands of the monomer and excimer can be observed simultaneously. Therefore, the metal coordination to the spacer ionophore can effectively alter the ratio between monomer emission and excimer emission so that the formation of excimer becomes a metal cation ratio probe (Figure 6.32). Several probes based on this mechanism have been reported by Kim's group for the detection of Cu^{2+}.

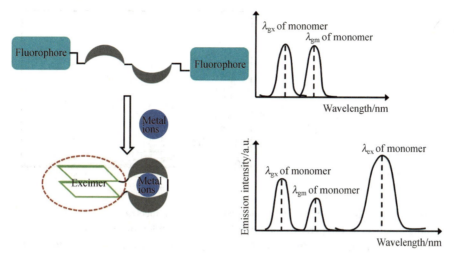

Figure 6.32 Excimer formation-based fluorescent sensing

These probes generally require highly p-delocalized planar systems such as pyrene as the monomer. Therefore, they normally display poor aqueous solubility. Moreover, the excimer formation is highly dependent on the distance resulting from metal coordination, which is difficult to predict. The two factors limit the practical application of this rationale to the design of ratiometric probes.

For obtaining the sensor for the sodium ions, a pair of dioxyanthracene molecules was included into a polyether ring. On their binding, the configuration of the molecule changes, and the excimer emission starts to be observed. Similar ideas were realized in the construction of sensors for other cations. A silver ion ratio fluorescence sensor based on pyrene functionalized heterocyclic receptor is proposed. It forms an intramolecular sandwich complex via self-assembly induced by silver ions. This results in a dramatic increase in fluorescence intensity of the excimer and a dramatic decrease in monomer fluorescence (Figure 6.33). The intensity ratio of excimer and monomer emissions (at 462 nm and 378 nm) is an ideal measure of Ag^+ ion concentrations. The application of molecular recognition constructs based on excimers is also popular in sensing neutral molecules, such as glucose.

Commonly, two pyrenyl groups are attached to the macromolecular sensor or label two molecules assembled on a sensor unit, but multiple binding can also be realized. Excimer is added into the cyclodextrin cavity to form reporting dye, which improves the performance of excimer sensor. Excimer-forming pyrene-conjugated oligonucleotides can be used to detect DNA and RNA sequences. Pyrene excimer probes can be applied for determination of nucleic acids in complex biological fluids.

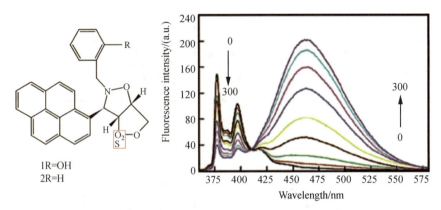

Figure 6.33 The structure and fluorescence emission spectra of pyrene derivative (1) forming excimers in the presence of silver ions [The addition of these ions in concentrations 0 μmol · L^{-1}, 4.0 μmol · L^{-1}, 10 μmol · L^{-1}, 15 μmol · L^{-1}, 20 μmol · L^{-1}, 40 μmol · L^{-1}, 75 μmol · L^{-1}, 150 μmol · L^{-1}, and 300 μmol · L^{-1} results in a decrease of intensity of the normal emission with typical bands at 378 nm and 397 nm (excitation 344 nm) and the appearance of a redshifted structureless maximum centered around 462 nm, typical of pyrene excimer.]

6.3.5 Configuration Changes or Conformation Transition Mechanism

The conformational change during the process of phosphorylation promotes the occurrence of FRET. Figure 6.34 displays a typical example. CFP and YFP are different-colored mutants of GFP derived from Aequorea victoria with mammalian codons and additional mutations. Upon phosphorylation of the substrate domain within phocus by a protein kinase, the adjacent phosphorylation recognition domain binds with the phosphorylated substrate domain, which changes the efficiency of FRET between the GFP mutants within phocus.

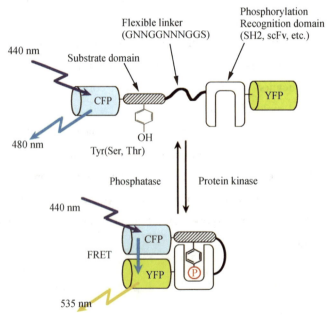

Figure 6.34 Principle of phocus for visualizing protein phosphorylation

As shown in Figure 6.35, on the left side of the compound is a kind of water-soluble salt, but the two pyrene rings are close to each other due to the strong hydrophobic interaction, so it is easy to see excimer emission of pyrene. It can be used to identify the adenosine phosphate anion. When adenosine phosphate is introduced, the excimer of pyrene is destroyed due to the insert of bases, indicating the disappearance of excimer fluorescence emission. This can be used to check whether the species have been captured or recognized. The molecular configuration change is the cause of such information.

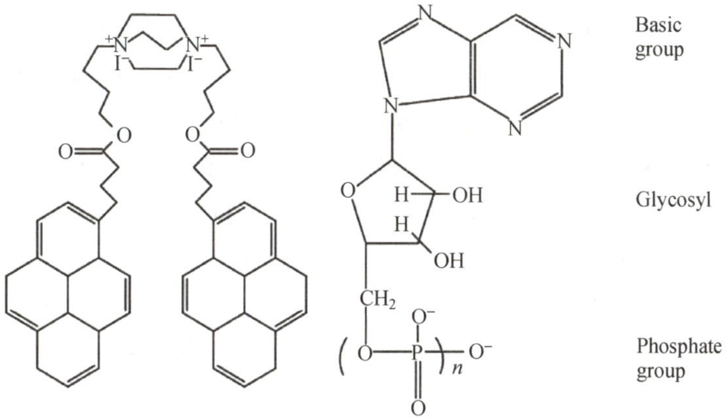

Figure 6.35 Configuration or conformation transition of adenosine monophosphate

The use of color changes upon the complexation of analytes is probably the simplest way of carrying out a chemical analysis because the color is immediately perceived by the naked eye. In spite of the use of many other much more sophisticated and precise alternatives, in some cases, the change of color is still a helpful tool. In the case of DNA testing, Figure 6.36 shows the

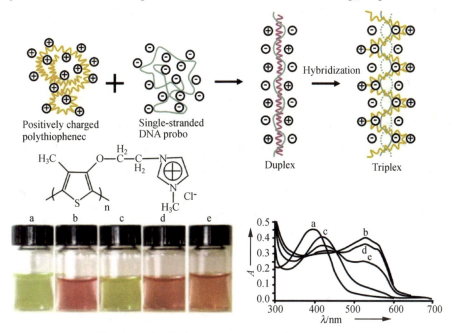

Figure 6.36 DNA detection by colorimetric method

detection by the colorimetric method. The detection of DNA hybridization is worked using the conformation transition of polythiophene, which is caused by the electrostatic interaction of water-soluble poly-thiophene and double-stranded DNA.

6.3.6 Excited-state Intramolecular Proton Transfer Mechanism

Very efficient could be the response when the same molecule contains both proton-donor and proton-acceptor groups in close proximity and connected by H-bond. Such a reaction is called the excited-state intramolecular proton transfer (ESIPT). It does not require a protic environment and may occur in any solvent, solid matrix, and even vacuum. The strict requirement is only on the structure of the molecule exhibiting ESIPT. The donor group of the proton is almost always a hydroxyl group, and the basic acceptor of a proton must be oxygen in the form of a heterocyclic nitrogen atom or a carbonyl group. The two groups, hydrogen bonding in the ground state, form a pathway for proton transfer in the excited state. The excitation leads to dramatic redistribution of electronic density in the fluorophore so that the proton donors become stronger donors, and acceptors become stronger acceptors. Hence appears the driving force for ESIPT.

In many systems, ESIPT is very fast and irreversible on the fluorescence lifetime scale, and therefore, it produces a single ESIPT band in the fluorescence spectrum, which lacks the switching ability needed for sensing. However, there are a number of dyes with a very attractive property of producing two bands in emission, one belonging to the initially excited LE state and the other to the product of ESIPT reaction. The dye was found in hydroxyphenyl benzothiazole, benzoxazole, and benzimidazole derivatives. Their two-band switching, dependent on pH and binding of ions, can potentially be explored in sensing.

ESIPT is a unique four-level photochemical process, with the electronic ground state of ESIPT fluorophores typically existing in an enol (E) form. Upon photoexcitation, the electronic charge of such molecules can be redistributed, resulting in greater acidity for the hydrogen bond donor group and increased basicity for the hydrogen bond acceptor within the E form. As a result, an extremely fast enol to keto phototautomerization ($k_{ESIPT} > 10^{12}$ s^{-1}) event takes place, with the excited state enol form (E^*) rapidly converting to its excited keto form (K^*). After decaying radiatively back to its electronic ground state, a reverse proton transfer (RPT) takes place to produce the original E form (Figure 6.37).

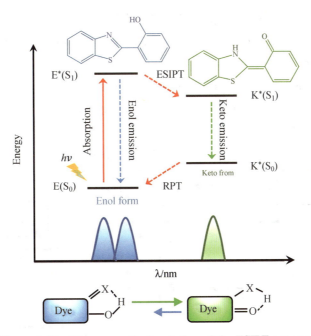

Figure 6.37 Diagrammatic description of the ESIPT process

Because of this rapid four-level photochemical process, ESIPT fluorophores are characterized by several particularly attractive features that make them useful as fluorescent probes and imaging agents. For instance, ESIPT fluorophores have an unusually large Stokes shift (about 200 nm) compared to traditional fluorophores (fluorescein, rhodamine, etc.). This helps avoid unwanted self-reabsorption and inner-filter effects. In addition, due to the transient characteristics of four-level photochemical process, the emission of ESIPT is very sensitive to its local environment. Consequently, the presence of polar and hydrogen bond donating solvents can result in inhibition of the ESIPT process, with the keto (K^*) emission often not being observed. Therefore, many ESIPT fluorescent probes require a large ratio of organic solvent or cetyltrimethylammonium bromide (CTAB), to create a sufficiently hydrophobic micellar environment to produce a fluorescence response. Due to the importance of the environment in ESIPT fluorescence, we have reported the measurement conditions in the text and for all figures and schemes. Most ESIPT fluorophores can be used for ratiometric sensing because they exhibit dual-emission spectra arising generated by the emission of excited enols (E^*) and excited ketones (K^*). This is particularly useful for fluorescence detection of biologically and/or environmentally important species because ratiometric probes provide direct information about the concentration of the target analyte without the need for calibration.

Fundamentals of Photophysics

The general strategy for developing ESIPT-based fluorescent probes is based on a design that involves blocking the hydrogen bond donor of the ESIPT fluorophore with a specific reactive unit that prevents the ESIPT process. As a result, since no exchangeable protons are available, only enol emission has been observed. However, exposure of a reactive unit in the probe to a specific analyte allows access to the keto, resulting in the ESIPT process being turned on (Figure 6.38).

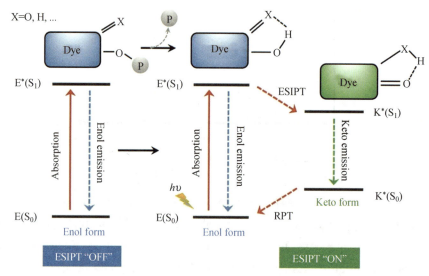

Figure 6.38　Diagrammatic representation of the most common strategy underlying the design of ESIPT-based fluorescent probes

An interesting and exciting development has been the combination of ESIPT with other fluorescence mechanisms. The systems produced as the result of this intellectual and chemical linking generally combine the advantageous attributes of these two mechanisms and overcome many of the limitations of ESIPT. Overall, ESIPT is an important sensing mechanism for the detection of biologically and/or environmentally important species.

6.4　Report

The chemical sensor should be a unit that can perform two important functions: providing the target (analyte) binding and reporting on this binding by generating an informative signal. Coupling these functions requires a functional linker or linking mechanism called transduction (Figure 6.39). The structure that is responsible for the generation of this signal is called a reporter. This chapter will discuss a variety of reporters and the mechanisms in the background of their operation, focusing on fluorescence reporting.

The report makes the information pass out by changing the fluorescence intensity (or other photophysical behavior) after getting the signal from the relay. The majority of fluorescent chemical sensors accomplish the report task mainly in the form of light emission, including fluorescence enhancement or fluorescence quenching. In this process, the most important thing is to synthesize new luminescent species and clarify the mechanism of their luminescence change.

Generally speaking, the light-emitting compounds used in the report are polycyclic aromatic compounds, intramolecular conjugated charge transfer compounds, and conjugated polymer. These typical compounds will be introduced in detail as follows.

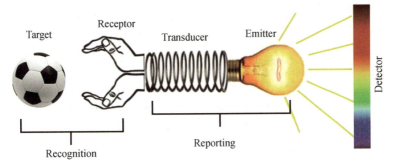

Figure 6.39 Basic functional elements of fluorescent chemical sensor and biosensor

6.4.1 Polycyclic Aromatic Compounds

As their name indicates, polycyclic aromatic hydrocarbons are aromatic hydrocarbons which contain more than one benzenoid (i.e. benzene-like) ring. The chemical group of polycyclic aromatic compounds, including the better-known subgroup of polycyclic aromatic hydrocarbons and the heterocyclic aromatic compounds, comprise several thousand individual compounds.

Hückel's "$4n + 2$" rule for aromaticity does not only apply to mono-cyclic compounds. Benzene rings may be joined together (fused) to give larger polycyclic aromatic compounds. A few examples are drawn in Figure 6.40, together with the approved numbering scheme for substituted derivatives. Among them, the peripheral carbon atoms (numbered in all but the last three examples) are all bonded to hydrogen atoms. Unlike benzene, all the C-C bond lengths in these fused ring aromatics are not the same, and there is some localization of the pi-electrons. The six benzene rings in coronene are fused in a planar ring; whereas the six rings in hexahelicene are not joined in a larger ring but assume a helical turn due to the crowding together of the terminal ring atoms (in the structure below, note that the top right and center-right rings are not attached to one another). This helical configuration renders the hexahelicene molecule chiral, and it has been resolved into stable enantiomers.

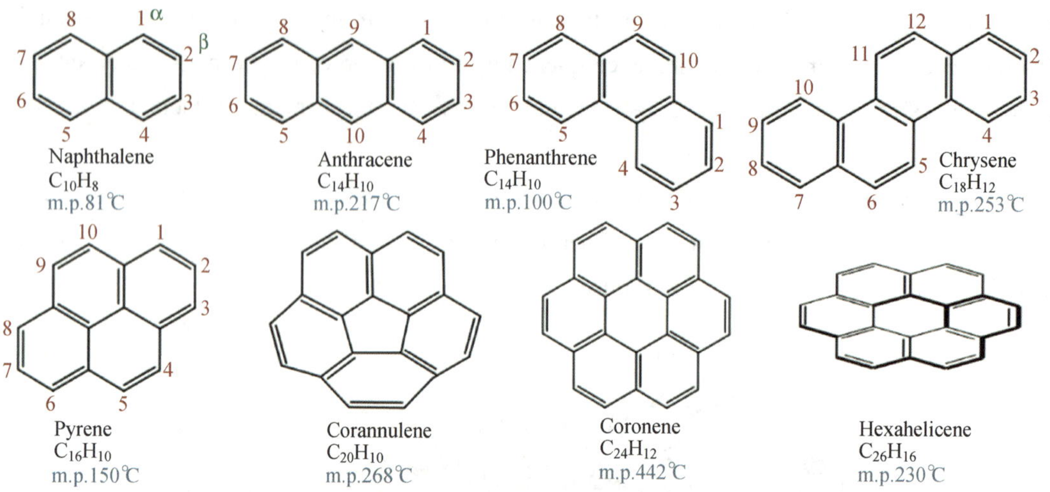

Fundamentals of Photophysics

Figure 6.40 Examples of polycyclic aromatic hydrocarbons

- Naphthalene $C_{10}H_8$ m.p.81℃
- Anthracene $C_{14}H_{10}$ m.p.217℃
- Phenanthrene $C_{14}H_{10}$ m.p.100℃
- Chrysene $C_{18}H_{12}$ m.p.253℃
- Pyrene $C_{16}H_{10}$ m.p.150℃
- Corannulene $C_{20}H_{10}$ m.p.268℃
- Coronene $C_{24}H_{12}$ m.p.442℃
- Hexahelicene $C_{26}H_{16}$ m.p.230℃

A majority of polycyclic aromatic hydrocarbons tend to assemble in an edge-to-face fashion in the crystalline state. Face-to-face contact is generally considered better for organic electronics as there is the possibility for increased π-π overlap between the faces of adjacent molecules. This would allow a significant π-overlap between molecules to facilitate the carrier pathway. This is not to say, however, that edge-to-face interactions have no π-π overlap, as it can be seen in the crystal packing found in pentacene, which has been shown to have excellent device properties.

The polycyclic aromatic compound is a non-conjugated system, which shall have the light-emitting part, as well as the electron donor (such as N, S, and O atoms on the lone pair electrons) and the acceptor. For instance, the intramolecular PET system causes fluorescence quenching (Figure 6.41).

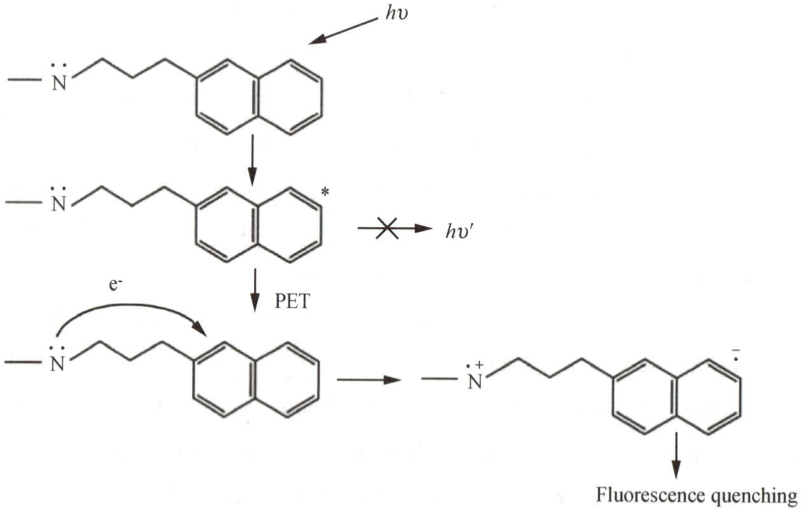

Figure 6.41 Fluorescence quenching caused by the intramolecular PET system

Recognition of tested species is often designed to interfere with the PET process. Several ways like protonation, coordination of metal cation, and formation of hydrogen bonds can stabilize the electrons at the electron-donating groups. Therefore, it can reduce the driving force of the PET process, making the fluorescence quenching process difficult. Finally, the excited aromatic groups will resume fluorescence or have fluorescence enhancement (Figure 6.42).

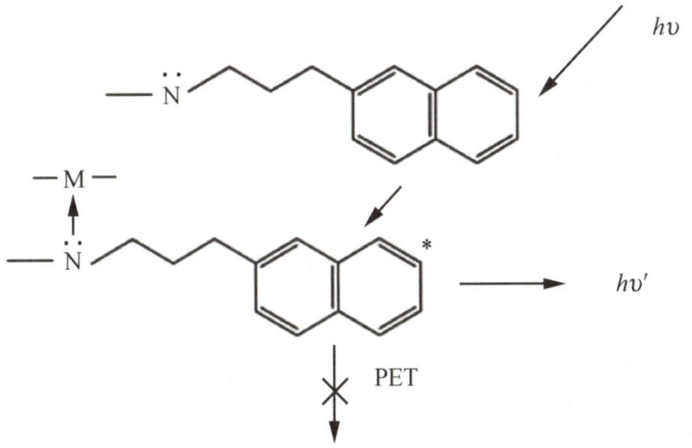

Figure 6.42 The excited aromatic groups which will resume fluorescence or have fluorescence enhancement

Calix[n]arenes are a class of polycyclic compounds that can allow (by proper chemical substitutions) achieving high-affinity binding of many small molecules. Calix[n]arenes are cyclic oligomers composed of $n = 4-6$ phenyl groups connected by methylene bridges that are obtained by phenol-formaldehyde condensation. They exist in a cup-like shape with a defined upper and lower rim and a central annulus. Consequently, their rigid conformation of skeleton forming a cavity together with some flexibility of side groups (allowing their flip-flop) enables them to act as host molecules for different small molecules and ions. Their basic structure resembling a vase (calix = vase) together with modifications with short peptides in its upper rim is presented in Figure 6.43. As can be seen, this beautiful structure resembles a basket of flowers.

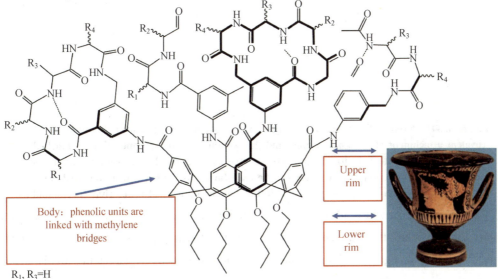

$R_1, R_3 = H$
$R_2, R_4 = CH_2COOH$

Figure 6.43 Calix[4]arene derivatives that mimic the properties of antibodies to recognize protein targets [R_1-R_2-R_3-R_4 could be H, CH_2COOH; $(CH_2)_4NH_2$. This arrangement provides the recognition pattern of negative and positive charges.]

Fundamentals of Photophysics

Based on the calixarene scaffold, a wide variety of structures can be realized by modified cations of both upper and lower rims. Examples of such derivatives are presented in Figure 6.44. Typically, various derivatives with differing selectivity for various guest molecules can be prepared by functionally modifying these sites. Mostly, they are small molecules and ions, but with properly modified cations, they can target the structural elements of large molecules, such as proteins. At these sites, fluorescent labeling can be achieved.

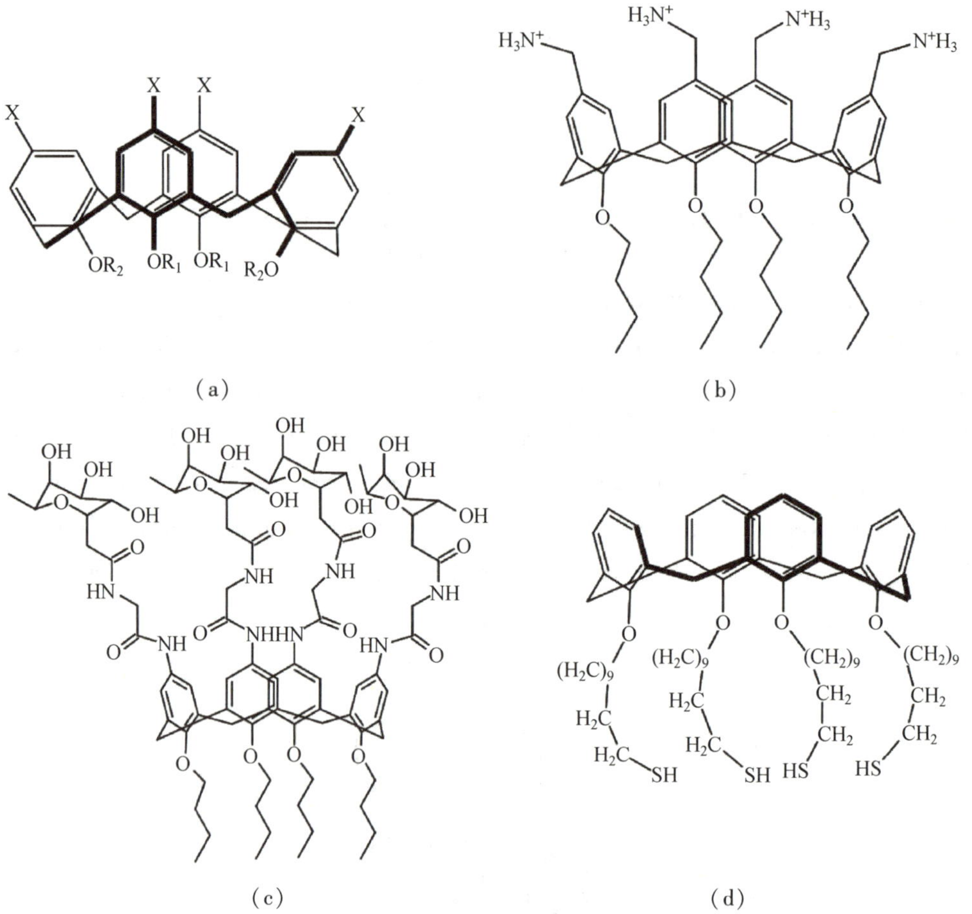

Figure 6.44 Parent calix[4]arene in a cone conformation. (a) The sites of substitutions with covalent attachment of recognition units and fluorescent dyes in the upper rim are marked X. In the lower rim, the substitution sites are denoted as R1 and R2. (b) Calix[4]arenes with amino groups at the upper rim. They can be used for convenient chemical-modified cations. (c) (d) Typical calix[4] arenes with substitutions at the upper and lower rims correspondingly

Calix[4]arene structure allows many possibilities for targeting proteins. Their upper rim displays four positions that can be used for covalent attachments. Moreover, the nature and symmetry of such recognition elements may be varied to selectively target the highly irregular surfaces comprised of charged, polar, and hydrophobic sites.

A series of synthetic receptors are prepared, in which four peptide loop domains are attached to a central calix[4]arene scaffold. First, each peptide loop is based on a cyclic hexapeptide, in which two residues have been replaced by a 3-aminomethyl benzoate dipeptide mimetic, which also contains a 5-amino substituent for anchoring the peptide to the scaffold. Then, through the attachment of various peptide loops, the authors made a series of calix[4] arenes expressing negatively and positively charged regions as well as hydrophobic regions and thus achieved binding to the complementary regions of several proteins. Finally, they demonstrated the binding to platelet-derived growth factor (PDGF) and inhibition of its interaction with its cell surface receptor PDGFR.

Fluorescence reporting on the binding of different targets to calix[4] arene structures is commonly introduced in two ways.

(1) Using the property of these molecules to bind different dyes at the site within the cavity, where there occurs the binding of many targets. Then the binding of the target molecule can displace the fluorescent dye. When the dye dissociates, it changes its fluorescence. The chelating properties of calix[4] arenes towards different dyes were reported by many researchers. In all these cases, the formation of the complex leads to dramatic fluorescence quenching, and this result suggests a convenient possibility of converting the receptor into a highly efficient sensor. This suggests using calix[4] arene derivatives in competitor substitution assays. Ample possibilities for modulating the target affinities by covalent modifications and a broad choice of responsive dyes with different affinities promise easy development of simple and efficient assays for many targets.

(2) Covalent conjugation of calix[4] arene side groups with fluorescent dyes. In this case, several mechanisms of fluorescence response can be realized. One of them can be easily applied in ion sensing and is based on induction/perturbation of photoinduced ICT. For instance, the tert-butylcalix[4] arene was synthesized either with one appended fluorophore and three ester groups or four appended fluorescent reporters. The dyes were 6-acyl-2-methoxynaphthalene derivatives, which contained an electron-donating substituent (methoxy group) conjugated to an electron-withdrawing substituent (carbonyl group). This is a typical arrangement for fluorescent reporters operating on the basis of the ICT principle.

6.4.2 Intramolecular Conjugated Charge Transfer Compounds

Intramolecular conjugated charge transfer compound is a conjugated system, it contains intramolecular exist "electron push" and "electron pull" two parts (Figure 6.45).

Fundamentals of Photophysics

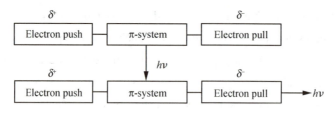

Figure 6.45 "Electron push" and "electron pull" parts

ICT is also an electron transfer in principle. However, the difference is that ICT occurs within the same electronic system or between the systems with high level of electronic conjugation between the partners, and it has its own characteristic features (Figure 6.46). The electronic states achieved in this reaction are not the "charge-separated" but "charge-polarized" states. Accordingly, they are still localized states with distinct energy minima.

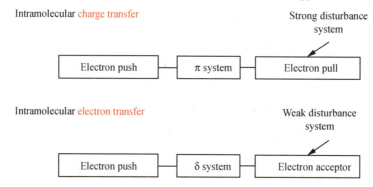

Figure 6.46 Comparison between ICT and intramolecular electron transfer

The two states of PET and ICT are easily distinguished by their absorption and emission spectra. In PET, the strong quenching occurs without spectral shifts. In contrast, the ICT states are often fluorescent but exhibit changes in intensities. In addition, their excitation and emission spectra may exhibit significant shifts that depend on the environment. This allows for providing the wavelength-ratio metric recording. In some cases, switching of intensity between two emission bands, normal (often called the locally excited, LE) and achieved in ICT reaction, can occur. This is even more attractive for ratio metric measurement.

Commonly, the ICT states are observed when the organic dye contains an electron-donating group (often a dialkylamino group) and an electron-withdrawing group (often carbonyl). If these groups are located on opposite sides of the molecule, electronic polarization is induced. Because an electron donor becomes in the excited state a stronger donor and an acceptor a stronger acceptor, an electronic polarization can be substantially increased in the excited state. Therefore, the created large dipole moment interacts with medium dipoles resulting in strong Stokes shifts.

The change in the medium conditions can produce the LE-ICT switching. Fluorescence from the normal LE state is commonly observed in low-polar solvents and cryogenic conditions, where the spectra may contain residuals of vibrational structure. As polarity and temperature increase, the ICT fluorescence emerges, with a broad and structureless long-wavelength shifted fluorescence band. Moreover, these shifts become larger because the solute-solvent dipole interactions are stronger in polar solvents. That is why the so-called "polarity probes" are, in fact, the ICT dyes. In addition, the ICT states are very sensitive to electric field effects and, therefore, to the presence of nearby charges.

This discussion clearly shows that the realization of ICT offers many possibilities. However, the application of this effect cannot be as broad as that of PET, and only organic dyes (and not all of them) can generate an efficient ICT emission. Nevertheless, within this family of dyes, one may find many possibilities for producing fluorescence reporter signals in sensing.

(1) Switching between LE and ICT emissions by the direct influence of target charge (Figure 6.47). This switching can be produced by binding an ion to either an electron donor or acceptor site. The processes occurring in the interaction of the ion-chelation group (as a receptor) with a cation are well-described. When an electron-donor group is attached to the receptor, the cation reduces the electron-donating character of this group. Hence, owing to the resulting reduction of electronic conjugation, a blue shift occurs in the absorption spectra together with the decrease in molar extinction. On the contrary, a cation interacting with the acceptor group enhances the electron-withdrawing character of this group; the absorption spectrum is thus redshifted, and the molar extinction is increased. The fluorescence spectra are, in principle, shifted in the same direction as those of absorption spectra.

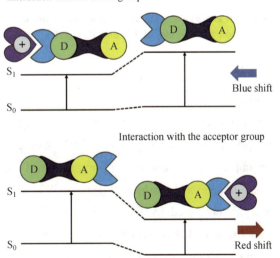

Figure 6.47 Spectral shifts of ICT sensors resulting from the interaction of a bound cation (+) with an electron-donating (D) or electron-withdrawing (A) group

(2) Coupling of ICT emission with dynamic variables. The ICT states can be stabilized and modulated by the dynamics of surrounding molecules and groups of atoms. Typically, the ICT state possesses a strongly increased dipole moment, and in the absence of strong interactions with surrounding dipoles, its energy may lie higher than that of the LE state. For this reason, the LE emission at shorter wavelengths will be observed in low-polar or highly viscous environments, where the strong dipole interactions are absent or cannot form by dipole rotations at the time of emission. In contrast, in highly polar environments, the surrounding dipoles tend to reorganize to their equilibrium configuration, stabilizing the ICT state. This makes the ICT emission to be energetically favorable and therefore intensive. Because of stronger interaction, a decrease of excited-state energy occurs, and the emission is shifted to longer wavelengths. Therefore, efficient transduction with the detectable change of reporter spectrum can be provided by the change in dynamics of molecules and groups of atoms at its location (Figure 6.48).

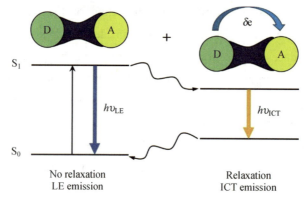

Figure 6.48 The simplified energy diagram showing the influence of molecular relaxations on the energies of LE and ICT states

Other aspects involve:
(1) Increased environment sensitivity of ICT emission.
(2) Quenching ICT states.
(3) Modulation of electronic conjugation within the dye exhibiting ICT emission.

The explained general principles of efficient usage of ICT mechanism in sensing above are best demonstrated in the example of designing the Zn^{2+} ion sensor of a new generation. The two compounds, whose formulas are presented below (Figure 6.49), one containing two protons (6.16) and the other one is substituted by fluorine atoms (2). Because these compounds generate strong shifts in absorption and fluorescence spectra on the zinc ion binding, this allows for precise ratiometric recording. Compound 6.17 demonstrates a 1∶1 ligand-Zn^{2+} binding mode and yields a dissociation constant of $K_d = 2.4$ μmol·L^{-1}. This compound is strongly asymmetric, which allows the ICT increase excitation, generating a strong dipole moment in the excited state. In this configuration, the metal ion binds to the acceptor rather than to the donor binding site so that the ion binding increases but does not decrease the charge-transfer character of the excited state together with substantial shifts of spectra to longer wavelengths. Instead of commonly observed quenching in polar media, such increase makes the Zn-bound state even more intensive in emission. This example demonstrates that the general principles are effective and can be applicable to a broad range of donor-acceptor fluorophores modified with tailored chelating sites for selective sensing of charged targets.

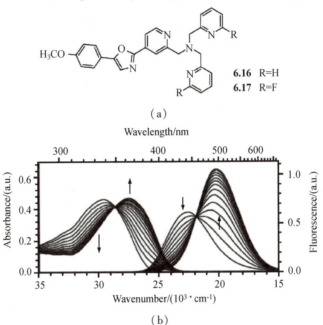

Figure 6.49 The ICT-based sensor for zinc ions: (a) the basic sensor molecule (1) and its fluorinated derivative (2); (b) absorption (left) and fluorescence emission (right) spectra for the titration of compound (2) with zinc ions in the micromolar concentration range in methanol

Another example of an intramolecular conjugated charge transfer compound is shown as follows. The acception of tested species in different positions of compounds can lead to the redistribution of intramolecular charge density and the charge transfer state-level change; these will be sensitively displayed on the fluorescence emission spectrum. As shown in Figure 6.50, the crown ether ring and ethylene polyamine in the compounds serve as the receptor, which can be used to identify the different metal ions.

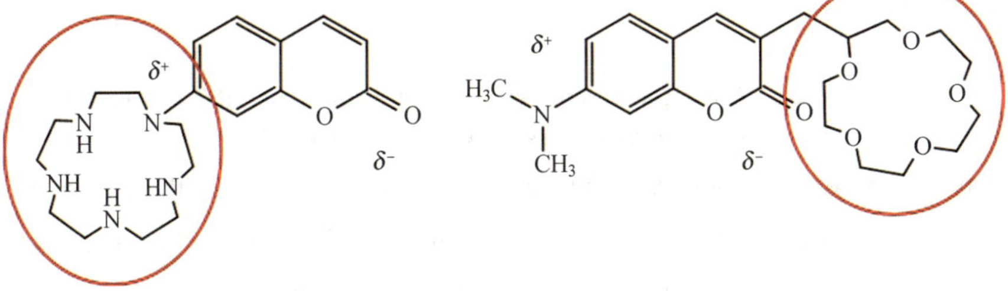

Figure 6.50 Crown ether ring and ethylene polyamine in the compounds

Fundamentals of Photophysics

6.4.3 Conjugated Polymers

Conjugated polymers are powerful fluorescent materials which are suitable for the applications of chemical sensors. Figure 6.51 shows the structures of some typical polymers that are applicable for the detection of analytes at low concentrations. These polymers include: poly (p-phenylene ethynylene, PPE), poly (p-phenylene vinylene, PPV), polyacetylene, and polyfluorene. Those polymers bearing ionizable pendant groups are water-soluble polyelectrolytes.

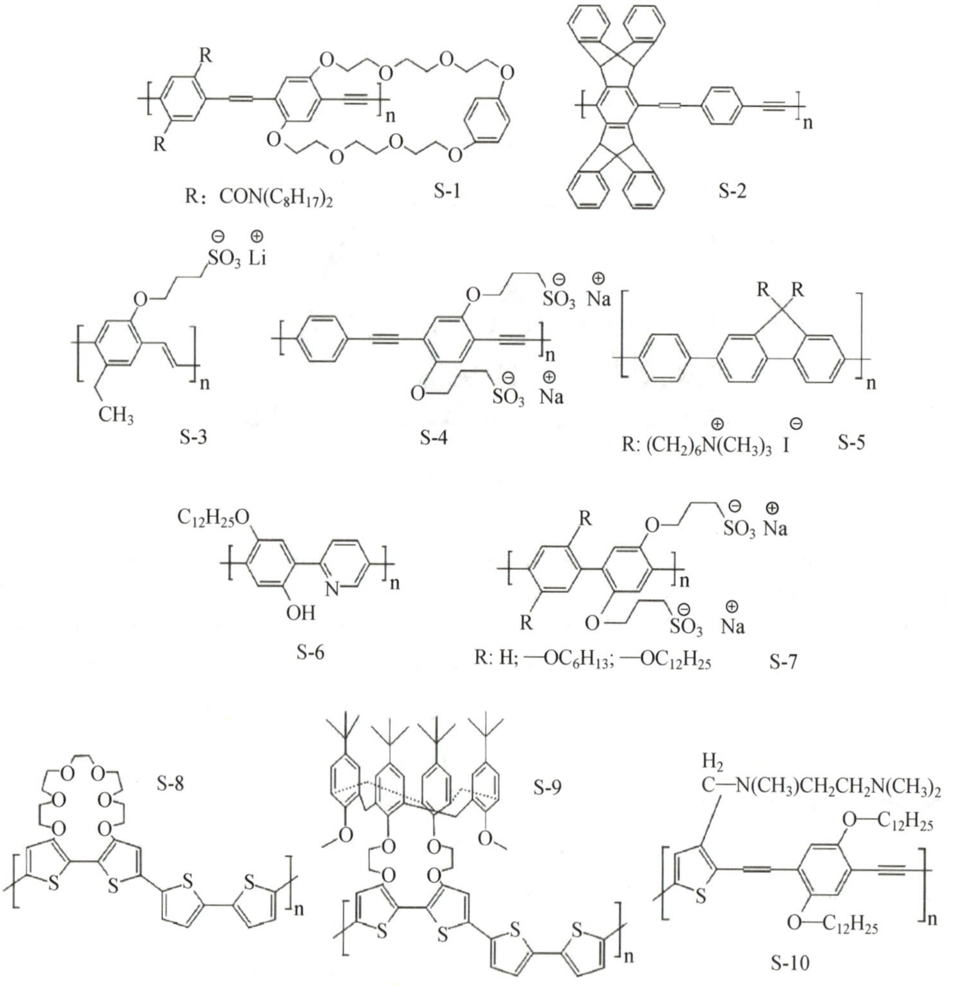

Figure 6.51 Chemical structures of typical conjugated polymers used as chemical sensors for organic compounds

The sensing ability of conjugated polymers depends on the fact that the noncovalent binding of extremely small amounts of analytes can quench their fluorescence (Figure 6.52). This phenomenon, known as super quenching, is due to the pronounced delocalization of excitons formed in conjugated polymers upon light absorption. Due to this delocalization, excitons can rapidly move along the polymer chain to quenching sites. This mode of action is called fluorescence turn-off sensing. On the other hand, fluorescence turn-on sensing is observed when an analyte is capable of selectively detaching a quencher previously non-covalently linked to the polymer.

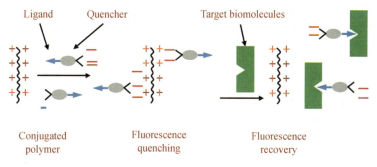

Figure 6.52　Conjugated polymer fluorescent sensors based on the electron transfer mechanism

Conjugated polymers used to interconnect receptor units results in greatly enhanced fluorescence-based chemosensory responses relative to single receptor sensory molecules. The origin of this effect is the facile energy migration, which is prone to occur throughout the polymer. A key feature is that the energy migration is a property of the collective system rather than a property of discrete units of the polymer.

Figure 6.53 shows a general schematic band diagram of the molecular wire enhancement methodology. Although the formation of pseudorotaxane involves the formation of charge-transfer complexes, the electron transfer quenching mechanism need not involve direct intermolecular interactions with the analyte. As a result, this mechanism affords considerable design flexibility. Analyte binding events that introduce local narrowing of the band structure may also be used to produce amplified responses.

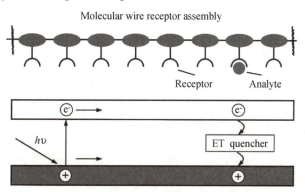

Figure 6.53　Schematic band diagram illustrating the mechanism by which the molecular wire receptor assembly can produce an enhancement in a fluorescence chemosensory response
[The horizontal dimension represents the position along the conjugated polymer shown schematically at the top. Excitations are created by the absorption of a photon ($h\nu$) and then migrate along the polymer backbone. Analyte binding produces a trapping site, whereby the excitation is effectively deactivated by electron transfer quenching.]

Polypyrrols are among the earliest examples of fluorescent conjugated polymer (FCP) sensors. Most common synthesis of this polymer is through oxidative polymerization. As early

Fundamentals of Photophysics

as the end of the 20th century, Chenthamarakshan and Ajayaghosh synthesized a pyrrole-derived oligosquaraine containing diethylene glycol monomethyl ether side chain, which is selective towards Li^+ with a 92% enhancement in fluorescence quantum yield with 40 μM of $LiClO_4$ (Figure 6.54).

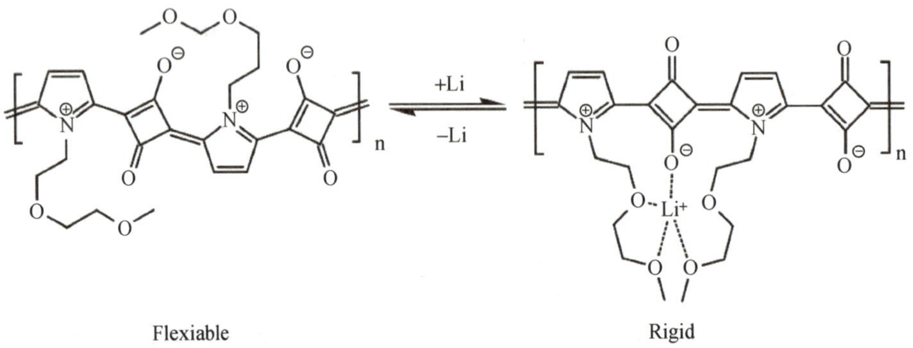

Figure 6.54 Shape changing upon addition of Li^+ of the pyrrole-derived oligosquaraine with diethylene glycol monomethyl ether side chain

PPV is a bright yellow fluorescent polymer. Its emission maxima at 551 nm and 520 nm are in the yellow-green region of the visible spectrum. Although the polymer itself is insoluble, modifications of the side chain have been developed to improve its solubility in organic solvent and aqueous solution for sensory applications. The bond angle of the ethylene group gives some flexibility to the polymer backbone; thus, sensing application through conformation change induced fluorescent change is possible.

Wang et al. presented a metal ion-sensitive polymer system which has been further exploited for years. In this approach, pseudo-conjugated, bipyridyl-phenylenevinylene-based polymers were prepared first; upon incorporating metal ions, such as Mn^{2+}, Zn^{2+}, Pd^{2+}, the polymer backbone would undergo a conformational change, extend the conjugation, so the photophysics of the polymer would change, in terms of the absorption and emission maximum red shift. However, this polymer system can only distinguish a group of metal cations but not a specific one, limiting its application (Figure 6.55).

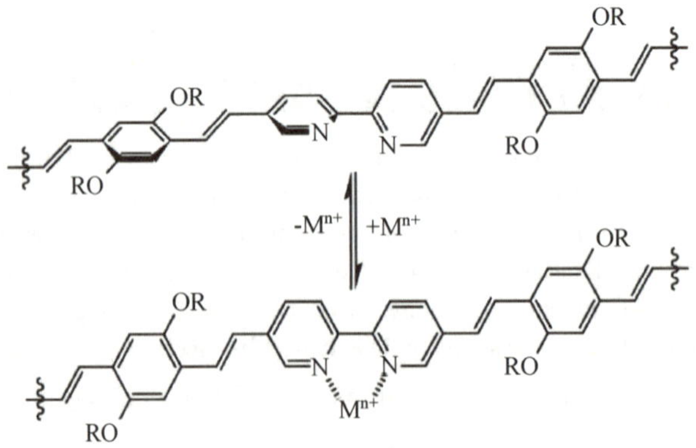

Figure 6.55 Poly(bipyridyl-phenylenevinylene) from Wang and Wasielewski

Polythiophenes (PTs) have been synthesized by electrochemical, chemical oxidation, and chemical coupling reactions. It has excellent properties, including low-cost synthesis, excellent environmental and thermal stability, mechanical strength, and magnetic and optical properties. In the application of sensors, PT can undergo a conformational change upon binding the analyte due to the flexibility of the carbon-carbon bond between thiophene rings. The twisting of the polymer backbone changes the length of conjugation, thus changing the maximum wavelength of absorption or emission. Marsella and Swager demonstrated a series of crown ether-containing PTs which show different l_{max} shifts upon binding with different alkaline metal cations (Figure 6.56).

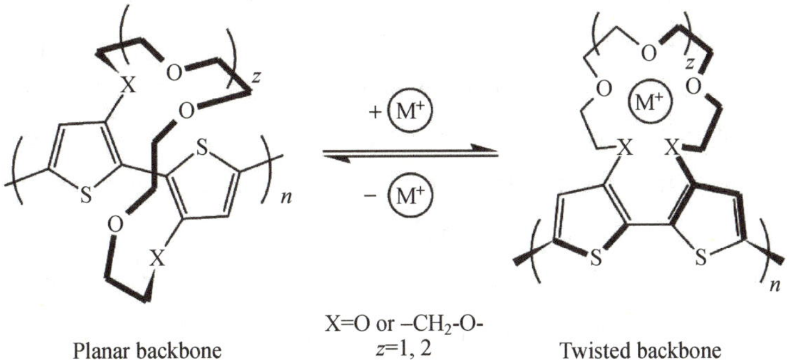

Figure 6.56 Scheme of crown ether pendent polythiophene switch from planar and twisted conformation upon binding

Recently, Guo et al. synthesized a series of novel PT derivative-based fluorescent sensor for various transition metal cations. The proposed sensing mechanism is the interaction between the metal cations and the ligand containing nitrogen or oxygen on the side chain of the PT. For instance, a tridentate nitrogen/oxygen-containing ligand was loaded on a thiophene ring followed by chemical oxidation polymerization. The polymer, poly[3-[2-(2-dimethylamino-thylamino)ethoxy]-4-methyl-thiophene], has a different response towards these metal cations: it shows a "turn-off" response towards Cu^{2+} in THF/Tris-HCl solution; a "turn-off" response to Cu^{2+} and Co^{2+} in MeCN/Tris-HCl solution; and a bathochromic shift on the maximum absorption from 420 nm to 472 nm when Cd^{2+} was added. The detection limit for Cu^{2+}, Co^{2+}, and Cd^{2+} were 2.0 nmol·L^{-1}, 2.5 nmol·L^{-1}, and 83 nmol·L^{-1}, respectively.

In addition, the polymer/Cu^{2+} complex in THF/Tris-HCl solution can be used as a "turn-on" sensor for neutral amino acids, homocysteine, and glutathione, with the fluorescent enhancement of 11.5-and 7.7-fold (Figure 6.57).

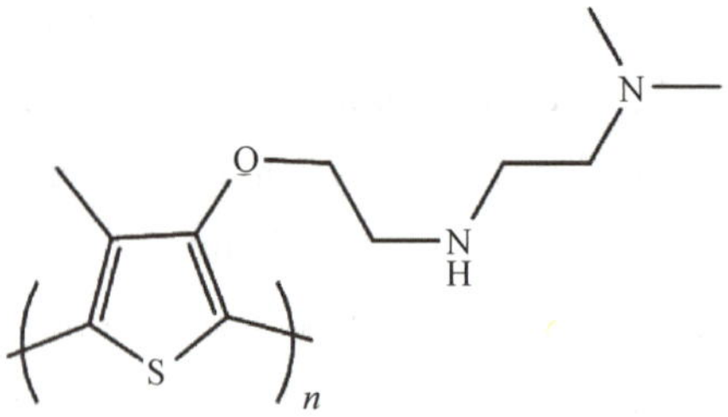

Fundamentals of Photophysics

Figure 6.57 Chemical structure of
poly[3-[2-(2-dimethylamino-thylamino)ethoxy]-4-methyl-thiophene] (PTMA)

Some research groups use a combination of these previous examples. The Jones Group has designed and synthesized a series of "turn-on" and "turn-off" sensors with pendant Lewis base receptor ligands on a poly[p-(phenyleneethynylene)-alt-(thienyleneethynylene)] (PPETE) backbone. These polymers have high fluorescent quantum yields and highly efficient quenching selectively towards certain transition metal cations. For example, for the terpyridine receptor polymer ttp-PPETE, a 5% emission quenching was observed under 4 nmol · L^{-1} Ni^{2+} concentration. For the N,N,N$_0$-trimethylethylene-diamino receptor polymer TMEDA-PPETE, 50% quenching of the fluorescent intensity was achieved upon 500 nmol · L^{-1} of Cu^{2+}, which correspond to only 10% load of the cations with respect to the receptor unit concentration. Recently, nonlinear conjugated polymers have been developed as a new class of chemosensors. The extended electron delocalization may improve the signal transduction of the sensors because the electron or energy communication is not limited to one dimension. Poly(phenyleneethynylene) with a bent structure was also utilized as a chemical sensor for various analytes. Zeng et al. reported a poly(m-phenyleneethynylene) with 2-thiohydantoin on the side chain, and this sensor can detect cuprous ion, hydrogen peroxide, and glucose. The bent structure of the polymer backbone provided a larger spatial angle for the receptor unit to rotate. In other words, the receptor unit is more exposed in the environment for the analyte to bind. Morisaki et al. reported an oligophenylene-layered polymer with nitrobenzene units on end, which showed PET energy transfer from the polymer to the terminal units. This can be a potential sensor system.

6.5 Fluorescence Chemical Sensor Application

6.5.1 Transition Metal Sensors

The toxicity of some metal ions has been well-known and which has raised environmental concerns widely. The Environmental Protection Agency (EPA) lists transition metal ions as priority pollutants. These pollutant metals include cadmium, chromium, copper, lead, mercury, nickel, silver, thallium, zinc, and so on. Traditionally, these trace metals in water are detected by complicated methods such as plasma-atomic emission spectrometry, plasma-mass spectrometry, or atomic absorption with furnace technique. However, the methods above are not very suitable for real-time monitoring of the environment. There is a growing need for portable polymer-based sensor devices for these metal pollutants. FCPs have high selectivity and sensitivity towards these cations; moreover, they are easy to synthesize and analyze. Therefore, many polymer sensors have been reported in the past two decades. In the design of the receptor unit, the characteristics of the cation, including the ionic diameter, charge, coordination number, and hardness, are taken into consideration.

Iron element is crucial in biological systems. The human body regulates the intake and reuse of iron tightly. Though deficiency of iron will lead to a variety of diseases, excessive iron is toxic. Iron ions commonly have two different oxidation states in nature: iron (II) and iron (III). Differentiation of these two ions poses a challenge in sensor research. Metallopolymer TMEDA-PPETE-Cu^{2+} has a 150-fold fluorescent "turn-on" response towards both Fe^{2+} and Fe^{3+}; however, this sensory system cannot distinguish between these two different oxidation states. An improved system from the same research group was reported. By changing the alkyl chain length and bite angle of the unconjugated diamino-based receptor, the resulting new oligomer called TMPDA-PPETE, can selectively differentiate Fe^{2+} and Fe^{3+} in the aqueous or biological system at nM concentrations (Figure 6.58).

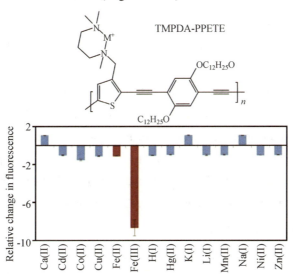

Figure 6.58 Fluorescence response of TMPDA-PPETE (chemical structure shown) to various 5 mM cations in room temperature solution

Wang's research group reported an anionic phosphonate-functionalized polyfluorene as a highly water-soluble Fe^{3+} chemosensory with 400-fold fluorescence quenching upon Fe^{3+} addition. Further, ultrathin multilayer films with alternating layers of the anionic phosphate-functionalized polyfluorene chemosensor and cationic poly (diallyldimethylamine) were found to be sensitive to Fe^{3+} with a detection limit of 10^{-7} mol·L^{-1}. Based on their initial researches, thin-film sensors of a derivative of the anionic phosphate-functionalized polyfluorene chemosensory prepared via covalent immobilization on a glass surface were fabricated. These films were able to detect Fe^{3+} with high sensitivity, selectivity, and reversibility with an improved limit detection of 8.4 ppb in THF solution and 0.14 ppm in aqueous solution (Figure 6.59).

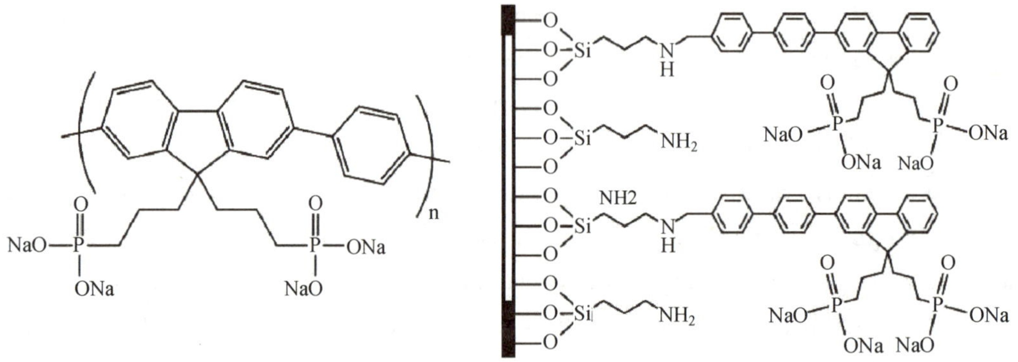

Figure 6.59 Chemical structures of the anionic phosphonate-functionalized polyfluorene (left) and phosphonate ester functionalized polyfluorene film sensor (right)

6.5.2 Fluorescent Sensors for Biological Analytes

FCPs have been developed widely with specific biological receptors to detect DNA, proteins, and various small biological molecules in the fields of biology, medicine, and life sciences. The challenges of DNA and protein sensing are to detect a specific sequence of DNA or a type of protein accurately. Different strategies were employed in the published literature: either the receptor unit contains the information towards a specific DNA or protein (such as the base sequence or the size and shape), or the FCP itself is "label-free"; however, it can report the detection by a "reporting fluorophore" using FRET, or the binding of the analyte will change the conformation or aggregation of the FCPs, resulting in a change in the fluorescence quantum yield or fluorescence lifetime. Due to the hydrophobic nature and rigid-rod-like backbone structure, it is important to introduce hydrophilic side chains or ionic pendant groups to CPs, which helps them with an increase of solubility in water.

i. ssDNA base sequence detection

Conjugated polymers permit the detection of DNA hybridization (pairing of complementary DNA single-strands, ssDNAs) and thus act as ssDNA sequence sensors. These sensors comprise an aqueous solution containing CP, a cationic conjugated polymer, and ssDNA-FL, a single-stranded DNA with a known base sequence labeled with a chromophore such as a fluorescein, FL. There is no interaction between CP and ssDNA. Irradiation with the light of a relatively short wavelength, which is not absorbed by FL, causes the fluorescence of CP. When an ssDNA with a specific base sequence is added, complementary to that of the probe ssDNA-FL, hybridization occurs. The double-strand thus formed becomes electrostatically linked to CP, thus allowing energy transfer from electronically excited CP^* to FL (Figure 6.60). The characteristic fluorescence of the FL groups generated in this way signals hybridization. Upon the addition of non-complementary ssDNA, the FL fluorescence is not observed. Relative to the CP^* emission, the FL emission spectrum is shifted to the long-wavelength region and can be reliably detected. Recent research on strand-specific DNA detection with cationic conjugated polymers has been concerned with their incorporation into DNA chips and microarrays.

$$CP^* + FL \longrightarrow CP + FL^*$$
$$FL^* \longrightarrow FL + h\nu$$

Figure 6.60 Energy transfer from an electronically excited conjugated polymer to fluorescein

ii. Imaging of intracellular mRNA

The localization and concentration of intracellular mRNA is a crucial factor in the control of protein synthesis. Control of c-fos mRNA is important because the c-fos protein is a transcription factor that participates in the control of the cell cycle and differentiation. The presence of c-fos mRNA was studied by using DNA oligomers that were expected to bind close to each other when hybridized with c-fos mRNA (Figure 6.61). The level of c-fos mRNA could be controlled by stimulation of the Cos cells. The donor-and acceptor-labeled oligomers were added to the cells via microinjection. The presence of the mRNA was found in the stimulated (left) but not in the unstimulated cells. These results show that RET imaging can be used to study the difficult problem of the regulation of mRNA in cells.

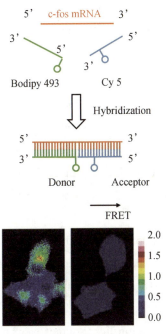

Figure 6.61 Intracellular detection of c-fos mRNA by RET ratio imaging

iii. Sensors for proteins

Pu and Liu reported that cationic and anionic poly (fluorenylene ethynylene-alt-benzothiadiazole) (PFEBT) could act as nonspecific "turn-on" sensor for proteins. In an aqueous solution, PFEBTs show a very weak fluorescence due to charge transfer of the excited states. However, the yellow fluorescence can be enhanced obviously by polymer aggregation upon complexation with proteins by electrostatic and hydrophobic interactions (Figure 6.62).

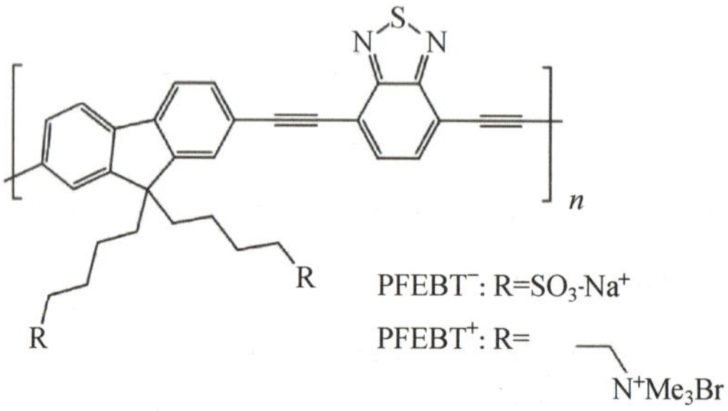

Figure 6.62 Chemical structure of cationic and anionic poly (fluoreyleneethynylene-alt-benzothiadiazole)

6.5.3 Water Quality Monitoring

Water quality monitoring is essential to prevent harm to human health and the aquatic ecosystem. Natural events and human activities can cause water pollution, resulting in unexpected poor water quality. By detecting and analyzing the parameters in the water, it is possible to identify the impacts and risks to human health and the ecosystem. Above technology can help to plan and implement appropriate management measures for water quality control.

i. Phytoplankton detection

Phytoplankton monitoring is one of the important tasks for supporting human health and environmental issues, especially for water quality control. Detecting and analyzing different groups of phytoplankton in water provides important information on aquatic ecological states and nutrient compositions. In addition, early detection of certain species of algae which cause harmful algal blooms (HABs) is essential to protect the water ecosystem and human health. Therefore, it is of great significance to take appropriate technology to realize the early detection of phytoplankton.

Due to different photopigment constituents, different phytoplankton groups exhibit unique fluorescence properties. Shin et al. developed handheld phytoplankton sensors using different excitation wavelengths (385 nm, 448 nm, and 590 nm) of LEDs to stimulate different species of phytoplankton selectively. A disposable PDMS chip and micro-vial were used for the sample holding and delivery purposes. Excitation light sources were placed underneath the sample, and the detector was installed on the opposite side, as shown in Figure 6.63, where the excitation light rays are projected to minimize the stray noise. Series of optical filters (dichroic and color filters) were used to block the remaining stray noise. The emission fluorescent lights from the green algae and cyanobacteria were 680 nm and 645 nm, respectively. It was successfully demonstrated to detect and differentiate different mixtures of green algae and cyanobacteria species using a multivariate algorithm. The limit of detection for green algae and cyanobacteria were 1 mg·L^{-1} and 4 mg·L^{-1}, respectively. [Figure 6.63 (a) and 6.63 (b)].

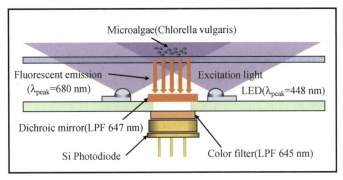

(a)

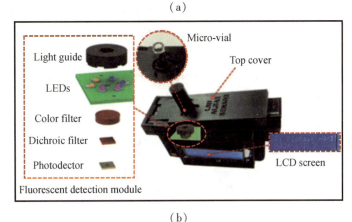

(b)

Figure 6.63 Portable fluorescence algae sensors: (a) a 3D-printed portable microalgal sensor with a disposable PDMS chip; (b) a handheld fluorescence platform for multiple phytoplankton detection (green algae and cyanobacteria)

ii. Dissolved organic matter detection

The increase in human pollution and the effects of climate change significantly affected the water quality. The pollutants in this natural water are closely related to the dissolved organic matter (DOM) concentration and its composition; therefore, it is important to understand the characteristics of DOM in the water. DOM is defined as any organic matter dissolved in the water that can pass through the water filter with a pore size of 0.2 μm. Since not all DOMs are light interactive, the light-absorbing portion, defining which as colored dissolved organic matter (CDOM), is measured by fluorescence-based detection. Due to the advantages that the fluorescence-based detection method offers, various studies have been conducted to concentrate on and develop an on-site detection of CDOM fluorescence to assist the water quality monitoring.

Fundamentals of Photophysics

Natural water typically has CDOM responsible for strongly absorbing light in the range from 250 nm to 450 nm and fluorescing from 400 nm to 450 nm. However, it is a challenge that utilizing blue light excitation for CDOM detection due to the presence of chlorophyll pigments in natural bodies of water; thus accurate assessment of the emission signal is essential. Lewis et al. studied an algorithm to accurately estimate the CDOM in the arctic ocean to compensate for the overly estimated CDOM when chlorophyll *a* present in the water. For remote sensing, satellite-based measurements were widely used to quickly assess large water areas such as the ocean. In addition, hyperspectral remote sensing was reported to achieve higher resolution than satellite-based sensing while covering a relatively larger area in the last decade and successfully differentiated many important multiple water quality parameters such as CDOM, chlorophyll *a*, diatoms, turbidity, and so on.

Blockstein et al. reported a portable fluorometer to measure chlorophyll pigments and CDOM concentration. Light-emitting diodes with 405 nm and 465 nm were utilized to selectively stimulate the CDOM and chlorophyll, respectively. Two custom fabricated thin optical filters were directly attached to a single sensor array to selectively measure two different fluorescence signals, as shown in Figure 6.64. The thin glass substrates were applied to absorb and decay the excitation lights. For demonstration, standard fluorescein dye was selected to simulate chlorophyll *a*. The sensor was tested while it is completely submerged under the water. The limit of detection for the fluorescein is $0.7 \text{ nmol} \cdot \text{L}^{-1}$.

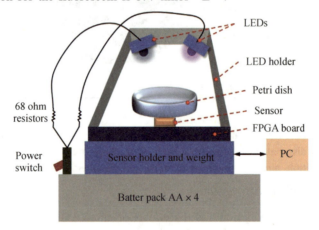

Figure 6.64　A miniature fluorometer to measure chlorophyll and CDOM concentration in the aquatic environment

iii. Heavy metal ion detection

Heavy metal pollution in water has become a strong threat to marine animals and humans since they can bioaccumulate in living organisms directly or through consumption. Among many metal ions, copper, lead, mercury, chromium, and cadmium are known to be highly toxic to humans. For instance, copper can cause liver damage, lead is known to damage the brain, and low doses of mercury exposure can cause severe damage to any animals' nervous system, including humans. Therefore, it is critical to detect those metal ions from the aqueous system rapidly.

Various methods have been developed for detecting different heavy metals such as electrochemical, spectroscopic, and optical detection. Although spectroscopic and electrochemical detection methods have been widely used, optical detection methods showed great potential for portable sensing platforms. One widely used optical detection method is colorimetric sensing which utilizes selective reagents and indicator dyes. They only react with desired target metal ions and absorb a specific wavelength of the color.

Guo et al. reported a carbon dot doped hydrogel waveguide for detecting Hg^{2+} in the water, as shown in Figure 6.65. Although the reported work is not a stand-alone system, it showed great potential to be a key element of the device and easily incorporated into a portable device. Polyethylene glycol diacrylate (PEGDA) was selected as the material for the waveguide, and the fluorescent carbon dots with 7.8 nm diameters were selected as the active material. The waveguide exhibited a peak absorption at 352 nm and fluorescence emission at 475 nm. The detection range was from 0 $\mu mol \cdot L^{-1}$ to 5 $\mu mol \cdot L^{-1}$, and the limit of detection was 4 $nmol \cdot L^{-1}$.

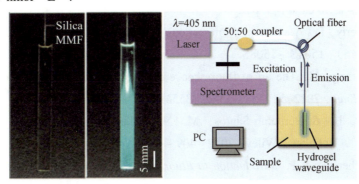

Figure 6.65　A carbon dot doped hydrogel waveguide for detecting Hg^{2+} in the water

Heavy metal detection in an aqueous environment is still challenging since it includes many elements and particles that can potentially interfere with the target metal ions, and it may lower the detection accuracy. Several strategies are taken to improve the detection performance of heavy metals in water. For example, cellulose filter papers can be used to pretreat the water before the test to eliminate the debris from the solution. These methods could minimize the interference from the dissolved matter in water and accurate characterization of fluorescence emission spectra of the sample solution (noise mainly from CDOM and chlorophylls).

References

[1] Chen, A., Wu, W. & Jones, W. E. *Comprehensive Supramolecular Chemistry II* [M]. Amsterdam: Elsevier, 2017: 179 – 195.

[2] National Research Council. *Expanding the Vision of Sensor Materials* [M]. Washington DC: The National Academies Press, 1995: 73 – 88.

[3] Shin, Y. H., Gutierrez-Wing, M. T. & Choi, J. W. Review: recent progress in portable fluorescence sensors [J]. *J. Electrochem. Soc.*, 2021,168(1): 017502.

[4] Basabe-Desmonts, L., Reinhoudt, D. N. & Crego-Calama, M. Design of fluorescent materials for chemical sensing [J]. *Chem. Soc. Rev.*, 2007,36(6): 993 – 1017.

[5] Smith, M. B. *March's Advanced Organic Chemistry: Reactions, Mechanisms, and Structure* [M]. Hoboken: Wiley, 2013: 96 – 104.

[6] The Cation-π Interaction [EB/OL]. [2022 – 03 – 27]. http://www.its.caltech.edu/~dadgrp/research/cation-pi.pdf.

[7] Dougherty, D. A. Cation-π interactions involving aromatic amino acids 1-4 [J]. *J. Nutr.*, 2007,137(6): 1504S – 1508S.

[8] Schäferling, M. The art of fluorescence imaging with chemical sensors [J]. *Angew. Chem. Int. Ed. Engl.*, 2012,51(15): 3532 – 3554.

[9] Valeur, B. & Berberan-Santos, M. N. Molecular *Fluorescence: Principles and Applications* [M]. Hoboken: Wiley, 2012: 409 – 478.

[10] Demchenko, A. P. *Introduction to Fluorescence Sensing* [M]. Heidelberg: Springer, 2015.

[11] Valeur, B. & Leray, I. *PCT (Photoinduced Charge Transfer) Fluorescent Molecular Sensors for Cation Recognition* [M]. Heidelberg: Springer, 2001: 187 – 207.

[12] Achten, C. & Andersson, J. T. Overview of polycyclic aromatic compounds (PAC) [J]. *Polycycl Aromat Compd.*, 2015,35(2-4): 177 – 186.

[13] Cho, D. M. *Partially Fluorinated Polycyclic Aromatic Compounds: Syhthesis and Supramolecular Behavior* [D]. Lexington: University of Kentucky, 2007.

[14] Zhou, Q. & Swager, T. M. Fluorescent chemosensors based on energy migration in conjugated polymers: the molecular wire approach to increased sensitivity [J]. *J. Am. Chem. Soc.*, 1995, 117: 12593 – 12602.

[15] Pu, K. Y. & LIU, B. Fluorescence turn-on responses of anionic and cationic conjugated polymers toward proteins: effect of electrostatic and hydrophobic interactions [J]. *J. Phys. Chem. B.*, 2010,114(9): 3077 – 3084.

Chapter 7

Photophysics in Life

7.1 Photosynthesis

Photosynthesis is a process used by plants and other organisms to convert light energy into chemical energy, which is released through cellular respiration to fuel the organism's activities. Some of this chemical energy is stored in carbohydrate molecules, such as sugars and starches, which are synthesized from carbon dioxide and water. Oxygen is also released as waste in most cases, storing three times more chemical energy than carbohydrates. Most plants, algae, and cyanobacteria perform photosynthesis; such organisms are known as photoautotrophs. Photosynthesis is primarily responsible for generating and maintaining the oxygen content of the earth's atmosphere and supplies most of the energy required for life on the earth.

Although photosynthesis is performed differently by different species, the process always begins when light energy is absorbed by proteins called reaction centers, which contain green chlorophyll (and other colored) pigments/chromophores. In plants, these proteins are kept in organelles called chloroplasts, which are most abundant in leaf cells. While in bacteria, they are embedded in the plasma membrane. In these light-dependent reactions, some energy is used to strip electrons from suitable substances (such as water) to produce oxygen. The hydrogen released by water splitting is used to generate two other compounds that serve as short-term stores of energy, enabling its transfer to drive other reactions. These compounds are reduced nicotinamide adenine dinucleotide phosphate (NADPH) and adenosine triphosphate (ATP), the "energy currency" of cells.

7.1.1 Natural Photosynthesis

The natural photosynthetic process basically involves the splitting of water by sunlight into oxygen, which is released into the atmosphere, and "hydrogen", which is not released in the atmosphere but instead is combined with carbon dioxide to produce various types of organic compounds (Figure 7.1). The burning of these compounds with oxygen, either by respiration (food) or combustion (fossil fuels, wood, and biomass), forms the original compounds (water and carbon dioxide) and releases the stored energy that originated from sunlight.

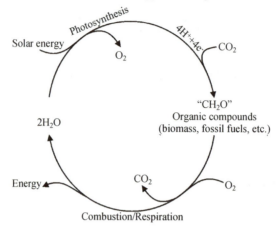

Figure 7.1 Schematic representation of natural photosynthesis, combustion, and respiration processes

In plants, algae, and cyanobacteria, sugars are synthesized through a subsequent series of light-independent reactions known as the Calvin cycle (Figure 7.2). In the Calvin cycle, atmospheric carbon dioxide is incorporated into existing organic carbon compounds such as ribulose bisphosphate (RuBP). Using the ATP and NADPH produced by the light-dependent reactions, the resulting compounds are then reduced and removed to form more carbohydrates such as glucose. In other bacteria, different mechanisms (such as the reverse Krebs cycle), are used to achieve the same goal.

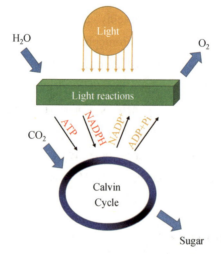

Figure 7.2 Photosynthesis changing sunlight into chemical energy, splitting water to liberate O_2, and fixing CO_2 into sugar

In photosynthetic bacteria, the proteins that collect light for photosynthesis are embedded in the cell membranes. In its simplest form, this involves the membrane around the cell itself. However, membranes may be tightly folded into cylindrical sheets called thylakoids or aggregated into round vesicles called intracytoplasmic membranes. These structures can fill most of the interior of the cell, giving the membrane a very large surface area, thus increasing the amount of light that the bacteria can absorb.

In plants and algae, photosynthesis occurs in organelles called chloroplasts (Figure 7.3). A typical plant cell contains approximately 10 to 100 chloroplasts. The chloroplast is surrounded by a membrane. This membrane consists of an inner phospholipid membrane, an outer phospholipid membrane, and an intermembrane space. Surrounded by the membrane is an aqueous fluid called the stroma. Embedded within the matrix are stacks of thylakoids (grana), which are the site of photosynthesis. The thylakoids are shown as flattened disks. The thylakoid itself is enclosed by the thylakoid membrane, and within the surrounded volume is a lumen or thylakoid space. Embedded in the thylakoid membrane are intact and peripheral membrane protein complexes of the photosynthetic system.

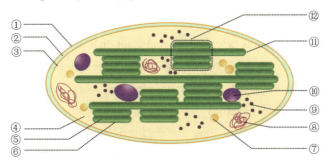

Figure 7.3 Schematic illustration of chloroplast ultrastructure: ① outer membrane; ② intermembrane space; ③ inner membrane (1 + 2 + 3: envelope); ④ stroma (aqueous fluid); ⑤ thylakoid lumen (inside of thylakoid); ⑥ thylakoid membrane; ⑦ plastoglobule (a drop of lipids); ⑧ plastidial DNA; ⑨ ribosome; ⑩ starch; ⑪ thylakoid (lamella); ⑫ granum (stack of thylakoids)

Plants mainly use the pigment chlorophyll to absorb light. The green part of the light spectrum is not absorbed but reflected, which is the reason why most plants have a green color. In addition to chlorophyll, plants also uses pigments such as carotenes and xanthophylls. Algae also use chlorophyll, but there are various other pigments such as phycocyanin, carotenes, and xanthophylls in green algae. Phycoerythrin in red algae (rhodophytes) and fucoxanthin in brown algae or diatoms result in a wide variety of colors.

Typically, plants convert light into chemical energy with a photosynthetic efficiency of 3% – 6%. Unconverted absorbed light dissipates primarily as heat, with a small fraction (1% – 2%) re-emitted as chlorophyll fluorescence at longer (redder) wavelengths. This fact allows measurement of the photoreaction of photosynthesis by using chlorophyll fluorometers.

The photosynthetic efficiency of actual plants varies from 0.1% to 8% with the conversion

Fundamentals of Photophysics

frequency of the light, light intensity, temperature, and proportion of carbon dioxide in the atmosphere. By contrast, solar panels convert light energy into electricity, with mass-produced panels having an efficiency of around 6% – 20%, while laboratory devices are more than 40% efficient. Scientists are studying photosynthesis in hopes of producing plants with higher yields.

The efficiency of both light and dark reactions can be measured, but the relationship between the two can be complicated. For example, the ATP and NADPH energy molecules produced by the light reaction can be used for carbon fixation or photorespiration in C_3 plants. Electrons may also flow to other electron sinks. For this reason, it is not uncommon for authors to distinguish between work done under non-photorespiratory conditions and those done under photorespiratory conditions.

7.1.2 Artificial Photosynthesis

Research into artificial photosynthesis as a renewable energy source has been going on for decades. This approach uses biomimetic technology to replicate the natural photosynthesis process, which uses abundant sunlight, water, and carbon dioxide resources to produce oxygen and energy-rich carbohydrates (Figure 7.4).

In general, a helpful fuel is a chemical reductant that can be stored, transported, and oxidized by oxygen in the air while releasing energy when needed. Thus, fuel production requires more than just energy. It requires an electron source and a material that can be chemically reduced with these electrons. If artificial photosynthesis is to contribute to filling society's energy needs, the electrons must be obtained from the oxidation of a low-cost and easily available compound, namely water, as occurs during the natural photosynthetic process. Water oxidation produces hydrogen ions, which are reduced to produce hydrogen gas, H_2, a useful fuel whose oxidation by air regenerates water. Alternatively, hydrogen can be used to reduce carbon dioxide to carbon-based fuels, which burn renewable carbon dioxide and water.

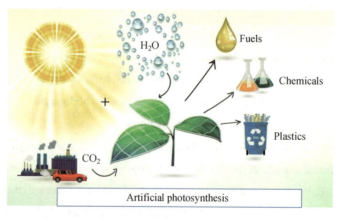

Figure 7.4 Schematic illustration of artificial photosynthesis

Therefore, research on artificial photosynthesis concentrates on using sunlight to split water into molecular hydrogen and molecular oxygen:

$$2H_2O + 4h\nu(\text{sunlight}) \rightarrow 2H_2 + O_2; \Delta G^0 = 4.92 \text{ eV} \quad (7\text{-}1)$$

and reducing carbon dioxide in aqueous solution to CO, ethanol, or hydrocarbons, such as methane formation shown in Equation 7-2:

$$CO_2 + 2H_2O + 8h\nu(\text{sunlight}) \rightarrow CH_4 + O_2; \Delta G^0 = 8.30 \text{ eV} \quad (7\text{-}2)$$

As carbon dioxide reduction is a challenging process from a kinetic viewpoint and can also be accomplished with molecular hydrogen, the attention of most scientists is focused on the photochemical water-splitting reaction.

While most people believe that if hydrogen could be generated from photochemical water splitting by using solar energy, both the energy and the environmental problems of our planet would be largely solved, but things are far more complicated than that. Storing the equivalent of the current energy demand would require splitting more than 10^{15} mol/year of water, which is about 100 times the scale of the most important chemical reaction performed in the chemical industry, namely, hydrogen fixation by the Haber-Bosch process. This comparison tells us that perhaps we are asking too much of ourselves and our planet. It is emphasized that the first step in solving the energy crisis should be to conserve energy and use it as efficiently as possible.

Of course, water splitting by sunlight is not a simple process; otherwise, it would already happen in nature. In fact, the electronic absorption spectrum of water does not overlap the emission spectrum of the sun, so that direct water dissociation by sunlight cannot take place. From the thermodynamic point of view, the most convenient process for splitting water by solar energy is that involving the evolution of molecular oxygen and molecular hydrogen from liquid water (Equation 7-3), whose low-energy thermodynamic threshold (1.23 eV), in principle, allows conversion of about 30% of the solar energy. Of course, a suitable light absorption sensitizer should be used. Furthermore, such a process involves two multi-electron transfer reactions (Equations 7-4 and 7-5), the second one involving four-electron oxidation of two water molecules coupled to the removal of four protons:

$$\frac{1}{2}H_2O \rightarrow \frac{1}{2}H_2 + \frac{1}{4}O_2; \Delta G^0 = 1.23 \text{ eV} \quad (7\text{-}3)$$

$$2H_2O + 2e^- \rightarrow H_2 + 2OH^-; E^0(\text{pH}=7) = -0.41 \text{ V}(\text{vs. NHE}) \quad (7\text{-}4)$$

$$2H_2O \rightarrow O_2 + 4H^+ + 4e^-; E^0(\text{pH}=7) = +0.82 \text{ V}(\text{vs. NHE}) \quad (7\text{-}5)$$

Since each proton can transfer only one electron in a photochemical process, two catalysts must be present in the system: one should collect electrons to produce molecular hydrogen, which is a two-electron reaction, and the other one should collect holes (positive charges) to produce molecular oxygen, which is a four-electron process. This means that oxygen evolution requires four successive photo-induced electron-transfer steps (threshold: 1.23 eV), coupled with proton transfer (proton-coupled electron-transfer, PCETs), as it happens in the OEC of the

Fundamentals of Photophysics

natural photosynthetic system.

Based on these considerations, a plausible artificial photosynthetic system should include the following essential features (Figure 7.5): an antenna for light-harvesting; an RC for charge separation; catalysts as one-to-multielectron interfaces between the charge-separated state and the products; a membrane to provide physical separation of the products.

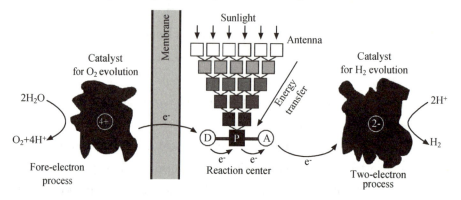

Figure 7.5 Schematic representation of the strategy for photochemical water splitting (artificial photosynthesis) (Five fundamental components can be recognized: an antenna for light harvesting, a charge-separation triad D-P-A, a catalyst for hydrogen evolution, a catalyst for oxygen evolution, and a membrane separating the reductive and the oxidative processes.)

7.2 Phototherapy

Light therapy is a medical treatment in which natural or artificial light is used to improve a health condition. Treatment could involve fluorescent light bulbs, halogen lights, sunlight, or light-emitting diodes. Phototherapy is also known as phototherapy and heliotherapy. The type of therapy and light used will vary by health condition (Figure 7.6).

Phototherapy has been used to treat diseases for over 3,500 years. In ancient India and Egypt, people used sunlight to treat skin diseases such as vitiligo. Modern phototherapy started with Niels Ryberg Finsen. He used sunlight and ultraviolet light to treat lupus vulgaris, a form of tuberculosis that affects the skin. Since then, the use of phototherapy has grown.

While ultraviolet light can be very harmful to the body, it can also help cure diseases. The skin has resident flora, particularly staphylococci bacteria, which can cause secondary bacterial infections in diseased skin. The ultraviolet light has bactericidal activity and, together with singlet oxygen formed by photosensitive mechanisms, can be helpful in destroying skin bacteria.

>>> **Chapter 7**   Photophysics in Life

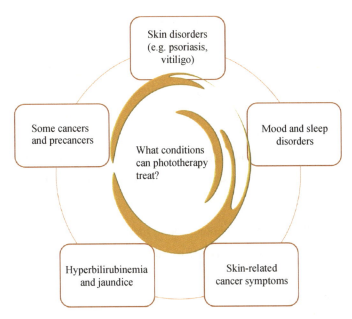

Figure 7.6 Schematic illustration of phototherapy

 7.2.1   Types of Treatment

i. Skin diseases

Phototherapy can be used to treat a variety of skin conditions, including Eczema, Psoriasis, Vitiligo, Itchy skin, and Cutaneous T-cell lymphoma. Treatment involves the use of ultraviolet light: a type of light present in sunlight, to slow the growth and inflammation of skin cells. Inflammation is one of the ways your immune system responds to infections, injuries, and foreign "invaders".

Ultraviolet B region (UVB) rays affect the outermost layers of the skin. Utraviolet A region (UVA) rays are slightly less intense but penetrate more deeply into the skin. These two types of ultraviolet light can be used in different ways. There are three main types of phototherapies used for skin diseases:

(1) Broadband UVB: Broadband UVB uses a wide range of spectrum of UVB rays. UVB rays exist in sunlight, but you cannot see them.

(2) Narrowband UVB: This involves using a smaller, more intense part of UVB to treat the skin condition. It's the most common type of light therapy used today.

(3) PUVA: Psoralen ultraviolet-A, or PUVA, combines UVA light with a chemical called psoralen, which comes from plants. Psoralen can be applied to your skin, or you can take it as a pill. It makes your skin more sensitive to light. PUVA has more side effects than some other phototherapy treatments. It is only used when other options cannot work.

ii. Mood and sleep disorders

Phototherapy is also used to treat mood and sleep disturbances. The seasonal affective disorder is a type of depression associated with certain seasons of the year. It usually begins in the fall and continues through winter. Phototherapy for SAD involves the use of a lightbox—a specially designed box that emits a steady, soft light. However, phototherapy used in this way has many side effects. They include headaches, fatigue or tiredness, insomnia, hyperactivity, and irritability.

Phototherapy can also help those with circadian rhythm sleep disorders, such as delayed sleep phase syndrome (DSPS). People with DSPS often can't fall asleep until the early hours of the morning or close to sunrise. Light therapy can help them shift to more normal sleep schedules.

iii. Cancers and precancers

It is a type of light therapy called photodynamic therapy that is used to treat certain types of cancers and precancers. It involves the use of a drug called a photosensitizer along with light. Photosensitizers are applied to the skin. When light strikes the skin, it interacts with the drug, producing oxygen that kills nearby cancer cells.

Photodynamic therapy is used to treat conditions such as: cancer of the esophagus, which is the tube that connects your mouth to your stomach; endobronchial cancer, a type of lung cancer; Barrett's esophagus, a precancerous condition usually caused by acid reflux.

Light therapy has some advantages over treatments like radiation and chemotherapy. For example, it usually does not have any long-term side effects. It leaves less scarring than surgery. And the cost of phototherapy is much lower than the other cancer treatment options. The downside is that it usually only works in areas on or under the skin, where light can reach. It also does not help much with cancers that have spread.

iv. For newborns

Phototherapy has been used to treat hyperbilirubinemia and jaundice for over 60 years. These conditions can cause a baby's skin, eyes, and body tissues to turn yellow. The yellow color comes from the excess bilirubin, a pigment produced when red blood cells break down. Light exposure lowers bilirubin levels in babies. It breaks down the bilirubin so the baby's body can properly clear it.

There are two main ways to treat jaundice with light therapy. The usual method is to place the baby under halogen spotlights or fluorescent lamps. During treatment, the baby's eyes are covered. Another technique is to use "biliblankets". The blankets have fiber-optic cables that shine blue light onto the baby's body. This method is most often used when babies are born early or other treatments have already been tried.

Compact fluorescent lamps and blue LED devices are also used to give babies phototherapy. They can stay close to the body because they do not generate much heat.

7.2.2 Photo-controlled Delivery

The ability to control the spatial and temporal distribution of a specific reaction is of the greatest importance in biology. This result may be achieved by means of compounds that are inert until irradiated with light of an appropriate wavelength. In the so-called cage compounds (Figure 7.7), a bioactive molecule is covalently bound ("caged") to a light-absorbing photolabile group, resulting in a biologically inactive species.

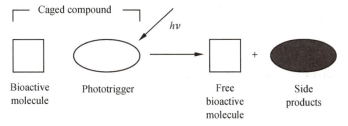

Figure 7.7 A cartoon representation of the photo-uncaging process

There are several possibilities for how the installed caging group can eliminate the biological functions of the molecule; most importantly, the steric demand of the caged group can inhibit the molecular interaction of the caged molecule with its biological partners. Upon light absorption, the photolabile group (also known as photo-trigger) undergoes a photoreaction that cleaves the species into the free bioactive molecule and products of the caging moiety.

Similar strategies have been used to cage simple drugs, such as acetylsalicylic, ibuprofen, and ketoprofen acids, as well as various biomolecules, including peptides, proteins, and nucleic acids, effectors that regulate gene expression, secondary messengers, neurotransmitters, and nucleotide cofactors.

Light-controlled delivery is quite important in platinum-based anticancer complexes because they are toxic to both healthy and cancerous tissues. The Pt^{2+} photocaged complex **7.1**, in which the Pt^{2+} ion is coordinated to a photoactive ligand via two amides and two pyridyl nitrogen atoms, is biologically inactive (Figure 7.8). Excitation with ultraviolet light, however, releases the metal ion that readily exchanges its ligands yielding complex **7.2**, which is toxic to human breast carcinoma cells.

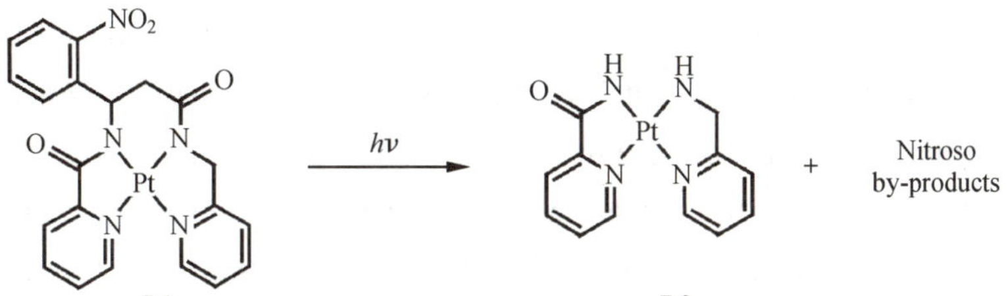

Figure 7.8 Pt^{2+} ion coordinated to a photoactive ligand through two amide and two pyridyl nitrogen atoms

Fundamentals of Photophysics

Light-controlled delivery of metal ions or ligands can be obtained from appropriate coordination compounds. The light-responsive functionality can either be a component of the encapsulated ligand or a property of the metal complex itself. For example, a metal ion encapsulated into a photocleavable cryptand can be ejected upon photoexcitation (Figure 7.9).

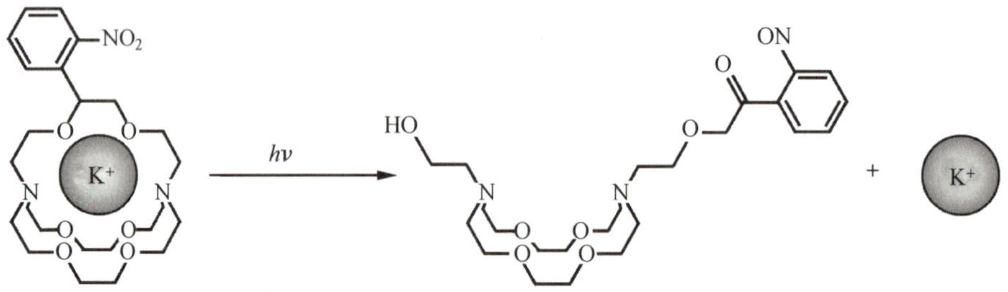

Figure 7.9 A metal ion encapsulated into a photocleavable cryptand ejected upon light excitation

Light-controlled ligand delivery is particularly important in the case of nitricoxide, NO, a bioregulatory molecule that plays critical role in cancer biology; it has been implicated in both tumor growth and suppression. In addition, NO is a γ-radiation sensitizer that may enhance the selective killing of neoplastic tissues. Complexes containing a metal nitrosyl bond (M-NO) can release NO upon photo-excitation. For example, an iron-sulfur nitrosyl cluster $Fe_2(\mu\text{-RS})_2(NO)_4$ bears a pendant chromophore that acts as a light-harvesting antenna, releasing NO upon continuous irradiation at 436 nm.

7.3 DNA Technology

7.3.1 DNA Sequencing

DNA consists of a linear string of nucleotides, or bases, for simplicity, represented by the first letters of their chemical names—A, T, C, and G. In the DNA double helix (Figure 7.10), the four chemical bases always combine with the same counterpart to form "base pairs". Adenine (A) always pairs with thymine (T); cytosine (C) always pairs with guanine (G). This pairing is the basis for the mechanism by which DNA molecules are copied when cells divide, and the pairing also underlies the methods by which most DNA sequencing experiments are performed. The human genome contains approximately 3 billion base pairs that spell out the instructions for making and maintaining a human being. The process of inferring the sequence of nucleotides in DNA is called DNA sequencing. Since the DNA sequence provides information that the cell uses to make RNA molecules and proteins, establishing DNA sequences is key to understanding how genomes work (Figure 7.11).

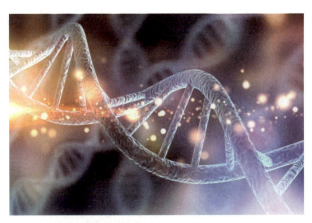

Figure 7.10 Schematic illustration of the DNA double helix structure

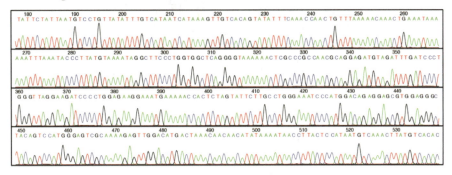

Figure 7.11 DNA sequence data from an automated sequencing machine

DNA sequencing essentially involves scientists taking samples of DNA from anywhere to determine how the four bases, thymine, adenine, guanine, and cytosine, are arranged in a DNA molecule, and this information can then be used further to determine what is known about that portion of DNA. Many things, including what organisms the DNA came from as well as where certain genes are located.

DNA sequencing first became practical in 1977. The original method involved chemically degrading DNA using conditions that were partially selective for one of the four bases. The DNA fragments were then separated by chromatography and detected by 32P autoradiography. In the same year, a method was available to generate fragments terminated in each of the four bases. This method used dideoxynucleotides to terminate enzymatic DNA synthesis. The fragments were again detected using radioactivity.

However, the use of radioactive tracers is problematic in cost, safety, and disposal. Beyond that, the use of radioactivity was not amenable to the degree of automation needed to sequence long DNA chains, chromosomes, or an entire genome. The use of fluorescence for DNA sequencing was first reported in 1986. Virtually all sequencing is currently done using fluorescence detection.

DNA sequencing uses a number of slightly different methods, but all the methods rely on dideoxynucleotide triphosphate (ddNTP) to terminate DNA synthesis. The basic idea of sequencing using ddNTP terminators is illustrated in Figures 7.12 and 7.13. In DNA, nucleotides are linked into a continuous strand via the 5' and 3' hydroxyl groups of the pentose sugar. DNA is replicated by the addition of nucleotides to the 3' hydroxyl group. To sequence DNA strands with unknown sequences, DNA polymerases are used for replication. Replication starts from a primer position with a known sequence. The most frequently used primer is the M13 sequence, which is 17 nucleotides long. In the example shown in Figure 7.13, a single fluorescent primer is used to initiate the reaction. The sample contains DNA polymerase and the four deoxynucleotide triphosphates.

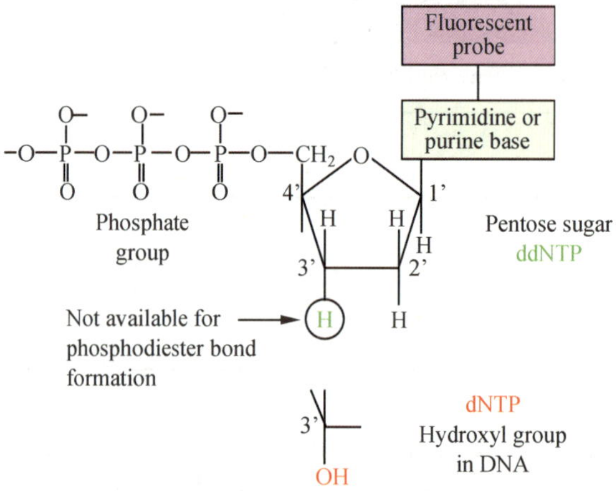

Figure 7.12 Schematics of a dideoxynucleotide triphosphate (ddNTP) (Fluorescent and nonfluorescent ddNTPs are used for DNA sequencing, depending on the method. The 2' group is hydrogen in DNA and is a hydroxyl group in RNA. In a ddNTP, the 3' hydroxyl group is not present, so the DNA chain cannot be continued.)

In a short period, DNA polymerase molecules are randomly distributed along the unknown sequence. The strands being synthesized have a sequence complementary to the unknown strand. The reaction mixture is divided into four portions, for each of the four bases. The DNA polymerase reaction is randomly terminated by adding one of the ddNTPs to each of the four parts of the reaction. The ddNTPs are added along the growing chain. The absence of a 3' hydroxyl group on the ddNTPs prevents further elongation and terminates the reaction. This results in a mixture of oligonucleotides of different lengths. Oligomers of different sizes are separated by polyacrylamide gel electrophoresis. It is worth noting that numerous fragments differing by just one base pair can be resolved: up to hundreds of bases. Each reaction mixture is electrophoresed in a separate lane. Each lane of the reaction mixture contains oligomers which are terminated with only one of the ddNTPs. The gels separate the DNA fragments according to size so that the sequence can be determined from the fluorescence of the separated DNA fragments.

>>> Chapter 7　Photophysics in Life

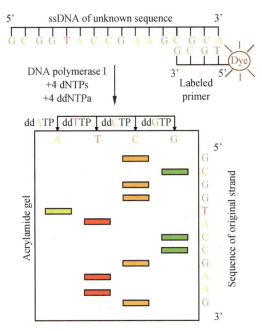

Figure 7.13　Four-lane DNA sequencing using nonfluorescent ddNTPs and a fluorescent primer for DNA synthesis

7.3.2　High-sensitivity DNA Stains

DNA is an important substance in the inheritance of life. The quantitative analysis and specific recognition of DNA molecules are significant to the development of genomics, virology, molecular biology, and other related disciplines. Due to the weak fluorescence of biomolecules, the fluorescence probe method is often used to detect biomolecules (Figure 7.14). The sensitivity of DNA fluorescence probes is an important factor affecting the detection results, so the development of more sensitive fluorescence probes and avoiding the interference of bioluminescence background has become a hot spot of current research.

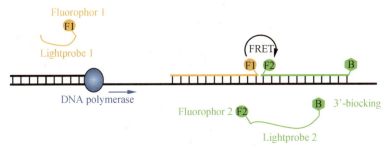

Figure 7.14　Schematic diagram of DNA fluorescence probe

Detection of DNA using stains has a long history, starting with the use of acridine dyes to stain chromatin. The situation was improved by the introduction of dyes like ethidium bromide (EB) and propidium iodide, which fluoresce weakly in water and more strongly when bound to

DNA. DNA can be detected by exposing the gels to EB. When EB is used, the gel typically contains micromolar concentrations of EB to ensure that DNA is bound to a large amount of EB. Because of the micromolar binding constants, sensitivity may be low due to the background of the free dyes.

There are now many greatly improved dyes with high affinity for DNA and little fluorescence in water. Some of these dyes are dimers of acridine or EB. The ethidium dimer was found to bind DNA 103 to 104 more strongly than the monomer. The homodimer of EB (Figure 7.15) was found to remain bound to DNA during electrophoresis. This result is surprising because the positively charged dye is expected to migrate from DNA in the opposite direction. This result indicates that the dyes do not dissociate from DNA on the timescale of electrophoresis. The DNA fragments can be stained prior to electrophoresis without maintaining a micromolar concentration of free dye. The DNA gels show little background fluorescence and allow the detection of DNA fragments with high sensitivity.

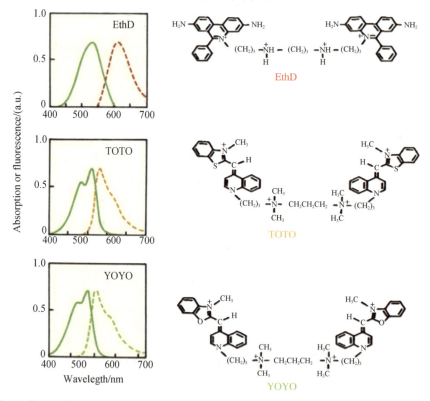

Figure 7.15 Chemical structures of high-affinity DNA dyes; absorption (dashed) and emission (solid) spectra of the dyes bound to DNA. [The relative enhancements of the fluorescence of the dyes on binding to DNA are (top to bottom) 35, 1,100, and 3,200.]

7.3.3 DNA Hybridization

DNA usually exists as a double-stranded molecule. The two strands bind to each other in a complementary fashion by a process called hybridization (Figure 7.16). DNA, naturally, as it is replicated, the new strand hybridizes with the old strand. In the laboratory, we can take advantage of hybridization by generating nucleic acid probes that we can use to screen cells for the presence or absence of certain DNA or RNA molecules.

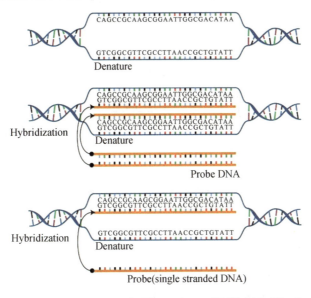

Figure 7.16 Schematic illustrations of DNA hybridization

Detection of DNA hybridization has a wide range of uses in molecular biology, genetics, and forensics. Hybridization takes place during polymerase chain reaction (PCR) and fluorescence in-situ hybridization. Various methods have been used to detect DNA hybridization by fluorescence. Several possible approaches are shown schematically in Figure 7.17. A common method is to detect an increase in RET when the complementary donor and acceptor labels hybridize (upper left). When these sequences are approached by hybridization, the presence of complementary DNA sequences can be detected by increased energy transfer. This can occur if the complementary strands are labeled with donors and acceptors.

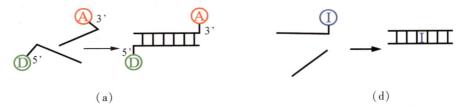

(a)   (d)

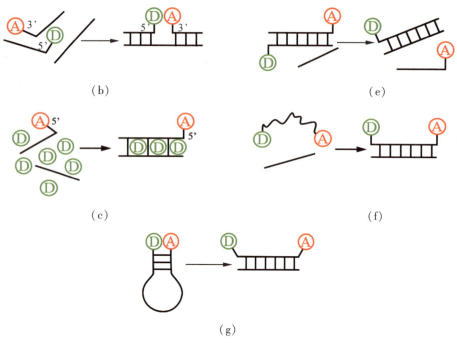

Figure 7.17 Methods to detect DNA hybridization by energy transfer (D, donor; A, acceptor or quencher; I, intercalating dye)

As shown in Figure 7.18, the DNA hybridization technique is performed by first heating DNA from two different sources of origin up to 86 degrees Celsius. Doing so unwinds the DNA strands, unraveling and separating them by breaking the hydrogen bonds between them. This process of separating the two DNA strands is called DNA denaturation or DNA melting.

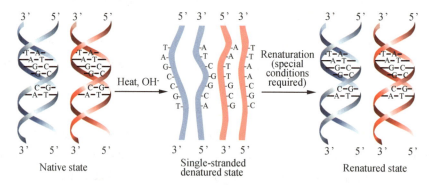

Figure 7.18 Schematic diagram showing the process of DNA hybridization where DNA is firstly denatured into single strands upon heating, followed by renaturing or reannealing back to double-stranded DNA upon gradual cooling

Once obtained, these single-stranded DNAs from both sources are mixed and gradually cooled. Gradual cooling affects the recombination of hydrogen bonds between similar DNA strands from both sources, bringing together complementary base pairs. This process is known as renaturation or the reannealing of DNA.

The greater the similarity of complementary base pairs in two sources, the greater number of hybrids are formed during renaturation. The percentage of hybrid formation during the DNA hybridization can be analyzed using radioactive or fluorescent probes or by UV spectrophotometry. The obtained hybrids are reheated, and the melting temperature is recorded. For the two closely related species, hybrid DNA would form a large number of complementary base pairs, while the melting temperature would be closer to 86 degrees Celsius to break hydrogen bonds. Hybrids formed with distantly related or dissimilar DNA will have fewer hydrogen bonds between fewer complementary bases, causing them to break easily at melting temperatures much lower than 86 degrees Celsius.

An example is shown in Figure 7.19, where the acridine dye is covalently attached to the 3' phosphate of an oligothymidylate.

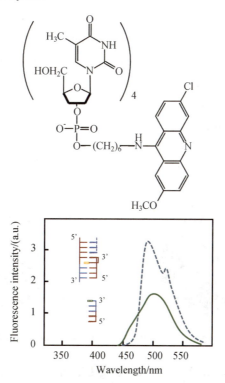

Figure 7.19 DNA hybridization detected by a covalently bound intercalating probe [Fluorescence intensity of the probe, an acridinedye covalently linked to the 3'-phosphate of an oligothymidylate (dashed); fluorescence intensity of the probe upon binding to a complementary adenine oligonucleotide (dotted).]

Acridine fluorescence increases approximately twofold upon binding to a complementary adenine oligonucleotide. Hybridization can be competitive where an increase in the amount of target DNA competes with the formation of donor-acceptor pairs. The acceptor can be either fluorescent or nonfluorescent, in which case the donor appears to be quenched. A competitive assay was performed with complementary DNA strands, where the opposite strands were labeled with fluorescein and rhodamine. Hybridization of the strands leads to quenching of the donor fluorescence. An increase in the amount of unlabeled DNA complementary to one of the labeled strands results in the displacement of the acceptor and an increase in donor fluorescence. Such arrays can be used in amplification reactions in which DNA is thermally denatured during each cycle.

7.4 Coffee Ring Effect

In physics, a "coffee ring" is a pattern left by a puddle of particle-laden liquid after evaporating. This phenomenon is named for the characteristic ring-like deposit along the perimeter of coffee spills. It is also commonly seen after sprinkling red wine. The mechanism behind the formation of these and similar rings is known as the coffee ring effect, or in some cases, the coffee stain effect, or simply ring stain (Figure 7.20).

This ubiquitous phenomenon occurs when the droplet contact line remains pinned during the drying process; the suspended particles tend to accumulate at the drop edge for capillary outflow to supplement local rapid solvent loss. Further investigation of this nonuniform redistribution process indicates that inner flows, including capillary flow and Marangoni flow, dynamics of the three-phase contact line, and particle-particle/particle-interface interaction will influence the final particle distribution.

Figure 7.20　Illustrations of the coffee ring effect

7.4.1　Mechanism of the Coffee Ring Effect

The coffee-ring pattern arises from the capillary flow caused by droplet evaporation: liquid evaporating from the edge is replenished by liquid from the interior. The resulting edge-ward flow can bring nearly any dispersed material to the edge. As a function of time, this process exhibits a "rush-hour" effect, that is, a rapid acceleration of the edge-ward flow during the final stage of the drying process.

Evaporation induces a Marangoni flow within the droplet (Figure 7.21). If the flow is strong, the particles are redistributed back to the center of the droplet. Thus, for particles to accumulate at the edges, the liquid must have a weak Marangoni flow, or something must happen to disrupt the flow. For example, surfactants can be added to reduce the liquid's surface tension gradient, thereby disrupting the induced flow. The Marangoni flow of the water is weak at the beginning and significantly reduced by natural surfactants.

Interaction of the particles suspended in a droplet with the free surface of the droplet is important for the creation of a coffee ring. As the droplet evaporates, the free surface collapses and traps suspended particles; eventually, all the particles are captured by the free surface and stay there until they move towards the edge of the droplet. This result implies that surfactants can be used to manipulate the motion of the solute particles by changing the surface tension of the droplet rather than trying to control the bulk flow inside the droplet.

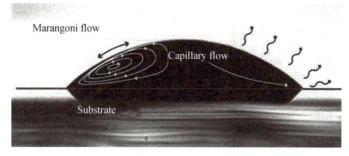

Figure 7.21 Streamline plots of the Marangoni flow and capillary flow

7.4.2 Suppression of the Coffee Ring Effect

The coffee-ring pattern is detrimental when a dried deposit needs to be applied evenly, such as in printed electronics. It can be inhibited by adding elongated particles, such as cellulose fibers, to the spherical particles that cause the coffee ring effect. The size and weight fraction of added particles may be smaller than those of the primary particles (Figure 7.22).

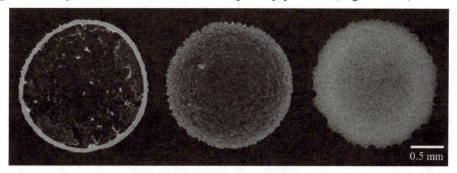

Figure 7.22 Stains produced by colloidal mixtures of polystyrene particles (diameter 1.4 μm) and cellulose fibers (diameter *ca.* 20 nm, length *ca.* 1 μm). [The polystyrene concentration is fixed at 0.1 wt%, and that of cellulose is 0 (left), 0.01 wt% (center), and 0.1 wt% (right).]

It has also been reported that controlling flow within droplet is an effective method to produce a uniform film, for example, exploiting solutal Marangoni flows which occurs during evaporation. Mixtures of low-and high-boiling point solvents were shown to suppress the coffee ring effect, changing the shape of a deposited solute from a ring-like to a dot-like shape.

Controlling the substrate temperature is an effective method to suppress the coffee ring. On a heated hydrophilic or hydrophobic substrate, a thinner ring of an internal deposit form, which is attributed to Marangoni convection. For example, the aqueous droplet in an open space at room temperature is shown in Figure 7.23 (a). In this case, a significant evaporation difference between droplet edge and droplet center is unavoidable, the droplet usually dries at a relatively slow evaporation rate, and most of the suspended particles are transported to the droplet edge by the strong capillary outflow. If the droplet dries rapidly enough, particles tend to aggregate at the gas-liquid interface rather than at the edge of the droplet and deposit directly inside [Figure 7.23 (b)]. Therefore, uniform deposition can be achieved by such a simple and universal evaporation kinetics control without changing the droplet composition or particle modification.

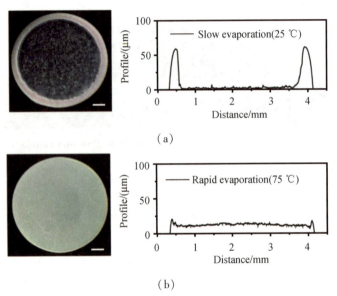

Figure 7.23 The suppressed coffee ring effect with increasing evaporation rate
[Images and morphology profiles of the final deposition by drying at room temperature (a, $T=25\ ℃$) and high temperature (b, $T=75\ ℃$) with constant humidity (50%). Most of the particles deposit at the drop edge, forming a ring-like fashion when drying at a slow evaporation rate. In contrast, a uniform deposition is formed with only a tiny part of the particle deposited on the edge when drying at a high temperature. The scale bar is 0.5 mm.]

Controlling the substrate wetting properties on smooth surfaces can prevent the pinning of the droplet contact line, which will suppress the coffee ring effect by reducing the number of particles deposited at the contact line. Droplets on superhydrophobic or liquid impregnated surfaces are less likely to have a pinned contact line and inhibit ring formation. Droplets that form an oil ring on the droplet contact line have high mobility to avoid ring formation on hydrophobic surfaces.

Alternating voltage electrowetting may suppress coffee stains without adding surface-active materials. Reverse particle motion may also reduce the coffee-ring effect due to the capillary force near the contact line. The reversal occurs when the capillary force exceeds the outward coffee-ring flow through the geometric constraints.

7.4.3 Example: Organic Crystal Growth

An example of suppressing the coffee ring effect is the use of a polymer blend solution to grow organic single crystals. The well-known coffee ring effect is responsible for the 3D growth of organic semiconductor single crystals (OSSCs), in which organic molecules tend to aggregate near the contact line due to the convection induced by evaporation. Therefore, to realize the 2D growth of the OSSCs, it is highly desirable to suppress the coffee ring effect to make the molecules evenly distributed during the crystal growth.

Researchers have demonstrated the wafer-scale growth of crystalline thin film composed of 2D organic crystals by suppressing the coffee ring effect using an organic semiconductor: polymer blend solution. The introduction of high-viscosity polystyrene (PS) in the organic solution can significantly reduce the evaporation-induced convective flow. Consequently, the 2D layer-by-layer molecular packing is greatly enhanced, enabling the uniform growth of high-quality, large-area 2D OSSCs, such as 2,7-didecylbenzothienobenzothiophene (C_{10}-BTBT, Figure 7.24).

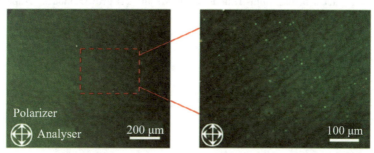

Figure 7.24 Polarized optical images of the large-size 2D C_{10}-BTBT crystal obtained from the C_{10}-BTBT:PS blends

The transport behavior of molecules in organic solution plays a crucial role in determining the growth mode of organic crystals. As shown in Figure 7.25 (a), for the C_{10}-BTBT solution, a high evaporation rate near the three-phase contact line will induce a convective flow for the solvent to supplement the evaporation loss. This flow brings the organic molecules from the fluid interior to the contact line, creating an undesirably uneven distribution of molecules. It is the outcome of the well-known coffee ring effect described by many workarounds. Therefore, it will induce a rapid aggregation of organic molecules at the contact line, leading to the 3D nucleation and, subsequently, 3D growth of the organic crystals [Figure 7.25 (b)]. To attenuate this unfavorable convective flow, the researchers consciously blended the high-viscosity PS with a molecular weight of about 2,000 kDa into the C_{10}-BTBT solution. Significantly, when the high-viscosity PS is used, the evaporation-induced convective flow will decrease 2.5 times in the blended solution compared to the C_{10}-BTBT solution. This can lead to a uniform distribution of organic molecules on the substrate [Figure 7.25 (c) and (d)], thus

Fundamentals of Photophysics

greatly facilitating the 2D growth of the organic crystals.

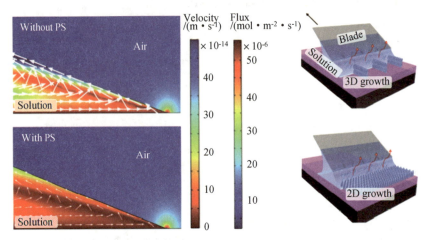

Figure 7.25 Simulated fluid flow around the three-phase contact line for (a) C_{10}-BTBT solution and (c) C_{10}-BTBT:PS blend solution (The convective velocity magnitude distribution and solvent evaporation flux are shown in the surface color. The arrows represent the flow directions, and the length of the arrows indicates the velocity of convective flow.) (b) and (d) are schematic illustrations of the 3D and 2D growth modes, respectively.

7.5 Sunscreen

7.5.1 Effects of Sunscreen

The alarming growth of skin cancer worldwide is the center topic of many studies. In the US alone, 1,000,000 new cases of skin cancer are reported each year, of which 10,000 cases ultimately lead to death. In countries such as Australia and South Africa, where the climate encourages the pursuit of outdoor activities, the incidence of skin cancer in the fair-skinned population is amongst the highest in the world. The causes of skin cancer are complex and unclear in the case of malignant melanoma. However, exposure to solar ultraviolet radiation has been implicated in malignant and nonmalignant melanoma. Therefore, the purpose of sunscreen is to protect against excessive ultraviolet radiation.

Sunscreen (Figure 7.26) is an essential part of a complete sun protection strategy. But sunscreen alone isn't enough to keep you safe in the sun. When used as directed, sunscreen is shown to reduce your risk of skin cancers and skin precancers. Regular daily use of sunscreen with SPF 15 can reduce your risk of developing squamous cell carcinoma (SCC) by about 40% and lower your melanoma risk by 50%.

Figure 7.26　Schematics of sunscreen

Solar ultraviolet radiation is divided into three regions according to the biological effects produced. The ultraviolet C region (UVC) includes the shortest wavelengths from 100 nm to 280 nm, but this region of the solar ultraviolet radiation is absorbed primarily by ozone and oxygen in the Earth's atmosphere and therefore does not reach the land surface. The intermediate range is the UVB from 280 nm to 320 nm, a portion of which cannot reach the Earth's surface due to absorption by stratospheric ozone. Since the absorption spectrum of DNA is spanned by the UVC and UVB, the wavelengths from 100 nm to 320 nm are most damaging to organisms. UVB is most significant in the induction of skin cancers under normal ultraviolet exposure conditions. The longer wavelength region, the UVA, from 320 nm to 400 nm, was once considered harmless because of its low energy but is now known to cause wrinkling, photo aging of the skin, dermatological photosensitivity, some erythema, and has been shown to contribute to melanoma induction.

The coverage surrounding the depletion of stratospheric ozone and increased rates of skin cancer has raised public awareness of the dangers of exposure to sunlight. In response to this problem, the use of sunscreen formulations to protect from the adverse effects of the sun is advocated. Sunscreen contains active ingredients that help prevent the sun's ultraviolet radiation from reaching your skin. As a result, the sunscreen industry is rapidly expanding, and protectants are now not only being incorporated into skin protection creams but also into almost every cosmetic product, including shampoos, moisturizers, and lipsticks, all in the hope of affording the consumer more effective protection.

7.5.2　Classification of Sunscreen

The protective function of a sunscreen formulation depends on its active ingredients. Sunscreen can be divided into physical sunscreen and chemical sunscreen according to the principle of sunscreen. Physical sunscreen ingredients, including titanium dioxide and zinc oxide

Fundamentals of Photophysics

minerals, block and scatter the rays before they penetrate your skin. Chemical sunscreen ingredients, such as avobenzone and octisalate, absorb ultraviolet rays before they can damage your skin. The schematic diagram of physical and chemical sunscreen is shown in Figure 7.27.

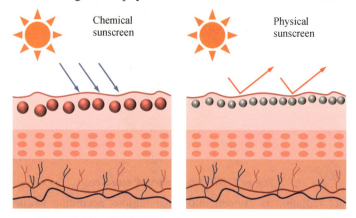

Figure 7.27 Physical sunscreen and chemical sunscreen

With the exception of a few miscellaneous compounds, the chemical absorbers currently used in the sunscreen industry can be classified into seven broad categories, as shown in Figure 7.28. These absorbers, in turn, can be subdivided into either UVA (benzophenones, anthranilates, and dibenzoyl methanes) or UVB (PABA derivatives, salicylates, cinnamates, and camphor derivatives) filters, depending upon which type of radiation they primarily absorb.

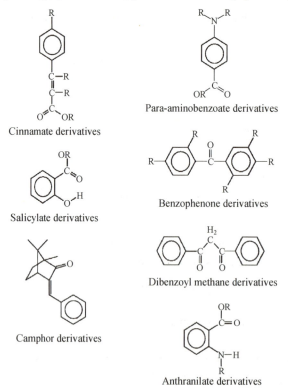

Figure 7.28 The seven major groups of chemical sunscreen filters currently used in the sunscreen industry

In general, these compounds usually contain an aromatic ring conjugated to a carbonyl group. Typically, an electron-releasing group such as an amine or methoxyl group is substituted in the ortho-or para-position of the aromatic ring. In short, therefore, these molecules contain conjugated systems that allow electron delocalization on the absorption of a photon of ultraviolet light. Thus, they achieve the purpose of absorbing any harmful ultraviolet radiation before it reaches the skin. However, we must be aware of the fact that any molecule with extra energy in its excited state is transient and must return to its ground state. During this process, various dissipative processes can take place, which can be a source of skin damage.

7.5.3 Precautions for Sunscreen Use

SPF stands for sun protection factor. Simply put, an SPF rating tells you how long you can stay in the sun without getting burned while wearing that sunscreen, as opposed to how long you can stay in the sun before you burn without wearing that sunscreen. For example, if it usually takes 15 minutes to burn without sunscreen and you apply an SPF 10, it will take 10 times longer (2.5 hours) to burn in the sun. The animation in Figure 7.29 shows how to protect yourself with sunscreen.

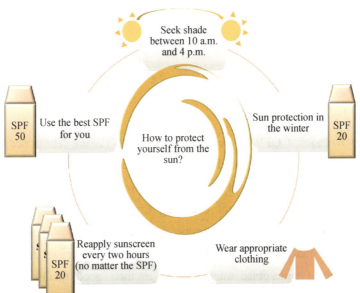

Figure 7.29 Several ways to protect yourself from the sun

The SPF number is determined through laboratory experiments that expose human subjects to a light spectrum meant to mimic the midday sun (when the sun's rays are most intense). Some subjects wear sunscreen, while others do not. A higher SPF doesn't mean great sun protection. It indicates that you will remain protected in the sun for a longer time. For example, both SPF 2 and SPF 30 products can protect the skin effectively. However, SPF 2 sunscreen requires to be used more frequently. To be safe, no matter what SPF you choose, it is a good idea to reapply sunscreen at least every two hours, as well as after swimming or sweating. In fact, the American Academy of Dermatology recommends the daily application of SPF 30 to all exposed skin.

Of course, one's total exposure to sunlight is also affected by one's occupation, clothing, lifestyle, etc. For instance, a teacher is exposed to less solar ultraviolet radiation than a bricklayer, cotton clothes afford more protection than polyester ones, and a swimmer or golfer is exposed to more ultraviolet radiation than a table tennis player.

Sunscreen safety is also a major concern. Take oxybenzone known as benzophenone-3 as an example. FDA approved the use of the chemical in the 1970s. Since then, studies conducted by the National Toxicology Program have shown that it can negatively affect the liver, kidney, and reproductive organs. Numerous studies have also suggested that it acts as an endocrine disrupter in laboratory animals.

Consumers looking to avoid these organic chemicals in sunscreens have few alternatives. They can choose a product that contains titanium dioxide or zinc oxide, both of which provide UVA protection. But most of these inorganically-based sunscreens on the market today contain metal oxides in nanoscale form, which also raises safety concerns. Therefore, in recent years, scientists have been working to conduct more research on nanomaterials about environment, health, and safety.

7.6 Organic Photovoltaics

Organic photovoltaics (OPVs) cells (Figure 7.30), are composed of carbon-rich (organic) compounds and can be tailored to enhance a specific function of the PV cell, such as band gap, transparency, or color. OPV cells are currently only about half as efficient as crystalline silicon cells and have shorter operating lifetimes but could be less expensive to manufacture in high volumes. They can also be applied to a variety of supporting materials, such as flexible plastic, making OPVs be able to serve a wide variety of uses.

OPVs have received widespread attention due to their good qualities, such as solution processability, tunable electronic properties, low-temperature manufacture, and cheap and light materials. While several other photovoltaic technologies have higher efficiencies, OPVs remain advantageous due to the low material toxicity, cost, and environmental impact. They have exceeded certified efficiencies of 13%, close to efficiency values obtained by low-cost commercial silicon solar cells.

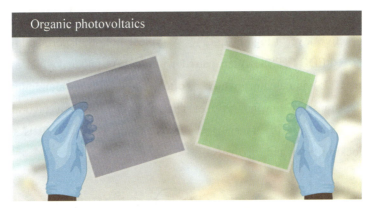

Figure 7.30 Schematic illustration of OPVs

An OPV cell is a solar cell where the absorbing layer is based on organic semiconductors (OSC) that are typically either polymers or small molecules. For organic materials to become conducting or semiconducting, a high level of conjugation (alternating single and double bonds) is required. Conjugation of the organic molecule results in the electrons associated with the double bonds becoming delocalized across the entire length of conjugation. These electrons have higher energies than other electrons in the molecule and are equivalent to valence electrons in inorganic semiconductor materials.

However, these electrons do not occupy a valence band in organic materials but are part of what is called the HOMO. Just like in inorganic semiconductors, there are unoccupied energy levels at higher energies. In organic materials, the first one is called the LUMO. Between the highest HOMO and LUMO of the OSC is an energy gap—often referred to as the band gap of the material. With increased conjugation, the band gap will become small enough for visible light to excite an electron from HOMO to LUMO.

7.6.1 Device Structure of an Organic Photovoltaic

The majority of OPVs used in modern research is solution-processed bulk heterojunction (BHJ) cells. Depending on the orientation of the electrodes, its architecture can be classified as conventional or inverted (Figure 7.31). There remains a small amount of work using planar bilayer junctions in research, but these will not be discussed here.

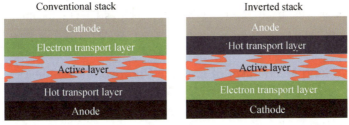

Figure 7.31 The stacks used in a conventional and inverted OPV cell, where the layers are not given to scale

Fundamentals of Photophysics

Charge-carrier transport is facilitated by hole-and electron-transporting interfacial layers on either side of the active layer (Figure 7.32). A typical hole-transporting layer (HTL) in a conventional stack is PEDOT:PSS, often paired with an ITO anode, while a typical electron-transporting layer (ETL) is calcium, which is often paired with an aluminum cathode. These layers promote the transport of one type of charge carrier through favorable energy level positioning while discouraging the transport of the other carrier. The HTL is sometimes known as the electron blocking layer and vice versa.

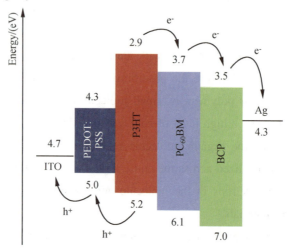

Figure 7.32 An approximation of charge carrier transport in the full stack of a conventional OPV, using the organic molecule BCP as an ETL (The smooth movement of energy levels facilitating transport is known as a band gap cascade.)

In the past few decades, the majority of acceptors used were derived from fullerene (normally in the form of PCBM). However, there has recently been a significant movement towards non-fullerene acceptors (NFAs), especially those based on small molecules. These have yielded higher efficiencies and stabilities than fullerene-based acceptors. As opposed to typical fullerene acceptors, which have poor light absorption in the visible regime, NFAs are typically designed to absorb highly, allowing exciton generation in both the donor and acceptor components of the active layer.

Donor OSCs vary more widely but are often polymer-based. Examples of highly performing donor materials include PBDB-T and PTB7. Donors are generally classified according to band gap and are known as wide band gap (>1.8 eV), such as P3HT; medium band gap (1.6 – 1.8 eV), such as PCDTBT; or narrow band gap (<1.6 eV), such as PTB.

7.6.2 Working Mechanism of Organic Photovoltaics

As with other PV technologies, the purpose of an OPV is to generate electricity from sunlight. This is achieved when the energy of light is equal to or greater than the band gap, leading to the absorption and excitation of an electron from the HOMO to the LUMO (Figure 7.33).

The excited electron will leave behind a positively-charged space known as a "hole". Due to the opposite charges of the hole and electron, they become attracted and form an electron-hole pair, also known as an "exciton". To remove the charged particles from the solar cell, we must separate the electron-hole pair, and this process is known as "exciton dissociation".

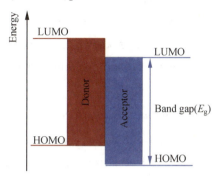

Figure 7.33 The donor-acceptor band gap offset typical in an OPV,
used to overcome exciton binding energy and facilitate dissociation

In a typical inorganic semiconductor, the attraction between the electron and hole (known as the exciton binding energy, E_b) is small enough to be overcome by thermal energy at room temperature (about 26 meV). This is due to a high dielectric constant-meaning. There is significant screening between the electron and hole, reducing the attraction between them. The ease in separating the electron and hole allows easy exciton dissociation.

In contrast, OSCs have low dielectric constants, leading to large E_b values in the range of 0.3 - 0.5 eV. As a result, exciton dissociation cannot be achieved by thermal energy alone in OSCs. To overcome this problem, we need at least two different OSCs within an OPV. The energy levels between the two different OSCs are offset, with the difference being greater than E_b, allowing exciton dissociation at the interface between them.

Depending on how the exciton dissociates, the OSCs are classified as either a donor or an acceptor (referring to whether the electron has been donated by a material or accepted by a material). In most OPVs, the donor will absorb the most light, and therefore the exciton will be generated on this material. At the interface with the acceptor, the exciton will dissociate. The electron will be donated to the acceptor material, which has a deeper HOMO and LUMO level, while the hole remains on the donor material.

The steps that govern the OPV function can be summarized as follows (details are shown in Figure 7.34).

i. Absorption of the light leading to exciton generation

Light with high enough energy levels can be absorbed by the OSC and excite electrons from the HOMO to the LUMO to form an exciton. If the energy of the light being absorbed is greater than the band gap, the electron will move to a higher energy level than the LUMO and decay down. This process is known as "thermalization", during which the energy is lost as heat.

Thermalization is a key energy loss mechanism in photovoltaics.

ii. Diffusion of the exciton to a donor-acceptor interface

Once formed, the exciton diffuses through the OSC component to the donor-acceptor interface, where the offset between LUMO levels will drive exciton dissociation. This must occur within a certain amount of time. If not, the excited electron will return to the empty energy state (known as the hole), a process known as "recombination". The time taken is known as the "exciton lifetime", which is often represented as the distance that the exciton can diffuse in this time (which is around 10 nm).

iii. Dissociation of the exciton across this interface

The electron at the interface will move to the acceptor material, and the hole will remain on the donor. These charge carriers will still be attracted, forming a charge-transfer state. When the distance between the pair increases, the attraction decreases. Eventually, the binding energy between them is overcome by thermal energy, and a charge-separated state is formed. While the electron-hole pair is still attracted in the charge-transfer state, recombination can occur across the interface between the two materials.

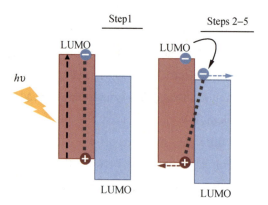

Figure 7.34 An approximation of the basic steps that govern OPV function under light illumination

iv. Charge carrier transport

The charge carriers will then diffuse to the appropriate electrodes (i.e. the holes to the anode and electrons to the cathode) through the relevant interfacial layers.

v. Charge carrier collection

At the electrodes, the charge carriers are collected and used to do work in the external circuit of the cell, producing a current.

7.6.3 Advantages and Disadvantages of Organic Photovoltaics

Nowadays, silicon solar panels are an industry standard, but these rigid and heavy blocks may be shunted aside by plastic rivals. Lightweight, flexible solar panels that could be printed and stuck onto buildings or placed in windows or cars, turning light into electricity in locations inaccessible to their heavier cousins. Organic means carbon-containing molecules, and OPVs can be considered as plastic solar cells. They offer advantages over silicon-based solar panels in many aspects.

Organic semiconductors are cheaper. They are lightweight and offer flexibility in their architecture, and in principle, they can be more environmentally friendly. In theory, plastic solar cells should also be easier to manufacture. The main difference between silicon technologies and OPV is that we are able to print it or coat it onto something as a thin film.

Plastic solar panels can weigh around 500 g per square meter, which is more than 40 times lighter than their silicon counterparts. Plastic panels can be attached to the fronts of buildings or placed on the roofs of buildings that might struggle to support standard solar panels safely. Organic solar cells are also much thinner than silicon solar cells, offering substantial savings on materials, which is good for the environment.

OPVs can be as thin as a few millimeters in thickness and can be placed onto plastic polyester films. Thin semitransparent OPVs can be fitted inside window panes, so that office windows could filter out some sunlight while turning it into electricity.

Solar cells based on the internal photovoltaic effect are probably the most elegant demonstration of renewable energy generation. Despite decades of research and development of thin-film photovoltaics, cells based on crystalline silicon have kept their standing of market leader. Although silicon solar cells have excellent properties, such as high efficiency and stability, and have also been scaled to a comparatively low cost, they have significant disadvantages, such as high-temperature processing with environmentally destructive materials, non-flexibility, and a large module weight (around $10 \text{ kg} \cdot \text{m}^{-2}$).

OPVs are arguably the most radical approach among the various alternative thin-film systems. The intrinsic properties of organic semiconductors allow the coating of plastic substrates in a roll-to-roll process at room temperature with extremely thin (around 100 nm) absorbers, mainly consisting of carbon. This results in flexible modules that are lightweight (around $500 \text{ g} \cdot \text{m}^{-2}$), display homogeneous surfaces with adjustable color and can be semitransparent. These unique properties enable many new applications that are not possible using conventional silicon photovoltaics. A prime example is building-integrated photovoltaics, in which light weight and transparency are key features; organic cells on facade elements or transparent cells on windows combine functions such as sun protection and energy generation.

OPV's great strength lies in the diversity of organic materials that can be designed and synthesized for the absorber, acceptor, and interfaces, but the scaling efficiency and lifetime to large area modules should be further improved.

7.6.4 The Future of Organic Photovoltaics

While a large majority of OPV literature remains focused on efficiency values, the main issues restricting the commercialization of OPVs are scalability and long-term stability. Indeed, some literature has suggested that the current efficiencies obtained could be competitive with other technologies if scaled appropriately. At present, there is little consideration of the synthetic complexity of materials or scalable deposition techniques, so these are likely to be an area of focus in the future, along with greener solvent systems.

OPVs have struggled with long-term stability, primarily due to damage from water and oxygen ingress. Improvements in this area are likely to be found with a better understanding of NFAs, as this group of materials has shown significant promise in terms of long-term stability.

NFAs are also likely to be critical in further improving efficiencies. Some studies have shown significantly less non-radiative recombination losses in NFAs compared to conventional fullerene acceptors, and some acceptors have shown the ability to work with very small LUMO offsets. The exact mechanisms of non-radiative recombination mitigation and exciton dissociation in these systems is still under discussion but will likely remain a significant area of investigation as OPV studies move forward.

Areas of particular interest in OPVs include:

i. NFAs

These are acceptors based on materials other than fullerene derivatives and are typically based on an acceptor-donor-acceptor (A-D-A) structure. The most efficient NFAs are those based on indacenodithiophene cores, such as ITIC and IT-2F.

ii. Singlet fission

In singlet fission, absorption of a high-energy photon generates a singlet exciton that is then converted into two triplet excitons, generating two excitons from a single photon. This can then theoretically overcome the Shockley-Queisser limit on efficiency.

iii. Ternary cells

In ternary OPVs, three OSCs are used typically in the active layer instead of two to improve the absorption of the cell in an attempt to boost efficiency. Ternary OPVs now have achieved efficiencies exceeding 14%, and more details can be found in a previous Ossila blog post.

iv. Molecular design

The majority of highest efficiency OPVs have been obtained by donor-acceptor pairs specifically tuned to give highly complementary energy levels by chemical modification. This energy level tuning by molecular design is likely to be an area of significant focus as OPVs move forward.

v. Scalable deposition techniques

There has been a recent focus on the manufacture of OPVs using more scalable techniques than spin coating, such as spray coating, blade coating, slot-die coating, and inkjet printing. This is likely to become increasingly relevant as OPVs move towards commercialization.

Given the advantages of an all-carbon, flexible, low-cost technology, OPVs could revolutionize renewable energy generation. However, it is still crucial that the efficiencies of organic solar cells can compete with silicon and other thin-film technologies. In hard numbers, this would mean lab efficiencies of well above 20% and module efficiencies of at least 15%. It is unlikely that pure optimization will be sufficient to reach such goals. There are several fundamental loss processes that require considerable basic research efforts to understand the limits of organic solar cells. The most important of these is voltage loss, which is related to the problem of exciton separation. Almost "ideal" solar cells, such as those based on GaAs, lose about 0.3 V compared with the voltage equivalent of the optical gap. Organic solar cells typically lose more than double this value. The key to reducing these losses is understanding the microscopic processes occurring at the photoactive heterojunction and finding a systematic way to optimize the gap at this junction. The potential for improvement is highlighted by the surprising observation that "diluted" organic solar cells, in which one of the components of the donor-acceptor system is present in only low concentration, still work well and have significantly reduced voltage losses. Moreover, subtle aspects of molecule alignment at the interface can lead to dramatic shifts (>0.5 eV) in the energy-level alignment. To optimize solar cells in which a 0.1 eV shift in band structure has a profound influence, researchers must understand and control this effect to achieve optimized cells.

The second key challenge is to increase the exciton diffusion length to a range in which all excitons are collected, even in regions where the absorption is weak. Here, it is essential to clarify whether the diffusion length is limited by intrinsic or extrinsic (that is, determined by impurities) factors. A basic understanding would then allow a directed search to be performed for materials with long diffusion lengths. Reports of very long diffusion lengths in single crystals indicate that the intrinsic potential for long-range energy transport is huge. Because it seems unlikely that single crystals will ever be of importance in real devices, the challenge is to realize thin films with large oriented domains and few defects that allow diffusion lengths at least several times longer than the absorption length. In that case, the organic solar cell community would happily throw the concept of the bulk heterojunction overboard in favor of planar structures, which are much simpler to deposit and allow better control of the interfaces.

7.7 Organic Light-emitting Diodes

OLED is an abbreviation for Organic Light-Emitting Diode. An OLED is similar to an LED; however, an OLED has an emissive electroluminescent layer of the film made up of organic molecules. Light is emitted when an electrical current travels through the organic molecules. It is basically a type of light-emitting diode or LED that has an emissive electroluminescent layer that acts as a film of organic compounds and is responsible for emitting light when an electric current is applied. Nowadays, OLEDs are extensively used for developing digital displays in several devices such as television, monitors, phones, portable handheld gaming devices, smartwatches, etc. An OLED prototype display is shown in Figure 7.35. OLEDs are also incorporated in solid-state lighting devices. An OLED is thinner and has a better display compared with a vacuum fluorescent display (VFD). Additionally, an OLED has a brighter, higher contrast display with faster response times, wider viewing angles, and less power consumption.

Figure 7.35 Photographs of OLED prototype displays

7.7.1 Device Structure of an Organic Light-emitting Diode

A general OLED comprises a sheet of organic materials deposited on a substrate, which is placed in between the cathode and the anode, as shown in Figure 7.36. The delocalization of π electrons is due to the conjugation over a portion of the whole molecule, resulting in the organic molecules becoming conductive electrically. These materials behave like organic semiconductors as their conductivity typically lies between insulators and conductors. In these materials, the role of the valence and conduction bands of inorganic semiconductors is performed by LUMO and HOMO.

Initially, polymer OLEDs were designed to have a single organic layer. However, nowadays, multilayer OLEDs can be developed with two or more layers to improve the efficacy of the device. Along with the number of layers, the material used for aiding charge injection at electrodes is also important in the final functioning of the device.

The conductive property of the material used decides whether there would be a more gradual electronic flow, or a charge blockage or resistance from traveling to the opposite electrode and being unexploited. The substance is chosen depending on material properties such as electrical conductivity, optical transparency, and chemical stability. Nowadays, OLEDs have a simple bilayer structure that comprises an emissive layer and a conductive layer. Based on the chemical structure of the material, the emitter can either be fluorescent or phosphorescent.

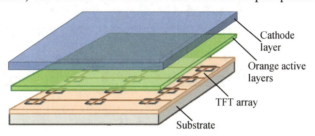

Figure 7.36　Device structure of a typical OLED

7.7.2　Working Mechanism of Organic Light-emitting Diodes

When the operation begins, a potential difference is applied across the OLED. The anode is kept at a higher potential with respect to the cathode (Figure 7.37). The material of the anode is based on material properties such as electrical conductivity, optical transparency, and chemical stability. The LUMO of the organic layer (at the cathode) receives the injected electrons, and the HOMO (at the anode) withdraws the electrons, or in other words, injects electron-hole pairs. In organic semiconductors, the holes are comparatively more mobile than the electrons. Therefore, the recombination of electrons and holes into an exciton occurs closer to the emissive layer.

This results in the decay of an excited state that leads to the emission of radiations having a wavelength ranging in the visible spectrum. The precise wavelength or frequency of the emitted radiation is determined by the band gap of the material, i.e. the difference in energy levels of the HOMO and LUMO. In the case of phosphorescent emitters, the excitons (singlets and triplets) decay radiatively. However, in the case of fluorescent emitters, the triplets do not emit any light. These fluorescent emitters possess a maximum intrinsic efficiency of 25% only. However, phosphorescent emitters (particularly of short wavelength) have a lower lifetime than fluorescent emitters.

The electron-hole fermions generated have a half-integer spin. Excitons can exist in singlet or triplet states based on the combination of different spins of electrons and holes. For each singlet exciton, three triplet excitons are formed. The triplet state decay (prevalent in phosphorescent) forbids spin and therefore increases the transition time span. Phosphorescent

Fundamentals of Photophysics

OLEDs facilitate intersystem crossing from both triplet and singlet states by using spin-orbit interactions. This improves internal efficiency. Nowadays, OLEDs are extensively used for developing digital displays in several devices such as television, monitors, phones, portable handheld gaming devices, smartwatches. OLEDs are also incorporated in solid-state lighting devices.

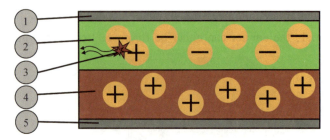

Figure 7.37 Schematics of an OLED: ① cathode; ② emissive layer; ③ emission of radiation; ④ conductive layer; ⑤ anode

7.7.3 Advantages and Disadvantages of Organic Light-emitting Diodes

Advantages of OLEDs include:

(1) OLEDs are biodegradable substances.

(2) OLEDs are comparatively lighter, thinner, and more elastic than the crystalline layers in liquid crystal displays or LEDs.

(3) OLEDs are very flexible; therefore, they can be easily folded and rolled up as required in roll-up displays inserted in certain fabrics these days. The reason behind this is that the substrate used in OLED is polymer rather than the glass used for an LED or an LCD.

(4) OLEDs are comparatively brighter than normal LEDs. The artificial contrast ratio of OLEDs is higher. This is due to the fact that the organic layers of OLEDs are much narrower than the analogous inorganic crystal layers of an LED. Moreover, the conductive and emissive layers of OLEDs do not use glass (which absorbs some portion of light) and can have a multi-layered design.

(5) Unlike an LCD, an OLED setup has no requirement for a backlight. This helps in reducing the energy or power consumption of an OLED device. LCDs need illumination to help in producing a visible image which necessitates more energy, whereas OLEDs are capable of generating their own light.

(6) The production process of an OLED is easier, and it can be processed into large thin sheets. Comparatively, it is much more difficult to produce such a large number of liquid crystal layers.

(7) OLEDs provide a wider viewing angle compared to LCDs. This is because an OLED pixel emits light directly. The pixel colors of an OLED are not shifted along with the change in the angle of observation from normal to a right angle.

(8) An OLED has a more rapid response time compared to an LCD.

Despite the advances achieved, there are some obstacles during the development of OLEDs:

(1) The lifespan of an OLED is lower than LCD. The green and red OLED films have longer lifespans of about 46,000 to 230,000 hours. However, the blue OLEDs have much shorter lifetimes, up to 13,000 – 14,000 hours approximately.

(2) The substances used for producing blue light in an OLED degrade more rapidly than the substances producing other colors, which causes a reduction in the overall luminescence of the OLED.

(3) OLEDs should not come in contact with water because it leads to instant degradation.

(4) OLEDs need about three times more power to display an image having a white background. The extensive use of white backgrounds can lead to reduced battery life in mobile phones and other devices.

(5) OLEDs are expensive. They cost around 10 to 20 times more than similar performing LEDs.

(6) There is a lack of a wide range of commercially available OLED products.

(7) OLEDs have a high capacitance that limits the device modulation bandwidth to about 100 kHz range.

(8) OLEDs have low light efficiency.

The differences between an OLED and an LED are listed in Table 7.1.

Table 7.1 Comparison between an OLED and an LED

OLED	LED
In the case of OLEDs, the emissive electroluminescent layer is made up of organic compounds.	In the case of LEDs, the emissive electroluminescent layer is made up of inorganic substances.
In OLED television, each pixel works individually.	In the case of LEDs, the emissive electroluminescent layer is made up of inorganic substances.
They have a lower light efficiency.	They have a higher light efficiency.
They can be thin and small due to their flexibility.	They are comparatively less flexible.
They do not use a backlight as they can produce their own light.	They cannot produce their own light and therefore use a backlight.
They are expensive.	They have comparatively lower manufacturing costs.
OLEDs do not require any kind of glass support.	LEDs require glass support.
They provide a wider viewing angle.	They have a comparatively lower angular range.

 ### 7.7.4 The Future of Organic Light-emitting Diodes

OLED technology may usher in a new era of large-area, transparent, flexible and low-energy display, and lighting products. The flexibility of OLEDs enables manufacturers to

Fundamentals of Photophysics

produce OLEDs using roll-to-roll manufacturing processes and allows for the production of flexible display and lighting products. OLEDs are commercially produced on rigid glass substrates mainly. However, first applications like watches or bent displays using flexible OLEDs have entered the market lately.

Developing sufficiently durable and flexible OLEDs will require better materials and further development of manufacturing tools and processes. Flexible plastic substrates need improved barrier layers to protect OLEDs from moisture and oxygen. Thin-film encapsulation also is needed to create thin and flexible metal-and glass-based OLEDs.

These advances ultimately may lead to very flexible OLED panels for display and lighting products, ensuring that any surface area (i.e. flat or curved) can host a light source. Recent demonstrations by display and lighting companies have already hinted at the potential of flexible OLED technology. Substantial development efforts are being invested in this area, and, if successful, flexible OLED panels may become commercially available as early as the last half of this decade.

7.8 Glow Sticks

Glow sticks (Figure 7.38) are plastic cylinders containing two liquids that temporarily create light when mixed together. Typically, the cylinders are about 10 cm to 13 cm long and less than 2.5 cm in diameter. Glow sticks are available in many colors and are often used for decoration or entertainment, such as at parties, concerts, and other nighttime events. They also have practical uses for camping, military or police operations, underwater activities, or certain emergencies. Thin glow sticks that are made of a more flexible plastic can take the form of necklaces, bracelets, or other shapes.

Regardless of their forms, glow sticks depend on a chemical process known as chemiluminescence to produce light. In chemiluminescence, a chemical reaction causes a release of energy. Electrons in the chemicals become excited and rise to a higher energy level. When the electrons drop back to their normal levels, they produce energy in the form of light.

The chemicals used to create the reaction in glow sticks are usually hydrogen peroxide and a mixture of phenyl oxalate ester and the fluorescent dye, or fluorophore, that gives the glow stick its color. Common colors of glow sticks include yellow, green, pink, blue, and orange. They also are available in red, white, yellow-green, and other shades and colors.

>>> **Chapter 7**　Photophysics in Life

Figure 7.38　**Schematics of glow sticks**

　　A glow stick's hydrogen peroxide is contained in a small glass or breakable plastic vial that floats within the mixture inside the stick. This is why the user of the glow stick must bend it to make it start glowing. When the stick bends, the vial breaks, the hydrogen peroxide is released, the chemical reaction begins, and the distinctive glow appears. The chemicals that are used might be toxic, so if the glow stick itself breaks, it should be thrown away, and the chemicals that might have leaked out should be washed off the user's skin and any other surfaces with which they came into contact. In addition to the color, the duration of the glow (usually several hours) also depends on the exact composition and quality of the chemicals inside. Some people say that a glow stick can be preserved by sticking it in a freezer. Indeed, cooling a glow stick will slow down the chemical reaction that is taking place inside it. The glow will not be as bright, but it will continue for a longer period of time.

　　Conversely, heating a glow stick, such as by placing it in a microwave, will speed up the chemical reaction. This will produce a brighter light. However, the glow will not last nearly as long because the reaction will use up all of the available hydrogen peroxide more quickly. Microwaving glow sticks might not be recommended by some manufacturers, and caution should always be used when it is being done.

7.8.1　Mechanism of Glow Sticks

　　The chemicals that react together in glow sticks to create light are kept separate until the right moment. The glow stick's outer plastic tube holds a solution of an oxalate ester and an electron-rich dye along with a glass vial filled with a hydrogen peroxide solution. The reaction will start when breaking the glass tube, and the hydrogen peroxide will be released. When the chemicals mix, the reaction takes several steps before releasing light.

　　First, the hydrogen peroxide and oxalate ester react to form a high-energy intermediate, but the precise nature of that intermediate is still something of a mystery. Many chemists believe it is the strained molecule 1,2-dioxetanedione. Despite 50 years of looking for it, there is no direct

Fundamentals of Photophysics

evidence of that compound.

Glow sticks to light up when oxalate esters react with hydrogen peroxide to form a high-energy intermediate (possibly 1,2-dioxetanedione, Figure 7.39). This intermediate reacts with dye, which moves to an excited state (indicated with Dye*) and then releases light as it relaxes.

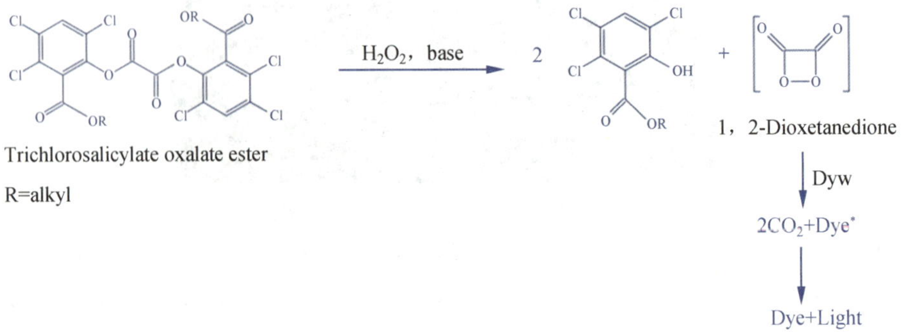

Figure 7.39 Chemical reactions involved in the glow stick

Although chemists are not certain of the precise structure of the high-energy intermediate, they know it is a good electron acceptor. It snags an electron from the dye and then breaks down into carbon dioxide and a negatively charged carbon dioxide radical anion. The dye, which has become a positively charged radical cation, takes back an electron from the CO_2 radical anion.

By taking back the electron, the dye gains excess energy. The molecule uses that energy to move into an excited state before dropping back down and emitting the energy as a photon of light, and finally the glow stick will glow.

7.8.2 History of Glow Sticks

The first glow stick was made by Dr. Edwin Chandross, a Brooklyn-born organic chemistry specialist in the 1960s. People have different theories as to why glow sticks were invented. It is widely assumed that they were made as emergency flares and for other recreational purposes. However, the scientist had nothing complex on his mind while inventing the glow stick. He was captivated by the idea that fireflies emit light and glow naturally. He only desired to imitate fireflies. Many other scientists have improved on Edwin's innovation throughout the years. Dr. Edwin Chandross was initially intrigued by chemiluminescence. His interest was piqued by the luminol experiment at the Massachusetts Institute of Technology. After graduation, he fumbled through countless experiments until he landed on one brilliant experiment that helped him discover the gateway to chemiluminescence. He reasoned that per oxalate and esters were the most important components. He wanted to put his idea to the test by developing a substance that, when mixed with hydrogen peroxide, would provide an active component. This demanded the usage of two components. One important component was chloride (a volatile oxalic acid derivative). After successfully identifying the initial ingredient, he went through a series of updated trials to find the optimal luminescence-producing mixture after synthesizing a test chemical that lit up gently.

Chandross published the reaction in the journal *Tetrahedron Letters* (1963), and it caught the attention of Michael M. Rauhut, who was manager of exploratory research at the chemical firm American Cyanamid. Rauhut and his colleague Laszlo J. Bollyky swapped the reactive oxalyl chloride that Chandross used for less reactive and longer-lasting oxalate esters, settling on an aromatic oxalate ester. American Cyanamid chemists also added a little salicylate base to catalyze the reaction, as well as experimenting with a rainbow of dyes. In modern glow sticks, rhodamine B produces a radiant red; 9,10-bis(phenylethynyl)anthracene gives a green glow, and 9,10-diphenylanthracene lights up with a blue hue.

7.8.3 Dangers of Opening Glow Sticks

Glow sticks are safe as long as precautions are followed and the chemicals are kept inside. Cutting open a glow stick can also cause the broken shards of glass to fall out. The packaging on glow sticks says they are nontoxic. However, the safety warnings on glow sticks warn us not to puncture or cut the plastic cover on the glow stick.

Although glow sticks contain no deadly dangerous chemicals, the chemicals should still be handled and treated with respect. Some glow products use a chemical called dibutyl phthalate. Other glow products contain a small glass vial inside the plastic tube that contains a mixture of hydrogen peroxide in phthalic ester. Outside of the glass vial is another chemical called phenyl oxalate ester. When the tube is cracked, the glass inside is broken, and the chemicals all mix together in a reaction that causes the glow.

Dibutyl phthalate is used to help make plastics soft and flexible. It is also used in glues, nail polish, leather, inks, and dyes.

Hydrogen peroxide is used as a cleaning agent. Over-the-counter hydrogen peroxide is diluted and not as strong as the hydrogen peroxide found in glow sticks. This hydrogen peroxide is corrosive to the skin, eyes, and respiratory tract. This is the type of hydrogen peroxide used in Steve's Elephant Toothpaste demonstration. It is not meant to be handled or mixed into other solutions.

Phthalic ester is a substance added to plastics to increase flexibility, durability, and transparency. Phthalates are being phased out in many products due to health concerns.

Phenyl oxalate ester is responsible for the luminescence in a glow stick. The reaction with hydrogen peroxide causes the liquid inside a glow stick to glow.

To sum up, these chemicals can sting and burn eyes, irritate and sting the skin, burn the mouth and throat if ingested. If the chemicals are ingested or spilled in the eyes or on the skin,

Fundamentals of Photophysics

it is recommended the area is rinsed with water, and the local poison control center is contacted.

7.9 Infrared Detection

Infrared detection devices (Figure 7.40) are sensors that can detect radiation in the infrared portion of the electromagnetic spectrum ($>10^{12}$ Hz to 5×10^{14} Hz). Usually, such devices form the information they gather into visible-light images for the benefit of human users. Alternatively, they may communicate directly with an automatic system, such as the guidance system of a missile.

Because all objects above absolute zero emit radiation in the infrared part of the electromagnetic spectrum, infrared detection provides a means of "seeing in the dark"—that is, forming images when the light in the visible portion of the spectrum ($>4.3 \times 10^{14}$ Hz to 7.5×10^{14} Hz) is scarce or absent. Because the warmer an object is, the more infrared radiation it emits. Infrared imaging is also helpful for detecting outstanding heat sources that may be invisible or hard to detect even when there is ample visible light (e.g. exhaust heat from ships, tanks, jets, or rockets). Many devices used by police, security, and military organizations, including user-wearable, gun-mounted, vehicle-mounted, missile-mounted, and orbital systems, exploit some form of infrared detection technology.

Figure 7.40 Photographs of common infrared detection devices

7.9.1 Principles of Infrared Detection

Infrared radiation (IR) light consists of electromagnetic radiation that is too low in frequency (i.e. too long in wavelength) to be perceived by the human eye yet is still too high in frequency to be classed as microwave radio. IR light that is just beyond the human visual limit ($>1.0 \times 10^{14}$ Hz to 4.0×10^{14} Hz) is termed near IR, while light farther from the visible spectrum is divided into middle IR, far IR, and extreme IR. Military and security systems utilize mostly near IR, and a narrow band in the far IR centered on 3.0×10^{13} Hz because the Earth's atmosphere happens to be transparent to IR radiation primarily in these two "windows".

All objects above absolute zero glow in the far IR, so no source of illumination is needed to image scenes using such radiation; to image scenes in near IR, illumination from a LED or filtered light bulb must be supplied. Howerer, near-IR images are still cheaper than passive far-IR imagers.

There are two designs for electronic IR imagers. The first is the scanner (Figure 7.41). In this design, light from a tiny portion of the scene to be imaged is focused by an optical and mechanical system on a small circuit element sensitive to photons in the desired IR frequency range. The intensity of the signal from the IR detector element is recorded, then the mechanical-optical system shifts its focus to a different scene fragment. The response of the IR detector element is again recorded, the view shifts again, and so forth, systematically covering the scene. Many scene-covering geometries have been employed by scanning imagers; the scanner may record horizontal or vertical lines (rasters), spiral outward from a central point, covers a series of radii, and so on.

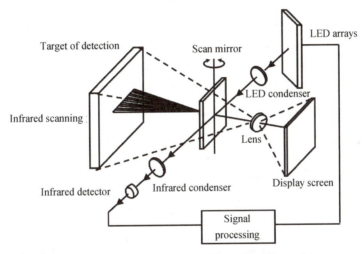

Figure 7.41　Schematic illustration of an IR scanner

The second basic type of IR imaging system is the "starer" (Figure 7.42). Such a system is said to "stare" because its optics do not move like a scanner's, scanning the scene a little bit at a time. Instead, they focus the image onto an extended focal plane. A flat (planar) array of tiny sensors is located in this plane, each equivalent to the single IR sensor employed in a scanning system. The system can record an entire image at once by measuring the IR response of all the elements in the flat array simultaneously (or rapidly). Image resolution in a staring scanner is limited by the number of elements in the array, whereas in a scanning system, it is limited by the size of the scanning dot.

Hybrid designs, in which partial or entire scan lines are sensed simultaneously by rows of sensors, have also been developed. Chemical films have also proved useful for IR imaging, but these are rarely used nowadays.

Fundamentals of Photophysics

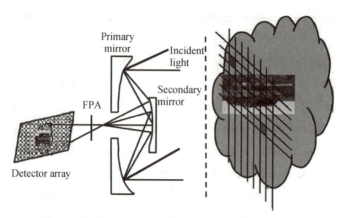

Figure 7.42 Schematic illustration of a "starer"

The earliest IR imagers built in the 1940s, 1950s, and 1960s were scanners. Starers were not technologically feasible until the early 1970s when large-scale circuit integration made possible the manufacture of focal-plane arrays with good resolution. As integrated-circuit technology has been refined, focal-plane arrays have become cheaper. Starers have many advantages, including greater reliability due to the absence of moving parts, quicker image acquisition, and freedom from internally-produced mechanical vibration.

Formerly, to keep the sensor from blinding itself with its IR radiation, both scanners and starers need to be cooled by liquid nitrogen. In recent years, however, uncooled IR imagers, both scanners and starers, have been increasing in quality and decreasing in price.

7.9.2 Applications of Infrared Detection

Aircraft, ground vehicles, surface ships, human beings, industrial facilities, rockets, and warheads entering the atmosphere are some of the objects of military interest that emit IR in telltale patterns. A wide variety of military IR systems have been developed to exploit these patterns. For example, heat-seeking missiles [Figure 7.43 (a)] that home in on the IR-bright gasses emitted by jet-aircraft engines have been commonplace since the 1950s. Heat-seeking missiles have also been developed for use against surface vehicles and ships. Also, starting in the early 1960s, military IR-imaging satellites [Figure 7.43 (b)] that have been observing the Earth detect the IR emissions of rocket and missile launches, and modern proposals for ballistic-missile defense depend heavily on space-and ground-based IR detectors that will track missiles and warheads as they arc through space. The US military is currently designing a new system of satellites dedicated to tracking missiles using IR imaging, the Space-Based Infrared System. According to the United States Space Force, this system will have a "unique capability to track missiles throughout their trajectory (not just during the hot boost phase when IR emissions from the missile are most intense), allowing the system to cue missile defense systems with accurate targeting data" effectively.

 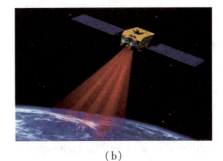

(a) (b)

Figure 7.43 (a) **Heat-seeking missile equipped on a military aircraft and** (b) **infrared detection by satellite.**

Various IR camera systems for "seeing in the dark" are also commonplace. These may be mounted on vehicles or stationary locations to allow nighttime surveillance of a fixed area. Night-vision systems worn on helmets and mounted on portable weapons usually do not operate by sensing IR. Instead, they amplify the visible light already present in a dark scene. Therefore, they are sometimes called "starlight" vision systems.

IR imaging is being investigated for use in the detection of landmines. Antipersonnel mines are typically buried only a few centimeters below the surface, so the heat radiation pattern of an area can, under some conditions, reveal their presence.

The security of a building or area of land from intruders is often enhanced by cameras that image the perimeter of the secure area and can be monitored by personnel in a central office. At night, such systems must be supplied with illumination or capable of IR imaging. Visible-light camera systems are cheaper and easier for human users to interpret. However, because excess illumination of an area by visible light (light pollution) is sometimes a concern, and because security forces may wish to keep an area under surveillance without making their presence known, IR systems are widely used for perimeter security and other surveillance tasks.

IR imaging has many other applications in police and security work besides surveillance. Aerial IR imaging can track vehicles, show which vehicles in a parking lot have arrived most recently, distinguish heated buildings, and locate buried structures (e.g. clandestine chemical laboratories) emitting heat through vents. IR images can be used to precisely determine the time of death of a person deceased for less than 15 hours or to detect document forgery by revealing subtle mechanical and chemical disturbances of the original paper and ink. The power consumption in a building can be estimated in real-time by observing the IR radiation emitted by the power transformer on the pole outside; modifications to walls or automobiles are often obvious in IR images; and IR images can reveal such visually inconspicuous features of crime scenes as use of cleaning solvents to remove blood, drag-marks across carpets, fresh paint, and explosives residues.

Fundamentals of Photophysics

7.10 Bionic Navigation Sensors

Polarization is a fundamental property of light, and light from any source is polarized to some degree.

Polarimetry is a valuable technique for obtaining polarization information about a wide range of sources and objects. Therefore, it is implemented in many different research fields, especially in the field of bionic polarization navigation. Bionic polarization navigation has received much attention in recent years due to its advantages, namely autonomy and no error accumulation. Bionic polarization navigation originated from biologists, who found that many insects in nature can perceive polarized light from the sky and use it to determine the heading angle for navigation (Figure 7.44). This method can assist the inertial navigation system in the case of satellite rejection by determining the heading angle of a carrier after measuring the azimuth angle of the solar meridian in the carrier coordinate frame.

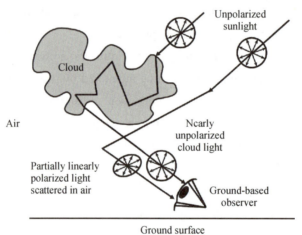

Figure 7.44 Schematic representation of the two components of light from cloudy sky regions reaching a ground-based observer [Unpolarized sunlight is scattered in the air and/or in a cloud. Direct cloud light is unpolarized (apart from the direction of rainbow scattering in water clouds), while light scattered in the air is partially linearly polarized.]

7.10.1 Mechanism of Polarization Navigation

As shown in Figure 7.45, the scattering of sunlight through the Earth's atmosphere produces a pattern of polarization across the sky. Solar radiation remains unpolarized until entering the atmosphere, where scattering interactions with atmospheric constituents cause a partially linear polarization pattern of the skylight. Along the solar and anti-solar meridians, the angle of the polarization is consistently perpendicular. However, as the Sun moves in the sky at an average speed of 15° per hour, the pattern of polarization is not constant, whereas its symmetry persists during the daytime.

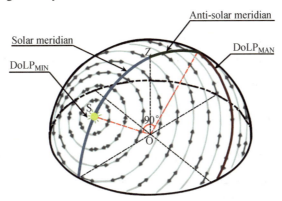

Figure 7.45 The linear pattern of polarization of skylight relative to the Sun (S) and an observer (O) [Gray arrows give the angle of polarization of the skylight. The degree of linear polarization (DoLP) is given by the thickness of the arrows. The DoLP is minimum around the Sun (approximately 10% of skylight is linearly polarized) and maximum along the red curve at 90° from the Sun (approximately 90%). The angle of polarization (AoP) is perpendicular to the solar (blue) and anti-solar (green) meridians.]

Particularly, insects are well-suited to the analysis of spatial orientation and animal navigation. Behavioral studies in honeybees, ants, and several other species show that insects largely navigate by two mechanisms. In familiar terrain, they use landmarks as guideposts for navigation routes and nest finding, as demonstrated elegantly in the classical study by Tinbergen (1932) on the bee wolf. In addition, probably more importantly, they rely on a vector-based mechanism of orientation, termed path integration. On their foraging trips, bees, ants, and other hymenopteran insects continually monitor their direction of travel using a sun-or polarized-light compass and estimate the distance to the nest through retinal image motion (in honeybees) or some other mechanism (ants).

Entomological studies have shown the existence of ommatidia which are sensitive to the polarization of skylight in the dorsal rim area (DRA) of insects' compound eyes. The optical structure of each ommatidium makes it sensitive to a single angle of polarization (AoP) and the corresponding orthogonal angle. The spectral sensitivity of polarization is in the ultraviolet range in most navigating insects, but some species are sensitive to other wavelengths, including the blue and green ones. During the past 40 years, several hypotheses have been put forward to explain the ultraviolet predominance in insects' DRA, but the most reasonable assumption seems to be that skylight polarization is clearly perceptible in the ultraviolet range even under canopies and clouds. The neural processing of polarized skylight in insects is thought to consist of three levels:

(1) The relevant information is first acquired by the ommatidia in the DRA, without any preference for any specific AoP.

(2) In the second phase, the information is transmitted to the optic lobe, where polarization neurons (POL-neurons) show high levels of synaptic activity in response to three specific orientations (10°, 60°, and 130°).

(3) Lastly, the central complex uses the neural responses delivered by the optic lobe to retrieve the insects' heading (POL-neurons in the central complex produce a uniform synaptic response to all possible angles).

7.10.2 Structure of Compound Eyes of Insects

The photoreceptor cells of insects are inherently sensitive to polarized light owing to the parallel orientation of the dichroic photoreceptor molecule, rhodopsin, in microvillar membranes. Photon absorption is maximal for light with an e-vector parallel to the microvillus axis. However, because polarization sensitivity interferes with the perception of color and brightness, it is reduced or actively suppressed in most parts of the eye and visual system by misalignment of microvillar orientation along the rhabdomere and, in addition, by the convergence of outputs from photoreceptor cells with different microvillar orientations. In many insect species, including locusts, ommatidia in a small dorsal margin of the compound eye, termed DRA, are particularly well-adapted for high polarization sensitivity. The DRA faces the sky, with optic axes directed upwards and slightly contralaterally, suggesting its role in the analysis of sky polarization. In DRA photoreceptor cells, microvilli are precisely aligned in parallel within the rhabdomeres. Self-screening is reduced by the short length of the rhabdoms.

Furthermore, in each ommatidium there are two blocks of photoreceptor cells with orthogonal microvillar orientations and thus sensitive to perpendicular e-vectors. The visual fields of DRA photoreceptors are often increased by degraded optics, reduction or lack of screening pigment between adjacent ommatidia, and enlarged cross-sectional area of the rhabdoms. Finally, the two sets of photoreceptor cells within each ommatidium are homochromatic, consistent with their exclusive specialization for polarized light detection.

The compound eye of the desert locust has a particularly prominent DRA, which shows all specializations described for other insects. Owing to its dark pigmentation in Schistocerca gregaria, the DRA can even be identified with the unaided eye [Figure 7.46 (a)]. As in many other insects, the locust DRA faces an area in the contralateral hemisphere of the sky with optical axes of ommatidia pointing 15° – 30° contralaterally. The locust DRA consists of about 400 ommatidia, and microvillar orientations are arranged in a fan-like pattern similar to that of the field cricket, the honeybee, and the desert ant [Figure 7.46 (a)]. Polarotaxis in tethered flying locusts is lost after painting the DRAs black [Figure 7.46 (b) and (c)], which indicates that polarization sensitivity is mediated exclusively by the DRA. Although the degree of sky polarization, especially under partly cloudy conditions and under the canopy, is highest in the ultraviolet, DRA photoreceptors in the locust (as in the cricket) are most sensitive in the blue.

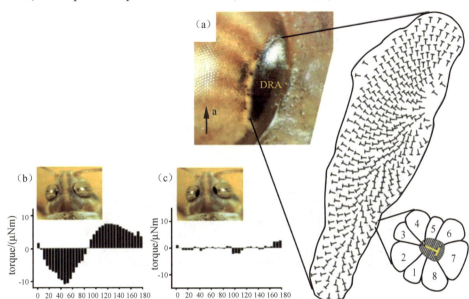

Figure 7.46 Sensory basis of polarization vision in the locust Schistocerca gregaria [(a) Polarization-sensitive DRA in the left compound eye of a locust, arrangement of ommatidia, and organization of a DRA ommatidium. The animal in (b) shows strong polarotaxis. After painting the DRAs of the eyes black (c), polarotaxis is abolished.]

7.10.3 Artificial Celestial Compass

An example of mimicking the polarization navigation behaviors of insects is shown in Figure 7.47. The polarization-dependent photodetector (PDP) is designed according to Lambert Law. The nano-gratings are mounted on the sensitive area of the photodetector, and the sensitive area is surrounded by 500 nm high electrodes. The isolation trench is designed to eliminate the crosstalk between neighboring PDPs. The polarizer that mounts on the sensitive area is a bilayer nanowire polarizer. The designed polarizer has the peak TM transmission close to 500 nm, and the downward trend after 500 nm is obvious, which is very helpful for polarization navigation.

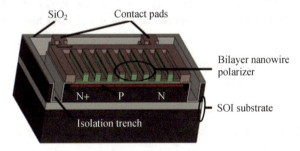

Figure 7.47 Schematic view of the PDP (The upper part is the bilayer nanowire polarizer, and the lower part is a photodetector in the silicon-on-insulator substrate.)

It has been discussed in the previous section that the polarized skylight is sensed by highly aligned polarization-opponent (POL-OP) units in the DRA of insects ommatidium. In each POL-OP unit, there are two photoreceptor channels with polarization directions perpendicular to each other. According to this special polarization-sensitive structure, six nano-gratings with different orientations are simultaneously fabricated on one chip. The six PDPs constitute three pairs of POL sensors (Figure 7.48, 0° and 90°, 60° and 150°, 120° and 30°), each pair of the POL-sensors and a log-ratio amplifier compose a polarization direction analyzer. The log-ratio amplifier receives input from the POL sensors and delivers the difference in their logarithmized signals.

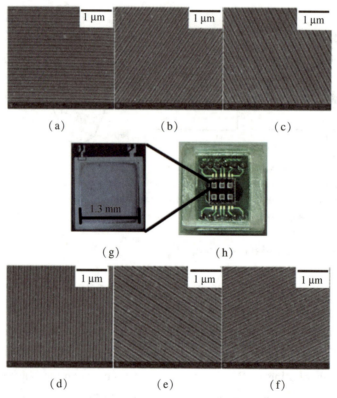

Figure 7.48 Optical and SEM images of the fabricated PDP
[(a) – (f): the SEM images of the grating in the sensitive area. The orientations are, correspondingly, 0°, 60°, 120°, 90°, 150°, and 30°; (g) enlarged optical image of the PDP;
(h) optical image of the PDPs in one chip (the contact pads of the PDPs have linked to the PCB pad).]

The final mobile robot platform for navigation is mainly composed of two driving wheels and two universal wheels, and a hardware platform, which agrees with the kinematics model of a two-wheel differential mobile robot. The overall hardware structure of the robot platform mainly includes the following modules:

(1) Processor module. The functions of the processor module include acquiring sensor data, implementing a path planning algorithm, and generating an execution layer control instruction.

(2) Polarization navigation sensor module. The polarization navigation sensor is used to

provide heading angle information, and its measurement accuracy is ± 0.3° in July.

(3) Velocity sensor module. The velocity of the mobile robot is detected by an incremental encoder with a resolution of 7 PPR/mm.

(4) Wireless communication module. The communication device used for the data transmission between the host computer and the processor is Bluetooth.

(5) Motor drive module. The motor drive module generates two pulse width modulation (PWM) waves to drive the left and right DC motors. By changing the duty cycle of the PWM wave, the control of acceleration and deceleration and start and stop of the mobile robot platform can be realized (Figure 7.49).

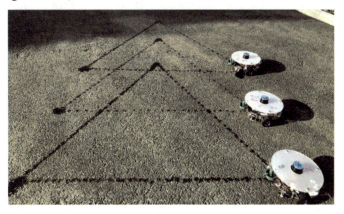

Figure 7.49　Triangular formation experiment of the bionic polarization navigation sensors

References

[1] Giorgiutti-Dauphiné, F. & Pauchard, L. Drying drops containing solutes: from hydrodynamical to mechanical instabilities [J]. *Eur. Phys. J. E.*, 2018,41(3): 32.

[2] Wang, W., et al. Controlled 2D growth of organic semiconductor crystals by suppressing "coffee-ring" effect [J]. *Nano. Res.*, 2020,13(9): 2478 – 2484.

[3] Martincigh, B. S., Allen, J. M. & Allen, S. K. *Sunscreens: The Molecules and Their Photochemistry* [M]. Heidelberg: Springer, 1997: 11 – 45.

[4] Leo, K. Organic photovoltaics [J]. *Nat. Rev. Mater.*, 2016,1: 16056 – 16057.

[5] Wan, Z. H., Zhao, K. C. & Chu, J. K. Robust azimuth measurement method based on polarimetric imaging for bionic polarization navigation [J]. *IEEE T. Instru. Meas.*, 2020,69(8): 5684 – 5692.

[6] Dupeyroux, J., Viollet, S. & Serres, J. R. An ant-inspired celestial compass applied to autonomous outdoor robot navigation [J]. *Robot. Auton. Syst.*, 2019,117: 40 – 56.